ENVIRONMENTAL BIOTECHNOLOGY

ENVIRONMENTAL BIOTECHNOLOGY

GEETHA BALI
RALLAPALLI RAMAMURTHI
S.B. SULLIA
AZIZ SHIRALIPOUR
SATISH KASTURY

A P H PUBLISHING CORPORATION
5 ANSARI ROAD, DARYA GANJ
NEW DELHI-110 002

Published by
S.B. Nangia
A P H Publishing Corporation
4435-36/7, Ansari Road, Darya Ganj
New Delhi-110002
☎ 23274050
Email : aphbooks@gmail.com

ISBN 81-7648-378-8

2026

₹ - 2595/-

Printed in India at
Efficient Offset Printers
New Delhi-110 035

Preface

The advent of new technologies has resulted in extensive exploitation of the environment causing serious environmental problems. It has become very important to adopt clean technologies for pollution prevention as well as modern technologies for remediation and management of environment. Biotechnology has emerged as a powerful and valuable tool for employing clean technologies and for sustainable environmental management.

The Center for Clean Environment Technology, Bangalore University is closely cooperating with the Center Natural Resources University of Florida and the Center for Environmental Studies, Florida Atlantic University, Boca Raton, USA in addressing the above issues. A number of workshops have been conducted by the above cooperating institutions in India as well as in the USA to discuss the latest developments in Biotechnology for management of environment. These workshops have facilitated intensive interaction between scientists and environment experts from India and the USA. The current publication is a selection of papers presented during three workshops hosted by the Center for Clean Environment Technology, Bangalore University. These workshops include Indo-US workshop on Applications of Biotechnology for Clean Environment and Energy in August 1998, Indo-US workshop on Waste Management and Role of Biotechnology in Waste Re-use in May 1999 and Biotechnology for Environmental Security in January 2001. The book also contains a few invited articles by renowned experts from different parts of the world. The articles discuss the latest technologies available for monitoring the environment, application of clean technologies for pollution prevention and biotechnology as a tool for solving some of the current problems.

Editors

Contents

List of Contributors

Abbhirami, R.
Biological Sciences Group
Birla Institute of Technology and Science
Pilani, Rajasthan, India

Anantha Murthy, K.S.
Department of Environmental Science,
Bangalore University, Bangalore, India

Anil Kush
Indo-American Hybrid Seeds (India) Pvt. Ltd.
Bangalore, India

Anuradha, S.
Biological Sciences Group
Birla Institute of Technology and Science
Pilani, Rajasthan, India

Arunkumar Nandini
Biological Sciences Group
Birla Institute of Technology and Science
Pilani, Rajasthan, India

Asha Syed
School for Forest Resources & Conservation,
University of Florida, Gainesville, Florida, USA.

Bhainsa, K.C.
Nuclear Agriculture & Biotechnology Division,
Bhabha Atomic Research Center, Mumbai. India

Cowan, P. E.
Landcare Research, Lincoln, New Zealand

Dasgupta, A.
Central Manufacturing Technology Institute,
Bangalore, India

Dev, D.V.
Department of Biotechnology, S.V. University,
Tirupati, India

Duckworth, J.A.
Landcare Research, Lincoln, New Zealand 8152

D'Souza, S.F.
Nuclear Agriculture and Biotechnology Division,
Bhabha Atomic Research Center, Mumbai, India

Frost Stan
Environmental Resources,
University of Salford,
School of Environment and Life Sciences,
Salford, UK

Geetha Bali
Center for Clean Environment Technology
Department of Microbiology and Biotechnology
Bangalore University, Bangalore, India

Halappa Gowda, T.P.
Department of Environmental Engineering,
SJC Engineering, Mysore, India

Harris, M.S.
Cooperative Research Center for Conservation and
Management of Marsupials,
Macquire University, NSW Australia 2109

Haughton Lee
Environmental Resources
University of Salford
School of Environment and Life Sciences
Salford, UK

Hegde, V.R.
Indian Resource Management Technology,
Bangalore

Hegde, V.S.
Indian Space Research Organization,
Bangalore, India

Hilda, A.
CAS in Botany,
University of Madras,
Guindy Campus, Chennai, India

Honne Gowda, H.
Indian Space Research Organization,
Bangalore, India

Jayaraman, V.
Indian Space Research Organization,
Bangalore, India

Kanchan Garg, B.A.
Department of Environmental Science,
Bangalore University, Bangalore, India

Kastury, S.
Florida Department of Environmental Protection
Tallahassee, Florida, USA

Kavitha, N.S.
CAS in Botany,
University of Madras,
Guindy Campus, Chennai, India

Keshava, N.
Department of Food Microbiology
Central Food Technological Research Institute
Mysore, India

Krishna Nand
Department of Food Microbiology
Central Food Technological Research Institute
Mysore, India

Komar, K.
Soil and Water Science Department,
University of Florida, Gainesville, Florida, USA

Latha, K.
CAS in Botany,
University of Madras,
Guindy Campus, Chennai, India

Ma, L.Q.
Soil and Water Science Department,
University of Florida, Gainesville, Florida, USA

Madyastha, K.M.
Bio Organic Chemistry Section,
Department of Organic Chemistry,
Indian Institute of Science, Bangalore, India

Mahesh, S.
Department of Environmental Engineering,
SJC Engineering , Mysore, India

Manja, K.S.
Division of Microbiology,
Defence Food Research Laboratory
Mysore, India

Mate, K.E.
Cooperative Research Center for Conservation and
Management of Marsupials,
Macquire University, NSW Australia

Melo, J.S.
Nuclear Agriculture and Biotechnology Division,
Bhabha Atomic Research Center, Mumbai, India

Molinia, F.C.
Landcare Research, Lincoln, New Zealand

Murali, H.S.
Division of Microbiology,
Defence Food Research Laboratory
Mysore, India

Mythreyi, S.
University of Iowa Medical Center
IOWA City, U.S.A.

Ogram Andrew, V.
Soil and Water Science Department
University of Florida, Gainesville, Florida, USA.

Pandya A. Gagan
Wyeth Lederle Vaccines
Pear River, New York 10965, USA

Pierro, T.
Center for Marine Structures and Geotechnique,
Department of Ocean Engineering, Florida Atlantic University,
Boca Raton, Florida, USA

Pringle Joseph
National Pollution Prevention Roundtable,
Washington D.C. USA

Rajalakshmi, N.
Department of Economics,
University of Madras,
Chepauk, Chennai, India

Ramamurthi Rallapalli
INSA Senior Scientist, Former Vice-Chancellor
S.V. University, Tirupati, India

Ramanujam, T.K.
Department of Chemical Engineering,
Indian Institute of Technology, Chennai, India.

Ray, S. K.
Biological Sciences Group
Birla Institute of Technology and Science
Pilani, Rajasthan, India

Ravishankar, S.T.
Department of Food Microbiology
Central Food Technological Research Institute
Mysore, India

Reddy, D.V.
Center for Marine Structures and Geotechnique,
Department of Ocean Engineering , Florida Atlantic University,
Boca Raton, Fl-33431. USA

Rockwood, D.L.
School for Forest Resources and Conservation,
University of Florida, Gainesville, Florida, USA

Rodger, J.C.
Cooperative Research Center for Conservation and Management of Marsupials,
Macquire University, NSW Australia

Roy Utpal
Biological Sciences Group
Birla Institute of Technology and Science
Pilani, Rajasthan, India

Shiralipour Aziz
Center for Natural Resources, Biomass Programs,
University of Florida, Gainesville,
Fl 32611-0230, USA

Shivayogimath, C.B.
Department of Chemical Engineering,
Indian Institute of Technology, Madras, Chennai, India

Shubha Maiya, P.
Department of Microbiology and Biotechnology
Bangalore University, Bangalore, India

Sidhu, K.S.
Cooperative Research Center for Conservation and Management of Marsupials.
Macquire University, NSW Australia

Srinivasa, P.C.
Department of Environmental Engineering, SJC Engineering, Mysore

Sulatha Dwarkanath
Bayer Corporation, Tarrytown,
NY 10591, USA

Sullia, S.B.
Department of Microbiology and Biotechnology
Bangalore University, Bangalore, India

Srinivasa, P.C.
Department of Environmental Engineering ,
SJC Engineering, Mysore, India

Sridevi Panchapakesan
Nuclear Agriculture and Biotechnology Division,
Bhabha Atomic Research Center, Mumbai, India

Varadaraj, M.C.
Department of food Microbiology
Central Food Technological Research Institute
Mysore, India

Varalakshmi, H.D.
Department of food Microbiology
Central Food Technological Research Institute
Mysore, India

Vijaya, T.
Department of Biotechnology, S.V. University,
Tirupati, India

Yemeniciler, A.
Center for Marine Structures and Geotechnique ,
Department of Ocean Engineering , Florida Atlantic University,
Boca Raton, Fl-33431. USA

1

Environmental Biotechnology : Concepts, Definitions and Criteria

GEETHA BALI AND RALLAPALLI RAMAMURTHI

INTRODUCTION

The conflict of interests between 'Development' and 'Degradation' from the point of view of ecosystems and environment has been debated, discussed and deliberated upon for several decades. Environment over the years has become a 'receptacle' for a myriad chemicals (Pesticides, plastics, dyes, petroleum hydrocarbons, heavy metals etc.) of truly anthropogenic origin threatening the very existence of soil and aquatic biota and causing irreparable damage to ecosystems. Innumerable commissions and reports have recognized the imperative need to evolve and implement a global action plan. While programmes relating education, awareness and community participation are long term initiatives expected to lead to minimize waste generation and optimize resource utilization, more urgent location specific effects to restore affected ecosystems have been contemplated utilizing physico-chemical and biotechnological methods. The former are found to be not cost effective and hence the emphasis has shifted to the concept of 'Bio-remediation' making effective

Centre for Clean Environment Technology Bangalore University, Jnana Bharati Bangalore, India.

use of microorganisms and other bio-systems. The aim of this paper is to present an assortment of biotechnological interventions to restore ecological balance. The applications of biotechnology include both diagnostics and remediation. These eco-friendly technologies could be used for waste minimization, waste conversion to resource, waste treatment to render the waste less toxic and for effective waste disposal.

CONCEPT OF BIOTECHNOLOGY

Biotechnology is highly multi-disciplinary involving many fields such as biology, microbiology, molecular biology, genetics, environmental science, chemistry including biochemistry, etc. Biotechnology can be defined as a series of technologies involving practical application of organisms or their cellular components for a variety of purposes including environmental management, industries, agriculture and food production and health care. Hence, it is considered under different sub-headings such as environmental biotechnology, agricultural biotechnology, medical biotechnology, industrial biotechnology etc.

Biotechnology is not new and was practiced thousands of years ago in making bread, cheese, wine, beer etc. The recent advances in molecular biology, immunology, genetics and microbiology as well as the development of powerful molecular techniques have added new dimensions, bringing biotechnology to the forefront to meet the needs of the modern society.

Biotechnology essentially involves the use of microbes, animal or plant cells or enzymes to synthesize, breakdown or transform materials (Smith, 1996). A differentiation is made between the 'traditional' biotechnology which refers to the conventional techniques that have been used for many centuries to produce beer, wine etc. and the 'new' biotechnology which embraces modern methods of recombinant DNA technology, cell fusion technologies along with modern developments in 'traditional' biotechnological processes. Today biotechnology enjoys a distinct and important status in the world market and is finding wide applications though problems exist in realizing its vast potential.

BIOREMEDIATION

The global environment is under great stress due to urbanization and industrialization as well as population pressure on the limited natural resources. The problems are compounded by drastic changes that have been taking place in the life style and habits of people. The environmental problems are diverse and sometimes specific with reference to time and space. The nature and the magnitude of the problems are ever changing, bringing new challenges and creating a constant need for evolving newer and more appropriate technologies. In this context, biotechnology has tremendous potential to cater to the need and holds hope for environmental protection and management. While some application such as for bioremediation are direct applications of biotechnology, there are many which are indirectly beneficial for the environment. Some aspects dealing with environmental remediation, pollution prevention and waste treatment will be emphasized in this article.

The problems of environment can be classified into the following subheads as most of the problems can be traced to one or more of the following either directly or indirectly.

1. Waste generation

a. Domestic waste (sewage, waste water, kitchen waste)
b. Industrial waste (by-products, effluents)
c. Agricultural waste (materials other than produce or materials left after separation/extraction of produce)
d. Food waste (food processing industries)
e. Wood waste

2. Use of chemicals for various purposes

a. Insecticides
b. Pesticides
c. Chemical fertilizers
d. Toxic products and by-products from chemical industries

3. Over exploitation of natural resources and loss

a. By conversion to non-recoverable forms, interference with biogeochemical cycles.

Waste generation is a side effect of consumption and production activities and tends to increase with economic advance. What is of concern is the increased presence of toxic chemicals such as halogen aliphatics, aromatics, polychlorinated biphenyls, polycyclic hydrocarbons etc. in the natural environment. These pollutants may reach air, water or soil and affect environment in several ways threatening the self regulating capacity of the biosphere. They may be present in high levels at the point of discharge or may remain low but can be highly toxic. The underground water sources are increasingly becoming contaminated. For example, the underground water sources have been permanently abandoned in the valley of the River Po in north Italy due to industrial pollution. Some substances may reach environment in small concentrations but may be subjected to biomagnification wherein their concentrations increase as they pass through food chain e.g. DDT. A number of reasons including our dependence on industries for economy, the extent of damage involved, as well as the extent of the area involved makes it difficult to totally prevent such pollution or adopt classical methods to remove the pollutants effectively and economically.

There are three main approaches in dealing with contaminated sites:

A. Identification of the problem
B. Assessment of the nature and degree of the hazard
C. Choice of remedial action.

Biotechnology has the potential to address the above problems with different degrees of success and its potential is ever increasing with the advent of more sophisticated molecular technologies.

BIOMONITORING AND BIOINDICATORS

Undesirable or harmful effects of air, water and soil pollution of ecosystem and biota can be monitored and predicted through the use of living systems and biological processes (Salanki et al., 1994). IUBS Commission has brought out a book of methods describing the use of bacteria, fungi, algae, fish, insect and other invertebrates from different ecosystems as test systems. Organisms such as rotifers, cladocerans, earthworms can be valuable as biomonitors to assess the health of the ecosystem.

Tributyltin (TBT) is a highly toxic compound widely used as an antifouling agent, a wood preservative, slimicide and as a biocide for industrial cooling systems, pulp and paper mills and breweries (Brisco et al., 1995). It has been shown to be highly toxic with idiosyncratic effects (Bryan and Gibbs, 1991). An *Escherichia coli* lux AB gene fusion library was used to screen for changes in gene expression on exposure to these compounds. It was found that light emission is significantly enhanced in the presence of TBT and dimethylsulfoxide (DMSO) in three clones. These clones have great potential as luminescent biosensors for detecting TBT and DMSO in the environment. Similarly, studies in the National Institute of Oceanography, Goa have shown that *Vibrio larveyi* and *Vibrio fisheri* from Arabian sea have excellent potential as biosensors based on similar principles. Immobilized enzymes can also be used for detecting trace quantities of chemical contaminants in soil and water. Portable paper and micro TLC methods have been developed for spot detection and separation of organophosphate (OP) residues, heavy metal compounds including copper, mercury, arsenic, chromium, cadmium and cyanides from surface deposits of grass, vegetables, soil and water samples. After a single step cleanup with hexane, OP residues are separated on TLC plates and sprayed with portable mammalian liver or egg white cholinesterase enzyme system followed by chromogenic substrate reagent and warmed with candle. The residues appear at nannogram levels as white chromograms with different Rf values. The heavy metal isolate after separation on Whatman No. 3 paper or TLC is sprayed with liver succinate dehydrogenase enzyme. The heavy metals appear as white chromograms and cyanides appear as pink spots at different Rf values (Dayananda *et al*, 1986; Usharani *et al*, 1984; Ramamurthi and Nandakumar, 1998).

IMPORTANCE OF MICROBES IN BIOREMEDIATION

Microorganisms play a pivotal role in biotechnology, especially in dealing with environment though we cannot exclude the use of higher organisms, especially plant and animal cell cultures in biotechnology. Microorganisms are primary fixers of photosynthetic energy and they act as systems for bringing about chemical changes in almost all types of natural and most synthetic organic molecules. Exploitation of the degradative capacities of microorganisms is the fundamental basis of organic pollutant bioremediation.

Knowledge of physiological characteristics, biochemical capabilities, ecology and genetic plasticity of the involved microbial species or consortia is an essential prerequisite for successful bioremediation of contaminated sites irrespective of the nature of the site or the choice of treatment system used. Successful bioremediation in the field requires multidisciplinary approach involving hydrologists, chemical, environmental and agricultural engineers, oceanographers, chemists as well as microbiologists, biochemists, ecologists and geneticists.

The development and application of cost effective, efficient and environmentally sound bioremediation technologies depends on the integration of many different disciplines as well as on advances in our understanding of the physicochemical, biological and ecological factors that govern microbial degradation of organic contaminants both *in situ* and *ex situ*. Considering the rapid advances over the past 10 years, the prospects for future successful bioremediation of contaminated soil, sediments and ground and surface waters are extremely promising.

Some of the factors making microbes invaluable in biotechnology include:

Their immense gene pool capable of unlimited synthetic and degradative processes.

Extremely rapid growth rates unmatched by higher organisms. Thus, immense quantities can be broken down under right conditions in short periods.

Availability of methodologies for selection of improved microorganisms from natural environmental pool as well as their modification (genetic engineering) to incorporate novel functional properties.

Microbes are also important catalytic agents in important geologic processes such as mineral formation, mineral degradation, sedimentation, weathering and geochemical cycling. Thus, they have a remarkable capacity to regulate orderly flow of materials because of their abilities to transform inorganic and organic materials. Microorganisms in soil and water will utilize any organic substance as source of energy and carbon by enzymatically breaking them into simple molecules that can be absorbed and utilized. Microbial population exists in a dynamic equilibrium which can be altered by modifying environmental conditions including nutrient availability. Not individual strains but a consortium of microbes will act on pollutants in nature.

Many biotechnological processes may be considered as having three components

(A) Obtain a suitable biological catalyst (microbes) for a specific process.
(B) Create the best possible environment for the catalyst to perform.
(C) Separate and purify the essential product from fermentation process.

Microbial ecology is the science that studies the interrelationship between microorganisms and their living and non-living environment. The decomposition of substances by microbial activity is called ***biodegradation***. Under suitable conditions, all natural organic compounds are biodegradable. They accumulate only when the environmental conditions are hostile to biodegradation.

Environmental biotechnology can be considered as the discipline that studies the application of biological systems and processes in waste treatment and management. Many successful biotechnological processes have been developed for water, soil, gas and solid waste treatments. To be employed suitably, the process should satisfy the following conditions:

A. The process should be efficient and economical
B. It should not lead to transformation of one problem to another
C. It should be environmentally safe.

However, there are a number of organic chemicals that are not easily biodegradable by microorganisms, e.g. lignin. Such chemicals are said to be recalcitrant. There are a number of synthetic compounds that are not formed by natural biosynthetic processes and are often ***recalcitrant***. These compounds are called ***xenobiotics*** and may contaminate the environment and have toxic influences. Since they are man-made, they are also called as anthropogenic compounds. These substances are resistant to biodegradation due to their structure and composition as well as due to environmental factors such as accessibility, concentration, presence of auxiliary carbon sources. E.g. phenol, chlorophenols, tetrachloroethane, dihydroxybenzene etc. Such compounds can be degraded anaerobically under special conditions. There has been marked increase in the production and use of xenobiotic chemicals during the last 50 years.

The use of microbes to remediate or restore the contaminated sites is termed ***bio-remediation***. This is different from their application to reduce or prevent the production of organic/inorganic contaminants at source. Bio-remediation specifically deals with restoration of environment already contaminated. Such a contamination might have taken place accidentally or incidentally as a result of production, storage, transport and use of organic and inorganic chemicals. Though the current trend is to encourage waste minimization, use of cleaner technologies and pollution prevention, bioremediation is indispensable because of the need to deal with the pollution that has already taken place and those that are inevitable. A number of cleanup strategies have been used in the past involving dispersal, collection, removal, landfill disposal and incineration. But these techniques simply dilute or sequester the contaminants or transfer them to another environmental medium. By contrast, bioremediation is a more effective and environmentally benign cleanup technology. This is so because it results in the partial or complete bioconversion of organic pollutants to microbial biomass and stable, usually nontoxic end products.

Bioremediation treatment technologies have been broadly divided into two categories based on whether biodegradation is stimulated *in situ* i.e. within the affected environment (soil, sediment, surface water, ground water etc.) or *ex situ* i.e. under contained conditions (compost pits, bio-reactors etc.). The latter usually involves removal of the contaminated material to the place of treatment with increased chances of exposure of the persons involved.

In Situ Bioremediation Techniques

In any natural setting, there is an existing microbial population which is ultimately responsible for the stabilization and treatment of waste products containing organic matter. This microbial population consists of a variety of microorganisms which invariably have minimum oxygen and nutrient requirements. Many of the problems encountered during handling of waste materials can be directly related to insufficient quantities of the two limiting factors namely oxygen and available nutrients. When the system is oxygen deficient, oxygen requiring bacteria do not thrive well and they give way to anaerobic bacteria. The natural respiration process of anaerobic bacteria can lead to generation of volatile organic compounds that are particularly noxious. In addition, there are bacteria that function with or without oxygen and they are called facultative anaerobes. The natural respiration process of these anaerobes involves the degradation of organic material into carbon-dioxide and water. Many treatment systems attempt to encourage the proliferation of these anaerobes in order to breakdown waste material without generating odours. Some waste treatment systems attempt to encourage the proliferation of aerobic bacteria thus eliminating the odor from decomposition by mechanically controlling the oxygen content of the waste material.

Availability of nutrients to the microbial populations within the waste material is another limiting factor determining the rate of breakdown of waste. They may be insufficient or may not be readily available. The nutrients which may act as limiting factors for the growth of microorganisms are amino acids, purines, pyrimidines and vitamins. These substances are often referred to as growth factors. By increasing the availability of nutrients to the microbial population in a readily available form, one can essentially control or accelerate the rate at which the microbes breakdown organic material.

Thus, *in situ* bioremediation involves stimulation of the degradative activities of endogenous microbial populations by the provision of nutrients and/or external electron receptors. Another way of enhancing microbial action is by the introduction of competent, exogenous microbes with or without nutritional enrichment. This procedure which is used to enhance microbial activity is called ***bioaugmentation.***

The addition of nutrients especially N and P has been tried extensively to treat marine environments and soil/aquifer systems contaminated with hydrocarbons. Nutrient cocktails have been added to marine and coastal environments. The addition of nutrients has also been employed for treating contaminated soils. For soil bioremediation, apart from the addition of nutrients, oxygen is provided through oxygen saturated water, air sparging or ***bioventing***. Water containing water oxygen and nutrients is added to the surface of the contaminated area and allowed to percolate through the soil and mix with the groundwater. Hydrogen peroxide has also been widely used as an alternate oxygen source in the percolation technology. In case of air-sparging, air is injected into water-saturated zone of soils and sediments. Bioventing involves supply of air to soil to be treated, using a combination of pumps and blowers. Using these devices, vacuum is applied to the target area, while continuously injecting low volumes of air. This procedure of bioventing is aimed at stimulating *in situ* aerobic degradation of contaminants, for its successful functioning, nutrients are often required to be added and adequate water levels must be present in the unsaturated zone to support microbial degradative activity. Bioventing has been extensively used for *in situ* remediation of unsaturated soils contaminated by petroleum hydrocarbons. Often bioventing is combined with another technique called soil vapour extraction for faster remediation of highly contaminated sites. In soil vapour extraction, much higher air flow rates are used to extract volatile hydrocarbons from petroleum-contaminated sites.

In situ bioremediation under anaerobic conditions may also be enhanced by the provision of electron acceptors, such as nitrate or sulfate. Degradation of organic pollutants in anaerobic microniches in soil, sediment and ground water environments is known to contribute significantly to overall *in situ* bioremediation rates. But *in situ* methanogenic bioremediation has not been studied much.

Ex Situ Bioremediation Techniques

These techniques are usually aerobic and involve treatment of contaminated soils or sediments using solid or slurry-phase systems. Solid phase systems include land farming (soil-treatment units), compost heaps and engineered biopiles. Land farming involves plowing, tilling and raking of the contaminated soil on site together with application of water, nutrients and microbial inocula if required. Care should be taken to prevent contamination of groundwater by leachate from the tilled zone. Land farm plots have also been used for the remediation *ex situ*, of heavily contaminated petroleum wastes. E.g. oily lagoon wastes.

Above-ground composting techniques involve physical removal of the contaminated soil or sediment to specially constructed platforms, lined pads, or compost sheds. Compost heaps consist of contaminated soil or sediment supplemented with composting material (straw, wood chips etc.) in order to enhance water-and-air-holding capacity and improve physical handling properties. Periodic mixing or turning is applied in order to ensure adequate aeration. In engineered biopiles, aeration is constantly supplied via a network of pipes inserted in the piles. Nutrients and microbial inocula are added, if required to compost heaps and biopiles. Composting techniques are used for remediation of highly contaminated sites and have proved successful for military sites contaminated with explosives such as TNT, RDX and Tetryl.

Slurry-phase bioremediation is a batch treatment technique in which excavated soils or sediments are mixed with water and treated in reactor vessels or in contained ponds or lagoons. In addition to the provision of adequate aeration and mixing, nutrients are added together with surfactants or dispersants as required. Temperature and pH are also adjusted to provide optimal or suitable environment in reactor vessel systems. Effective bioremediation has been obtained with slurry-phase systems for soils and sediments contaminated with a wide range of organic compounds, including pesticides, petroleum hydrocarbons, pentachlorophenol, polychlorophenol, polychlorinated biphenyls (PCBs) creosote, coal tars, wood preserving wastes and so on.

PRODUCTION OF ETHANOL FROM LIGNOCELLULOSIC MATERIALS USING CLOSTRIDIUM THERMOCELLUM

According to TIFAC of DST about 380 million mt of biomass were generated from 12 major crops during 1989-90 and paddy residues constituted 50 per cent of this. Conventional conversion of cellulosic biomass to ethanol is done by the step acid/enzymatic hydrolysis followed by yeast fermentation. The single step process is carried out either by monoculture or by co-culture fermentations. For effective fermentation of hemicellulose fraction, *Clostridium thermocellum* has been found to be very useful since it produces cellulase as well as xylanase systems. *Clostridium thermocellum* is an anaerobic, thermophilic (60°C), cellulolytic and ethanologenic bacterium and 1/3 of the genome is cellulase genes. Cloning is being attempted (Seenayya et al., 1998). Vijaya and Dev (1998) demonstrated that mesophilic *C. tertium* could efficiently degrade lignocellulosic material to yield 0.22 gm of ethanol per gm of substrate consumed at room temperature which is naturally a more economic proposition. A new strain SS 19 has been isolated shown to be highly efficient in the conversion of cellulose to ethanol in a single step.

Biodegradation of pentachlorophenol in treated wood

PCP is a wood preservative. Wood leachate poses environmental toxicity problems. Even in India PCP was used for catamarans. The toxic dangers to aquatic life and ultimate risk to humans are matters of great concern. The wood waste needs to be recycled and the presence of PCP is a constraint. Microbial degradation of PCP is a good possibility.

While PCP degradation by microbe has been known the same in wood leachate has never been tried. Arthrobacter can mineralize PCP into chloride and CO_2. Arthrobacter is able to use PCP as the sole C source. Microtox tests have proved reduced toxicity.

These above-ground reactors are also used for bioremediation of groundwater contaminated with volatile organics. Now many companies market microbial inocula that are said to significantly increase biodegradation of pollutants.

FACTORS INFLUENCING BIOREMEDIATION

A number of environmental factors influence in situ bioremediation rates. Some of them can be controlled or modified while some are difficult to control or modify.

Hydrogen Ion concentration (pH)

Majority of bacteria exhibit growth optima at or near neutral pH. Many soils are acidic throughout the world. Treatment commonly known as liming involves addition of finely ground agricultural limestone, calcium hydroxide, calcium carbonate, magnesium carbonate etc. during tilling and mixing of the upper layers. Treatment may affect solubility, bioavailability and chemical form of the organic pollutant and of soil macro- and micronutrients. Fungi are generally more resistant to acidic soils than soil and aquifer bacteria. Increase in soil pH decreases the availability of Ca, Mg, Na, K, ammonia, nitrogen and phosphorus. Decrease in soil pH-decreases the availability of nitrate and chloride.

Temperature

Temperature affects (a) bacterial metabolism (b) growth rates (c) soil matrix and (d) physico-chemical state of the contaminants. Generally *in situ* bioremediation is carried under mesophilic conditions (20° to 40°C)

Even for laboratory studies, bacteria with potential remediation value have focussed on mesophilic species because of the ease of cultivation and relatively short doubling times. In compost heaps or biopiles the temperature in the center of the soil/sediment may reach 70°C or higher during the initial active phase and thermophilic bacteria can be seen performing under such conditions. *In situ* technology for soils in tropical countries where the soil temperature may exceed 50°C merits further investigation. Even at 0.2°C to 8.3°C, purging of contaminated ground water, enrichment with nutrients and hydrogen peroxide have been carried successfully. Even modest increase in temperature can significantly raise bioremediation rates. A number of techniques are used to increase *in situ* soil remediation applications. These include use of mulches, plastic covers, vegetation cover to moderate fluctuations in soil temperature. In cold climate, steam may be injected to raise the soil temperature or heating return water. Temperature of compost heaps can be increased by irrigation with heated waters. For very cold environment, above ground liquid and slurry bioreactors where the temperature can be optimized are the only choices.

Water content and Geological characters

Water content especially water availability influences bioremediation rates. Water in soils or sediments may not be available to microorganisms because it is adsorbed to solid substances or tied up as water of hydration to dissolved solutes. This can be solved by irrigating the contaminated soil, compost heaps and biopiles. In general, *In situ* degradation rates are enhanced when the soil is granular or porous with a high permeability and uniform mineralogy. Rocky condition, low permeability, complex mineralogy and water logged or arid conditions are not favourable.

Nutrient availability

Nutrients are generally supplemented in both *in situ* and *ex situ* bioremediation of soils, sediments, ground and surface waters. Nutrient requirement depends on the nature of contaminants and the extent to which polluted site has been subjected to agricultural use. Remediation of petroleum hydrocarbon contaminated sites needs typically nitrogen, phosphorus. Not much is known about the importance of other macronutrients and trace elements.

External electron availability

Oxygen is used as an electron acceptor to increase bioremediation activity. A number of anaerobic bacteria can break down a variety of aliphatic and aromatic organic compounds of both natural and anthropogenic origin wholly or partially by denitrifying bacteria by sulfate, iron, and molybdenum reducers and by methanogenic consortia. Efforts are being made to use anaerobic bacteria for breaking down petroleum contaminated ground water in oil refinery sites in the presence of nitrates.

Benzene, toluene, ethyl benzene, xylene and chlorinated aliphatic and aromatic compounds have been successfully removed. Methanogenic bacteria can degrade chlorinated ethenes such as tetrachloroethane, trichloroethane, dichloroethane, perchloroethylene, carbon tetrachloride, chloroform, tetrachloromethane, alkylbenzenes, and a number of chlorinated aromatic compounds.

Bioavailability of pollutants

Anthropogenic organic polymers such as polystyrene PVC etc. are highly recalcitrant because of their insolubility and the lack of extracellular microbial enzymes capable of catalyzing depolymerization. However, non-polymer-degrading bacteria and actinomycetes are able to degrade oligomeric polystyrene fragments and low molecular weight fragments of lignin resulting from fungal attacks on the lignin polymer.

Low water solubility and a tendency to adsorb to particulate matter in soils and sediments are factors that can severely limit *in situ* biodegradation. Of many organic contaminants, polychlorinated and policyclic aromatic hydrocarbons are highly a polar and have low solubility in water and tend to adsorb strongly to organic matter in soils and sediments. Consequently, the rate limiting step is not the activity of microorganisms but the slow rate of desorption and dissolution of contaminants in the water phase. Improvements of biodegradation rates have been achieved by the addition of surfactants (either biosurfactants or synthetic detergents) to the contaminated zone. Biosurfactants produced by microorganisms within soils and sediments have also been shown to enhance biodegradation rates. Another promising approach is to add biodegradable solvents to assist desorption and dissolution rates with consequent increase in the biodegradation of the adsorbed pollutants.

Co-metabolism

Co-metabolism is a process whereby microorganisms involved in the metabolism of a growth promoting substrate also transform other organic contaminants which can be called as co-substrates which are not growth supporting if these co-substrates are provided as the only sources of carbon and energy. This is one of the most important mechanisms involved in the transformation of chlorinated organic aliphatic and aromatic contaminants by microbes. Such co-metabolic transformation of organic pollutants is an important process in both aerobic and anaerobic environments: e.g. Bacterial transformation of DDT, (dichlorodiphenyltrichloroethane) PCBs (polychlorinated biphenyls) and TCE (trichloroethylene). Similarly, microbes can sequentially remove halogen atoms from polluting halogenated compounds wherein halogen atoms are replaced by hydrogen under anaerobic conditions. Here halogen atoms serve as hydrogen acceptors and hence dehalogenation involves co-metabolism and provision of a growth promoting substance.

Gene expression

The ability of indigenous microorganisms to degrade organic pollutants is dependent on expression of the genes encoding the required enzymes. These genes may not express if these substances are available in very low concentrations. This can be overcome by adding substances that are structurally related to the organic pollutants which will act as inducers. Similarly, the presence of alternate carbon or energy source may repress the expression of the degradative enzyme needed to transform the target pollutant. For example, addition of glucose or amino acids to aquifer samples contaminated with toulene, ethylene dibromide, phenol and p-nitrophenol inhibits the degradation of these contaminants because the microbes will prefer the more easily degradable substrate.

Bioaugmentation

Where degradative microbes do not exist or where the process is too slow, microbial inoculants may be added to enhance bioremediation rates. This techniques is known as bioaugmentation and may involve (a) addition of natural isolates of bacteria or (b) genetically engineered organisms (GEMs).

There are rigid rules governing the release of GEMs as there is concern about their potential negative impact on the environment. The genetic patterns have evolved over several years and they are relatively stable. It is believed that altered genomes have greater instability and increase the chances of mutations some of which may not be safe. Bioaugmentation has met with varying degrees of success.

With the above background on the principles and techniques involved in the application of biotechnology for environment, a few well studied examples are given below

ENVIRONMENT WASTE TREATMENT TECHNOLOGIES

Waste water and sewage treatment

Biologic treatment of waste waters and sewage is the largest biotechnological industry going by volume and is one of the earliest. Waste water from sources such as homes, hospitals, schools and commercial buildings contain food residues, human excreta, paper, soap, detergent, dirt, cloth and other debris. This mixture is called sanitary or domestic sewage. Waste water which is often joined by the run-of from rain and industrial effluents in our country, flows through a network of street sewer pipes and microbial action takes place using the nutrients as the source of energy. High-energy organic substances are degraded to low-energy compounds utilizing oxygen.. The sewage reaches treatment plants. The biological disposal of organic waste is achieved in many ways through out the world. The general practice is to treat the sewage in three stages known as primary, secondary and tertiary treatments.

Primary treatment is the first stage where coarse particles are removed. A grinding mechanism may be used to reduce the remaining objects into smaller objects. The sewage then flows into a series of settling tanks where particles settle down. During the second phase called as ***secondary treatment***, the dissolved organic mater is allowed to be degraded or oxidized by microorganisms in a highly aerated condition with effective mixing to ensure contact of microorganisms with the substrate and air. The microorganisms multiply and form a biomass or sludge that can be removed and dumped. The sludge can be alternatively passed to an anaerobic digester (bioreactor) that will reduce volume of solids, the odour and the number of pathogenic microbes. Another way of treatment is to allow the dilute organic waste to pass through percolating or trickling filter bioreactors where the liquid flows over a series of surfaces which may be stones, gravel, plastic sheets etc. on which the microbes multiply and remove organic matter, utilizing it for their growth. Sometimes there is excessive microbial growth which reduces the efficiency of the system and they have to be cleared. Complications occur due to toxic action of industrial chemicals on microbes. There may be foam formation from detergents. However, controlled use of microorganisms has helped in the control of waterborne diseases such as typhoid, cholera dysentery etc.

Landfill technologies

Disposal of solid wastes is becoming an increasingly difficult and growing problem. While part of the matter is non-biodegradable including plastics etc. most of it is biodegradable. The practice is to deposit the waste in low lying sites, compress the waste and cover by a layer of soil. They can be a source of smell, and a health hazard if improperly maintained. Toxic heavy metals, hazardous pollutants and products of anaerobic decomposition may seep from the site to underground aquifers polluting urban water supplies. This can be controlled to some extant in a number of ways. The land fills are required to be air and watertight. Choosing a proper site making sure there are no water bodies in the vicinity and at a level lower than the landfill, covering the bottom of the landfill with a protective material such as geosynthetic liners while preparing the land fills can be helpful. Earlier they used to be "dumping yards" while now in advanced countries they are maintained and monitored as large bioreactor vessels. The latest technology is to bioaugment to promote microbial action. Methane production takes place and gas vents are installed to prevent explosion. The leachates are monitored and even treated.

Composting

It is an aerobic microbially driven process which converts soil organic wastes into a stable, sanitary, humus like material which is reduced in bulk and can be safely returned to environment for soil improvement or as substrate for mushroom production . It is not only a means of safely treating solid wastes but a method of recycling of organic matter derived from domestic, agriculture and food industry wastes. Composting is carried in a packed bed of solid organic particles in which the indigenous microbes will grow and reproduce. Free access to air is an essential requirement. The starting material may be arranged in piles, or covered tunnels or rotating bioreactors. Some form of pre-treatment of the waste may be required such as particle size reduction by shredding or grinding. The biological process results in oxidation of mixed organic substrates with oxygen to produce carbon-di-oxide, water and other organic by-products. After the composting process, the final product most often needs to be left for a variable time periods to stabilize. Successful composting requires optimization of growth conditions for the microorganisms. There can be rapid heat build-up by microbial reactions which may impair microbial activity. The temperature needs to be kept below 55°C. Moisture level should be 45 to 50%. Moisture level less than 40 % is too dry to support microbial activity and above 60% is too moist and may clog inter-particle spaces restricting aeration. Some solid materials are degraded very slowly e.g. lignin present in straw and wood materials. Such materials may even hamper degradation of other more easily degradable materials.

For large scale commercial composting, aerated pile system is used wherein the process takes place in closed buildings to facilitate odour emissions. Forced aeration with regular turning is necessary to create good compost condition.. Some such facilities can handle up to 60,000 tonnes per year. In Tunnel composting, composting is performed in closed plastic tunnels 30 to 50 m long, 4–6 m in width and height. Such systems are used for treating sewage sludge and domestic wastes. Some such composting plants can handle about 10,000 tonnes per year.

Another method is to use rotating drums of various sizes. Rotating drum system is especially suitable for domestic organic waste with high water content. Small drums can be used for garden waste and can be recycled readily. Waste gas outlets emitting sulphur and nitrogen compounds create problems of odour for which scrubbers could be installed or filtration could be done. For bio-filtration, a fixed bed or mass of organic material such as mature compost or microbially embedded wood chips is used. The gases pass through the mixture and the resulting biochemical activity can greatly reduce the offensive chemical smells. Germany, known for its extensive composting programme, produces more than a million tonne compost per year.

Bioremediation of oil contamination

Large areas of earth surface including land and water are contaminated with oil-derived compounds and toxic chemicals. More than 2 million tonnes of oil are estimated to enter the sea each year. 50% from industrial effluents, sewage and river overflows and rest from non-tanker shipping and natural seepage from below sea floor. Only about 18% comes from refineries, off-shore operations and tanker activities. Oil contamination is easy to detect unlike other pollutants. Most oils are relatively less toxic to environment but can affect bird and aquatic animal life very seriously.

Three main types of bioremediation technologies are currently being developed for treatment of oil spills: (a) addition of nutrients to open water oil slicks. (b) addition of microbes to oiled shorelines and (c) addition of nutrients and /or microbes to open water slicks. Just adding oil to an environment will stimulate growth of indigenous microbes as oil acts as a source of carbon. But there is a lag which can be several days to several weeks, before they can effectively degrade oil. During this period, certain fractions initially toxic to microbes undergo weathering. On March 24, about 11 million gallons of Prudhoe Bay crude oil were spilled into Prince William Sound, Alaska from the tanker *Exxon Valdez*. The spilled oil spread over 350 miles of shoreline. About 15 to 20 percent was lost by initial weathering due to volatilization. Mostly aliphatic hydrocarbons of less than 12 carbon atoms, aromatic hydrocarbons of less than 12 carbon atoms and aromatic hydrocarbons such as benzene, toluene, xylene and methyl-substituted naphthalenes were lost this way. The weathered oil was black in colour. Initial tests were conducted following which bioremediation was successfully undertaken to overcome this problem.

Phenol Biodegradation. Phenolic compounds are normally found in the effluents of petrochemical, textile, tannery industries and also from the discharges of units employing coal gasification in the range of 6-200 mg/l (Abramowicz, 1990; Aust, 1990; Buswell, 1975; Ellis, 1977; Knoll and Winter, 1987). Shashirekha and Subramanian (1997) have demonstrated that marine cyanobacteria do have great potential to degrade phenolics. *Phormidium valderianum* BDU 30501 has been found to grow well in phenol concentration of 50 mg/l. Salinity tolerance of the cyanobacterium provides additional advantage. The bacterium is found to remove 38 mg/l in seven days.

Bio-leaching

The biological reactions involved in metal leaching and extraction of metals involves oxidation of mineral sulphides. Many bacteria, fungi, yeasts, algae and even protozoa are capable of carrying out specific reactions. Many minerals exist in close association with substances like sulphur which has to be oxidized to free the metal. The bacterium *Thiobacillus ferrooxidans* is widely used for oxidizing sulphur and iron converting sulphur into sulphuric acid and converting iron sulphide to iron sulphate. The crushed ore is repeatedly washed with a bio-leaching solution containing microbes and some essential nutrients such as phosphate and ammonia. The leached liquor contains the metal which can be separated. In the USA about 10% of total copper produced is obtained in this way. About 30,000 tonnes of copper per year is believed to be extracted in this fashion world over. Bio-leaching is also practiced for obtaining uranium, an important energy source, from low-grade ore which would be uneconomical for extraction by conventional processes. Currently, sulphur coal is not used due to the production of the pollutant sulphur dioxide. But with bio-leaching, the sulphur coal can be made rid of sulphur-containing pyrite. Microbes are capable of releasing petroleum products from oil shelf and tar sands. Similarly microbes can be used for extraction of gold and silver from refractory ores which do not yield these metals with conventional treatment processes. Pre-treatment with microbes enables conversion and extraction with the usual procedure.

Because of their ability to accumulate metals from dilute solutions, microbes such as bacteria, yeasts and moulds can be used for extracting metals from industrial effluents as well as waste dumps to recover these metals economically and also to prevent pollution of environment. This is often more economical than regular mining operation. Bio-leaching can also produce refined material. Unlike smelting and other procedures used for metal extraction, bio-leaching does not pollute air and with care, ground water contamination can also be prevented. Gold recovery in South Africa is estimated to be 94% efficient using bio-oxidation against 90% through other methods. Municipal and industrial waste waters are rich in phosphorous. This promotes eutrophication of receiving water bodies. Phosphorus bacteria can be used to remove phosphorous.

Biological control

During the last 5 decades, pesticides have been extensively used all over the world which contributed for the dramatic increase in crop production. Third world uses one sixth of the total pesticides produced in the world. About 90,000 metric tonnes of pesticides is used in India of which about 40% is believed to be used in Andhra Pradesh alone. However, pesticide residues in food and their accumulation in the environment are causing concern. Many of the pesticides have a cumulative effect. Also many species of insects and other pests are developing resistance to the known pesticides. Some of the pesticides are recalcitrant or are degraded very slowly in the environment. More than 25 million agricultural workers are believed to suffer from pesticide poisoning globally.

Use of natural enemies: One of the current trends is to use natural enemies of the concern causing pests and weeds. For example natural enemies of insect pests could be other insects. Such insect biocontrol agents could be other predatory insects, parasites or hyperparasites or even birds. Similarly, aquatic and terrestrial weeds could be kept under control employing phytophagous insects. One recent example of successful biocontrol programme is the control of the noxious weed parthenium using a Mexican beetle *Zygogramma bicolorata* (Jayanth & Geetha Bali, 1994). The weed is believed to have originated from Mexico. The weed is under control in the presence of its natural enemies in its natural habitat. But when introduced into a new environment, it gains the advantage of not having its natural enemies and can grow uncontrolled. Introduction of one of its natural enemies, a beetle that occurs in its natural habitat, after carefully examining the safety of introducing them under quarantine, has resulted in checking the growth of the weed in Bangalore and surrounding areas. Similar studies have enabled the use of hymenopteran hyperparasites *Nasolynx thymus* for the control of Uzi fly a serious dipteran parasite of the silkworm *Bombyx mori* (Jhothi et al., 1998) There have been many such successful programmes undertaken in Australia, USA and other countries.

Use of BMCs and naturally occurring compounds

A promising method of environmentally safe insect pest management is the use of naturally occurring organic compounds that influence chemosensory behaviour as attractants, stimulants, arrestants, antifeedants or repellants These naturally occurring BMCs differ from conventional pesticides by the fact that they do not kill the insect pest but modify its behaviour. They score over pesticides in that (1) They are highly selective, interfering specifically with the behaviour of the target insect. (2) They are less hazardous for the non-target organisms. (3) Increased use of such compounds decreases the selection pressure by pesticides, decreasing the chances for development of pesticide resistance. A number of chemicals which are biologically active, isolated from plants, animals including pests themselves can be potential control agents (Geetha Bali, 2001)

Use of microorganisms

Bacillus thuringiensis: A number of microorganisms have been employed to control pests and diseases. The most widely used biocontrol agent is *Bacillus thuringiensis*, which are spore forming bacteria which are highly toxic to insects and more importantly they are specific in their action on insects. They have been used against lepidopterans, a class of insects to which most of the agricultural insect pests belong. Though the bacillus kills these insects, the action occurs over a period of few days and is not instantaneous. Now efforts are made to express lethal genes which will have more drastic and immediate action on pests. These genes have also been incorporated into a number of plant species making the plants pest resistant, e.g. Bt cotton. Similarly viruses can also be used to deliver such genes which code for substances toxic to insects. In Malaysia, baculoviruses have been used to protect coconut plantation from rhinoceros beetle.

Use of white rot fungi: White rot fungi such a *Phanaerochaete chrysosporium* are widely used for degradation of lignocellulose materials. These organisms which can degrade complex molecules such as lignin can degrade many of the most dangerous industrial pollutants such as PCBs. Symbiotic fungi which colonize the roots of a number of vascular plants enhance uptake of nutrients especially phosphorus and also increase disease resistance and tolerance to environmental stress. Efforts are made to inoculate trees with selected species of mycorrhizae. A number of nitrogen fixing microorganisms are being used instead of chemical fertilizers.

Vermicomposting

Solid waste which includes domestic and agricultural waste can be used for vermicomposting wherein specific species of earthworms identified for the purpose are introduced. These earthworms ingest the material along with soil, partially breaking down the organic matter and combining the matter with soil and excreting the ingested food as nitrogen rich substance which can be used as manure. More recent studies have shown that earthworms can also be used to treat effluents from industries including distilleries very economically.

BIOSORPTION

BIOSORPTION QUALITIES OF THE CYANOBACTERIUM MICROCYSTIS

Green alga *Chlorella*, blue green alga *Anabaena*, *Microcystis* marine algae and mosses have been tested for their biosorption qualities with the intent of using them for metal removal from aquatic ecosystems. Experiments conducted by Rai et al. (1998) in Varanasi proved that natural grown *Microsystis* compared to laboratory grown have greater efficiency (14% for Ni and 9% for Cd++ biosorption). Affinity is more for Cd++ and at higher concentration there is multilayer binding.

Use of water hyacinth : Now water hyacinth is being increasingly used to absorb impurities from water. It effectively removes the pollutants while producing plant biomass which can be converted into feed, fertilizer and energy. Water hyacinth has been used to make alcohol and other chemicals in Japan which are marketed. But encouraging water hyacinth in open ponds can be detrimental to the water bodies resulting in total drying up. When used, the plant is grown in separate artificial isolation tanks and the plants are periodically harvested.

Biodegradable plastics

There has been a tremendous increase in the use of petrochemical based polymers and plastics. But there is a growing concern on their accumulation in the environment. While efforts are made to discourage the use of plastics, efforts are made to develop alternate materials which are completely biodegradable. Polyhydroxybutyrate-polyhydroxyvalerate (PHBV) is one such biodegradable plastic that has been developed. A number of bacteria produce polyhydroxybutyrate (PHB) in nature . The addition of polyhydroxyvalerate makes PHB less brittle and more thermostable. The bacterial strain *Alcaligenes eutrophus* can be used to make PHBV by fermentation in the presence of glucose and propionic acid. Finally PHBV can be recovered from aqueous extraction. A number of microorganisms can degrade PHBV, breaking it into HB and HV which is used as a carbon source for growth. PHBV can be used in place of plastic for packaging etc. Poly-beta-hydroxybutyrate, produced by a number of microorganisms, can be used for making films, fibres and sheets and even bottles in place of industrial polymers. Efforts are also made to clone the PHB gene of *Alcaligenes eutrophus* into *E. coli* as the latter can grow and multiply rapidly and produce greater biomass. Similar efforts are made to express this gene in plants.

Treatment of waste gases

More recently, technologies have been developed for cleaning air using microbial degradation processes. With biofiltration, polluted air is led through a filter which consists of natural packing material such as compost. This packing material houses a wide variety of micro-organisms which have the potential of degrading a large number of organic compounds. Such biofilters are used in sewage treatment plants to process polluted air which causes odour. They find application in industries where emission of odour is a problem. Considerable research in this area has enabled combination of high conversion rates with high gas flow to make the system efficient. The packing material has been improved to make it more durable and long lasting. The pressure drop of the filter bed is also decreased to facilitate high gas velocities (400m/h). In the USA the cost works out to be around 0.25 to 1.25 US$ /1000 m^3 waste gas treated which is affordable.

However, treatment of xenobiotics needs additional research. Research has been carried on the treatability of polluted air from pharmaceutical companies. The air pollutants include substances like acetone, ethanol, isopropanol and methylene chloride. In the first step, acetone can be completely degraded and in the second step, ethanol and isopropanol can be treated. Some substances like methylene chloride can not be degraded. A third step has to be introduced inoculating the filters with a cultures of *Hyphomicrobium* sp, to degrade this compound. Accumulation of degradation products in biofilters can also cause problems. For example, accumulation of chlorides completely arrests degradation of methylene chloride which necessitates replacement of filters. The rate of degradation of recalcitrant xenobiotics can be very low. To overcome the problems due to xenobiotics, trickling filters have been developed. In a trickling filter, the gas flow is moving in the water phase through a bed of inert packing material with a developed biolayer which facilitates better conversion.

Limitations in the use of microbes for biodegradation

However, there are certain limitations in the use of microbes as bio-degrading agents. The major limitations include:

1. While the indigenous degradative microbes are fully adapted to the specific environment to be treated, the newly introduced or foreign microbes must be able to survive in the new environment and compete with the established indigenous microbes.
2. Added inocula must remain in close contact with the pollutants and in aqueous atmosphere avoid dilutions.
3. Harmful substances may be produced as a result of bacterial action.

Environmental sustainability

The ecosystem has to be protected from the adverse environmental effects associated with increased urbanization and industrialization. To achieve this, creative management of effluents and emissions, reducing waste generation and developing clean technologies are necessary. There is an increased awareness that rather than remediation following pollution, tackling of the problem at the source and pollution prevention should be the practice. Biotechnology will be valuable in evolving techniques for improving many types of existing biological and chemical engineering processes which are currently releasing pollutants into the environment. Development of integrated bioprocess designs for high quality bioprocesses which are efficient, controllable and clean has been the current approach. This has led to a new technology concept called "Eco-Tech" which is defined as "The use of a whole range of biodiversity in a holistic and low-invasive way in order to achieve benefits for humankind obeying ecological principles" and Biotechnology will play a central role in this concept of Eco-Tech..

REFERENCES

Abramowicz, D.A (1990) Aerobic and anaerobic degradation of PCBs : a review Crit.Rev.Biotechnol, 10 : 241-251.

Amit Nair (2000) Pesticides in India, Toxic Link, Delhi, Mumbai, Chennai.

Aust S.D. (1990) Degradation of environmental pollutants by Phaenerochaete chrysosporium Microb.Ecol., 20 :197-209.

Bryan, G.W. and Gibbs, P.E. (1991) "Impact of low concentrations of tributyl tin (TBT) on marine organisms" in Metal Ecotoxicology, concepts and applications, M.C. Newman and A.W. Melntosh (Eds) Lewis publishers Inc., Michigan.

Bryan, G.W., Gibbs, P.E., Hugget, R.J., Curtis, A., Bailey, D.S. and Dauer, D.M. (1989) Mar. Poll. Bull. 10 : 458-462.

Briscol, S.F., Caroline Diorio and Du Bow, M.S (1995) Luminescent Biosensors for the detection of Tritutyl tin and Dimethyl sulfoxide and the elucidation of their mechanism of toxicity in M. Moo-Young M., Anderson, W.A. and A.M. Chakavarthy (Eds.) Environmental Biotechnology : Principals and Applications (1995) Kluwer Academic Publishers Dordrecht/Boston/ London...645-655.

Busswell, J.A. (1975) Metabolism of phenols and cresols by *Bacillus stearothermophilus* J.Bacteriol. 124:1077-1083.

Dayananda, S., Usha Rani, A., Ramamurthi, R. and Nandakumar, N.V., (1986) Fenitrothion degradation in freshwater Poll. Res 5(1) : 19-21.

Ellis B.E. (1977) Degradation of phenolic compounds by fresh water algae. Plat Sci. Lett. 8:213-216.

Forster, C.F. (1985) Biotechnology and Waste Water Treatment, Cambridge studies in Biotechnology, Cambridge University Press, Cambridge.

Gary S. Sayler, Robert Fox and James W. Blackburn (eds.) (1991), Environmental Biotechnology for Waste Treatment, Plenum Press, New York.

Glaser, J.A. (1990) White Rot Fungus Applications to Hazardous Waste Treatment, USEPA Risk Reduction Engineering Laboratory, Cincinnati, Ohio.

Geetha Bali (2001) Role of peptides in the regulation of Insect behaviour and Eco-friendly strategies for Insect Pest Control, Sectional Presidential address, 88th ISC, 1-51.

Greenshields, R. (ed.) (1993) Industrialized Biotechnology International, Sterling Publications Ltd. Hong Kong.

Jayanth, K.P. and Geetha Bali (1994) Biological Control of *P. hysterophorus* by the Mexican beetle *Z. bicolorata* in Bangalore, India. FAO Plant Protection Bull. Rome. 42 : 207-213.

John E. Smith, (1996) Biotechnology, Cambridge University Press.

Jyothi, H.K., G. Veeranna and Geetha Bali (1998) Low temperature storage of the parasitoids of visibly *E. bombycis* L. a pest of silkworm *Bombyx mori* L. J. Biol. Control 12(1) : 31-35.

Knoll Gand Winter J. (1987) Anaerobic degradation of phenol in sewage sludge Appl.Microbiol. 32 : 192-194.

Peaprna, I., Kamali, K., Robinson, C.W., Mortimer, W.P. and Allen, D.G. (1995) Biodegradation of Pentachlorophenol in aqueous extract from PCP-treated wood by Arthrobactor sp in M.Moo-Young et al. (Eds.) Environmental Biotechnology: Principals and Applications. 320-331 Kluwer Academic Publishers, Netherlands.

Ramamurthi, R. and Nandakumar, N.V. (1998) Standardization and optimization of biodetector strips for pesticides and heavy metals DBT project Report.

Rai, L.C., Sarita Singh and Subhashnee Pradhan (1998) Biotechnological potential of naturally occurring and laboratory grown *Microcystis* in biosorption of Ni_2^+ and Cd_2^+ Curr. Sci. 74(5) 461-464.

Salanki, J., Jefferey, D., Hughes, G.M. (Eds) (1994) Biological Monitoring of Environments, Manual of Methods. Willingfood.

Seenayya, G., Gopal Reddy, Sai Ram, M., Swamy, M.V. and Sudha Rani, K. (2000) Production of Ethanol from Lignocellulosic materials using *Clostridium thermocellum* in Bioenergy W.H. Smith (Eds) Science Publishers. Enfield (NH), Plymouth (UK).

Usha Rani, A., Dayananda, S., Ramamurthi, R. and Nandakumar, N.V. (1984) Methyl Parathion degradation in soil. Geobios New Report 3, 185-187.

Vijaya, T. and Dav, D.V. (2000) Cellulose conversion to Ethanol by a mesophilic cellulolytic bacterium. In Ramamurthi, R. and Satish Kastury and Wayne H Smith (Eds) Bioenergy (2000) p. 117. Science Publishers. Inc., Enfield (NH) Plymouth (UK).

Shashirekha, S., Uma, L. and Subramanyam, G. (1997) J. Indus. Micro and Biotech 19, 130-133.

2

Characterization of Contaminated Sites by Molecular Diagnostics

ANDREW V. OGRAM*

INTRODUCTION

Bioremediation is frequently an attractive strategy for decontaminating surface and subsurface sites contaminated with organic pollutants, and may provide a relatively low cost and efficient means of environmental decontamination in many cases. This strategy may also be used as a companion to chemical remediation in sites contaminated with high levels of compounds that are toxic to bacteria. In this case, bioremediation is applied once the bulk of the contaminant is sufficiently decreased to levels that are not toxic to the degrading organisms. A prerequisite for any remediation strategy is characterization of the site with regard to various factors that may effect bioremediation, particularly characterization of indigenous bacterial communities (Hetizer and Sayler, 1993).

Knowledge of the types, concentrations, and activities of biodegradative bacteria and of the processes that control their activities is important in at least two regards:

* Soil and Water Science Department and Microbiology and Cell Science Department University of Florida, Gainesville, FL 32611-0290 USA.

- as part of site characterization for determining appropriate remediation strategies; and
- as part of monitoring the progress and effectiveness of bioremediation.

Although rarely used, concentrations of specific bacteria may also serve as indicators of remediation "endpoints;" i.e., as indicators of remediation sufficient to declare a site "clean." Identification and enumeration of individual bacterial species are usually deemed too time consuming and laborious for routine analysis, and is usually dispensed with in favor of evaluation of biodegradative potential of samples in laboratory microcosms.

Among the limitations of site characterization by bench top microcosms are: lack of sensitivity; inaccuracy due to an inability to cultivate most soil bacteria in the laboratory; deviation from what is actually occurring in the field; and cumbersome experimental setups. In addition, many sites are contaminated with mixtures of compounds, thereby greatly increasing the numbers of analyses required for sufficient site characterization. A relatively new approach, molecular diagnostics (Sayler and Layton, 1990), provides an alternative to traditional lab microcosms and has great potential to provide important information on resident microbial communities that would be unavailable from traditional lab microcosm studies.

Molecular diagnostics utilizes information regarding the structures and activities of resident microbial communities contained in the community's nucleic acid to characterize the biodegradative potential of a site. Much of the information required for accurate site characterization is encoded in the microbial community's DNA, and analyzing DNA directly isolated from samples may access much of this information. Utilizing DNA as an indicator molecule in this way bypasses the need for laboratory cultivation, and therefore bypasses much of the bias and uncertainty associated with laboratory incubations. In addition, molecular diagnostics can be a very rapid and accurate means of simultaneously screening numerous samples, thereby increasing the efficiency with which samples may be processed. This review will focus on the advantages and limitations of molecular diagnostics as a means of site characterization and monitoring for bioremediation applications.

BACKGROUND

Applications of relatively simple molecular genetic techniques for characterization of contaminated soil samples provides useful information regarding bacterial community structure and activity, and may yield much more accurate estimations of biodegradative potential than standard lab microcosm would provide (Ogram and Sayler, 1988). Some advantages of molecular genetic applications to site characterization are:

- rapid characterization of numerous samples;
- simultaneous characterization of multiple biodegradative pathways from mixed-waste sites;
- information regarding the specific biochemical pathway most likely to dominate remediation process;
- provide rapid assessment of the progress of biostimulation strategies; and
- rapid quantitative assessment of indicators of remediation endpoints.

The most commonly applied molecular genetic approach to site characterization is so-called "molecular diagnostics," referring to the use of cloned genes of interest ("gene probes") to measure concentrations of similar genes in microbial DNA or RNA directly extracted from the soil sample. Gene probes cloned from bacterial pathways for metabolism of the pollutant of interest may be used against a large number of samples in a relatively short period of time, simultaneously providing information on the specific pathways present and the concentrations of bacteria possessing the genes. Selected examples of gene probes are listed in Table 2.1. Information regarding specific pathway(s) involved in the bioremediation process will increase understanding of specific environmental factors that may be manipulated for optimization of the bioremediation process. The approach may be quantitative, and the relative concentrations of specific genes in a given sample is related to the potential for that sample to biodegrade the compound of interest. This information will aid scientists and engineers in developing and monitoring remediation strategies appropriate for the site and the types of bacteria available for remediation.

Table 2.1 : Examples of Gene Probes used in Molecular Diagnostics

Target Compound	*Target Enzyme*	*Target Gene*	*Host Strain*	*Reference*
Toluene, TCE	Toluene 4-monoxygenase	*tom A*	*Burkholderia cepacia G4*	Shields et al., 1989
Toluene, TCE	Toluene dioxygenase	*todC1C2BA*	*Pseudomonas putida F1*	Zylstra and Gibson, 1989
Toluene, TCE	Toluene-2-monooxygenase	*tmoABCDE*	*Pseudomonas mendocina KR1*	Yen et al., 1991
BTEX compounds	Catechol 2,3 dioxygenase	*xylE*	*Pseudomonas putida mt-2*	Assinder and Williams, 1990
TCE	Soluble methane monooxygenase	*mmoB*	*Methylosinus trichosporium OB3b*	Tsien and Hanson, 1992.
Naphthalene	Naphthalene dioxygenase	*nahAcd*	*Pseudomonas putida G7*	Simon et al., 1993.

Potential versus Activity

The potential for biodegradation to occur is indicated by the concentration of the gene of interest within the community DNA, and this information may be accessed by using DNA as the target for gene probes. This approach does not yield information regarding the *activity* of these genes, or serve as an indication of in situ biodegradation rates. The activity of the genes is expressed as RNA, and using RNA as a target for gene probes may, in some cases, correlate with biodegradation rates. Virtually all studies attempting to correlate biodegradation rates with mRNA concentrations have involved correlating *nahA* (naphthalene dioxygenase) mRNA concentrations with naphthalene biodegradation rates (Fleming et al., 1993; Sanseverino et al., 1993). Using RNA as a target in such studies may be limited for some genes due to the very high turnover rate of some mRNAs, but *nahA* transcripts have correlated well with biodegradation rates in a number of soils.

Bioavailability

An additional aspect of molecular diagnostics is the application of more sophisticated techniques to determine the bioavailability of the contaminant at a particular site, and the *in situ* activity of the degrading organisms. Hydrophobic molecules such as naphthalene may be tightly sorbed to soil particles, thereby decreasing their availability to the degrading bacteria. In many cases, lack of bioavailability is a greater limitation to bioremediation than low concentrations of biodegradative bacteria. The relative bioavailability is dependent on specific soil properties, and hence should be evaluated for each site. Bioluminescent reporter genes under the control of promoters in biodegradative pathways (such as for *nahA*) have been used to indicate the relative bioavailability of naphthalene in a given soil (King et al., 1990; Sanseverino et al., 1993). In essence, light is emitted when the nahA gene is induced (i.e., when naphthalene is taken up by the reporter strain and naphthalene dioxygenase is induced). The amount of light emitted is proportional to the availability of the substrate.

Limitations

As with any microbial method, molecular diagnostics has limitations, although most of these are being overcome with time. The greatest limitation to date is that the gene of interest must be known. Soils harbor a great diversity of bacteria (perhaps as many as 10,000 distinct species), most of which have never been cultivated in the laboratory. Laboratory cultivation is an obvious prerequisite for studying the genetics of biodegradation, and it is likely that we currently know very little of the diversity of genes involved in biodegradation of many organic pollutants. This lack of information limits the number of gene probes available for use, and in many cases molecular diagnostics may underestimate both the numbers of bacteria involved in bioremediation and the types of biodegradative pathways involved.

It is impossible to over-emphasize the great diversity of soil bacteria and our relative ignorance of the genetic and biochemical diversity present in the soil. A recent example of the failure of molecular diagnostics to accurately describe the biodegradative potential of a site involved a dominant and undescribed toluene pathway (Ogram and Scow, unpublished observations). In this case, DNA was isolated from soil that rapidly mineralized toluene, and analyzed with gene probes specific to all five toluene pathways known at that time (1996). No positive signals were observed with any of the five gene probes against the soil DNA, indicating that these genes were not likely to represent the dominant toluene degrading bacteria, if they were present at all in the samples. Toluene metabolizing bacteria were isolated from the soil, and none of these harbored genes similar to the five known pathways, indicating the presence of a novel toluene genetic pathway. The dominant toluene degrading bacterial strain in this soil was later identified as a *Rhodococcus* sp., a gram positive species, that shares relatively low similarity at the DNA level to all cloned genes involved in toluene metabolism. Virtually all that is known of the genetics and biochemistry of biodegradation was learned from gram negative species such as *Pseudomonas*. These species grow rapidly in standard enrichment cultures and frequently out grow other species that may be equally or more important in the soil. Until more is known of the diversity of biodegradative bacteria in soil, the general applicability of molecular diagnostics will be limited.

EXPERIMENTAL APPROACHES

Molecular diagnostics (with the exception of reporter genes) applies very basic procedures from molecular genetics to measure concentrations of a specific gene within a soil sample. The concentrations of these genes are related to the degradation potential of the target compound by the resident microbial community; the greater the number of genes, the greater the number of biodegradative bacteria, and hence, the greater the potential biodegradation rates. The construction and application of bioluminescent reporter strains is complicated and is beyond the scope of this review. The interested reader is encouraged to read more detailed reviews on this subject (Ogram, 1998; Ogram and Bezdicek, 1994). The general experimental approach may be divided into three general steps: (1) isolation of DNA; (2) hybridization; and (3) detection. A general flow chart is presented in Fig. 2.1.

Isolation of DNA

Numerous protocols for isolation of DNA from soils have been published over the last 19 years, although most follow a general scheme. Microbial cells are first lysed in the presence of the soil, followed by separation of the DNA from the soil particles. Extraction of DNA from soils frequently co-extracts humic substances that must be removed. Various strategies may be applied to purification of DNA from humic substances, and have included selective precipitations, electrophoretic methods, and chromatographic separations. The method of choice is frequently dependent on the individual sample due to the great variability in humic substances in soils. General strategies to aid in choosing a specific method are available (Ogram, 1998).

Hybridization

Hybridization is the process by which the gene probe combines with similar genes in target DNA. Target DNA is typically fixed to a nylon filter, such that DNA from many samples may be analyzed simultaneously. DNA is double stranded, with complementary base pairs (A-T; C-G) forming the genetic sequence of each strand. The gene probe is mixed with the target DNA under conditions that promote specific coupling of the two pieces of DNA (probe and target). The probe is typically labeled with either radioactive phosphorus (^{32}P) or with a non-radioactive fluorophore, such that the hybrid is labeled. The concentration of the label on the filter is proportional to the concentration of target genes within the community.

Detection

Once hybridization is complete, the concentration of the label associated with the nylon filter (e.g. the target DNA) must be determined. If a radioactive probe is used, the signals are usually detected by exposure to X-ray film; the density of the exposure is proportional to the concentration of the radioactive probe. If the probe is labeled by a fluorophore, detection systems specific to the fluorophore are used, and are described by the vendor of the labeling system.

APPLICATIONS

Molecular diagnostics has been successfully applied to characterize a number of sites, including those contaminated with a variety of priority pollutants such as fuel oil (Guo et al., 1997; Sanseverino et al., 1993; Stapleton and Sayler, 1998), chlorinated aliphatic compounds (perchloroethylene and trichloroethylene) (Brockman et al., 1994), and polychlorinated biphenyls (Layton et al., 1994). A few examples are described below.

Characterization of a Fuel Oil Contaminated Site

Molecular diagnostics is particularly well suited for characterization of sites contaminated with simple aromatic hydrocarbons such as toluene and naphthalene due to the relatively large body of knowledge available on the genetics and biochemistry of these compounds. Several pathways for toluene metabolism have been identified, and genes are available from these pathways for use as gene probes. Simple aromatics such as toluene and naphthalene are common constituents of fuel oil, and applications of molecular diagnostics are particularly effective in characterizing sites contaminated with fuel oil.

Leaking under ground storage tanks beneath gasoline stations (petrol pumps) are common sources of petroleum and diesel fuel contamination in soils. Typical components of these fuel oils are the so-called BTEX (benzene, toluene, ethyl benzene, and xylenes) compounds and naphthalene. BTEX and naphthalene are easily utilizable carbon sources for many soil bacteria, and one might expect that bacterial species capable of metabolizing these compounds would be enriched in contaminated soils. The presence and concentration of aromatic hydrocarbon-metabolizing bacteria would likely be an indicator of the potential success of a bioremediation approach, and the likelihood that so-called intrinsic bioremediation (e.g. natural attenuation) would be effective. In addition, background levels of these bacteria (or their genes) may be useful in defining indicators of remediation endpoints; indicators that remediation has reduced the levels of the contaminant to concentrations at which no ecological damage is likely.

In a recent study (Guo et al., 1997), molecular diagnostics indicated the potential success for bioremediation and natural attenuation of fuel oil contamination resulting from a leaking underground storage tank. Concentrations of genes specific to naphthalene, particularly *nahA* (encoding naphthalene dioxygenase) and *xylE* (encoding catechol 2,3-dioxygenase) were significantly enriched in regions of the contaminated zone, suggesting natural attenuation was proceeding in these regions. This indicates that *xylE* would be a useful biological indicator of fuel oil contamination, or as an indicator of a remediation endpoint. Concentrations of *xylE* decreased to background levels at high concentrations of the contaminant, however, indicating that high concentrations of the contaminant may be toxic to the hosts of these genes. A chemical or physical remediation approach may be required for highly contaminated zones, whereas natural attenuation or bioremediation would be appropriate in regions of the site that contain lower concentrations of the contaminant.

Monitoring Analog Enrichment for Remediation of PCE and TCE Contamination

Perchloroethylene (PCE) and trichloroethylene (TCE) are among the most significant ground water pollutants in the United

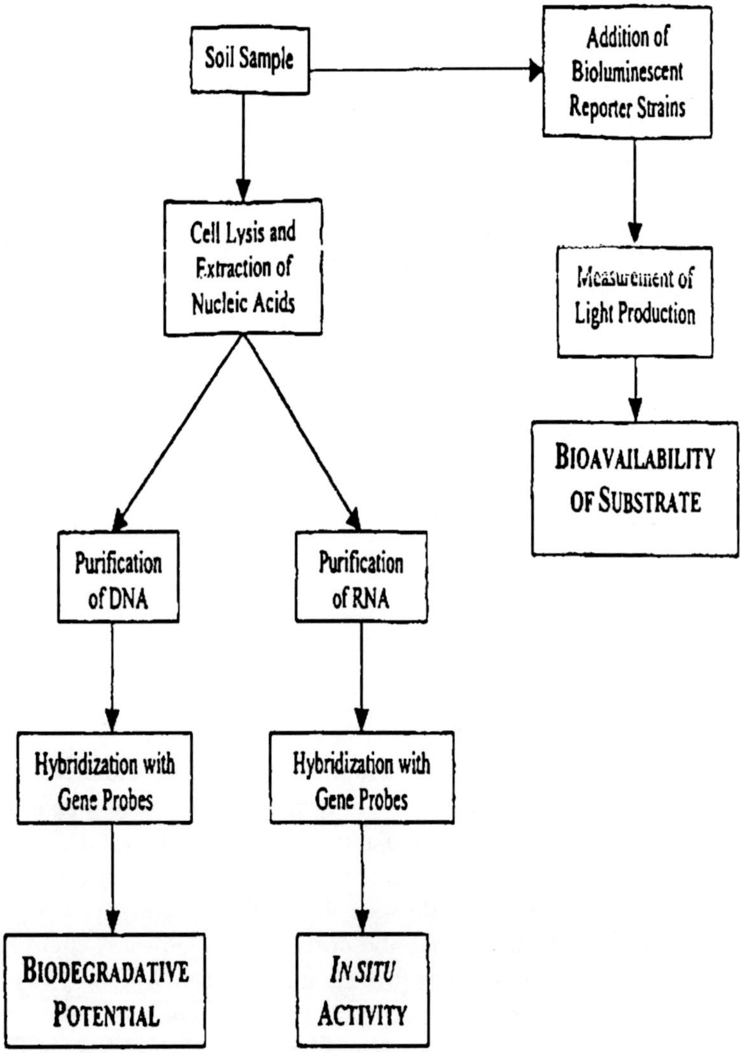

Fig. 2.1 : Flow chart of sample characterization by molecular diagnostics

States, and represent a challenge for remediation. Relatively few species of microorganisms are capable of utilizing chlorinated aliphatic compounds such as TCE as sole sources of carbon and energy, making direct stimulation or enrichment of TCE biodegradation difficult. Methanotrophic bacteria are among the most efficient at degrading these compounds, but do so through co-metabolism mediated by a monooxygenase (methane monooxygenase) exhibiting broad substrate specificity. One strategy for remediating groundwater contaminated with these compounds is by analog enrichment, or enrichment of the co-metabolizing bacteria and induction of the responsible enzymes by addition of the primary substrate for the enzyme. In this case, methane induces the responsible enzyme and serves as a growth substrate for methanotrophs, such that addition of methane to a contaminated aquifer (with sufficient oxygen and other nutrients) should stimulate the biodegradation of TCE.

The most ambitious application of methane enrichment to remediate a TCE contaminated aquifer was conducted by the US Department of Energy beneath their Savannah River Site in Aiken, South Carolina. Horizontal wells were installed to deliver mixtures of methane, air, and nutrients to the zone of contamination, and molecular diagnostics was used in conjunction with traditional methods such as most probable number analysis to monitor enrichment of methanotrophs (Brockman et al., 1994). In addition to gene probes directed toward soluble methane monooxygenase (sMMO), probes for other oxygenases capable of co-metabolizing TCE were applied. Increasing levels of the gene encoding sMMO were observed as the methane injection campaign proceeded, mirroring increases in MPN methanotrophs. In addition to increases in genes specific to methanotrophs, increases were also observed in genes encoding toluene monooxygenase (also active against TCE), indicating general stimulation of the microbial community. Concentrations in sMMO gene concentrations reflected observed decreases in the concentration of TCE, suggesting successful remediation. As the methane enrichment proceeded, a decrease in the disappearance rate of TCE was noted; this reduction in the rate of removal corresponded with sharp declines in the concentration of sMMO genes and toluene monooxygenase genes. The reason for this decline in rates is not known.

In this case, molecular diagnostics accurately monitored enrichment of methanotrophs during methane enrichment, a trend that was confirmed by more traditional analyses. In addition to monitoring enrichments of methanotrophs, molecular diagnostics indicated general enrichment of the microbial community such that additional (non-methanotrophic) TCE-co-metabolizing bacteria were enriched.

SUMMARY

Molecular diagnostics holds great promise as a tool for evaluating the biodegradative potential of individual sites due to the ability to rapidly characterize numerous samples within a contaminated site. As with any method, however, molecular diagnostics must be approached cautiously due to inherent limitations. Current limitations are largely due to a lack of knowledge of the diversity of environmentally relevant bacterial strains, but this limitation will be overcome with time. Molecular diagnostics are best applied toward characterization of potential degradation of compounds exhibiting simple pathways with low genetic diversities, such as those for toluene, naphthalene, and TCE, although a diversity of genetic pathways for these compounds has been observed between different soils. Advanced methods, such as bioluminescent reporter genes, are currently being evaluated, and are likely to become indispensable tools of remediation specialists in the near future.

REFERENCES

Assinder, S.J., and P.A. Williams. 1990. The TOL plasmids: Determinants of the catabolism of toluene and the xylenes. Adv. Micro. Physiol. 31 : 1-69.

Brockman, F., W. Sun, A. Ogram, W. Payne, and D. Workman. 1994. Baseline characterization and remediation-induced changes in TCE degradative potential using enrichment techniques and DNA probe analysis. In: Bioremediation of chlorinated and polycyclic aromatic hydrocarbon compounds (Hinchee, R.E., et al. Eds). Lewis Publishers, Boca Raton, FL.

Fleming, J., J. Sanseverino, and G. Sayler. 1993. Quantitative relationship between naphthalene catabolic gene frequency and expression in predicting PAH degradation in soils at town gas manufacturing sites. Environ. Sci. Technol. 27 : 1068-1074.

Guo, C., W. Sun, J.B. Harsh, and A. Ogram. 1997. Hybridization analysis of microbial DNA from fuel oil-contaminated and non-contaminated soil. Micro. Ecol. 34 : 178-187.

Heitzer, A., and G. Sayler. 1993. Monitoring the efficacy of bioremediation. Trends Biotechnol. 11 : 334-343.

King, J., P. DiGrazia, B. Applegate, R. Burlage, J. Sanseverino, P. Dunbar, F. Larimer, and G. Sayler. 1990. Rapid, sensitive bioluminescent reporter technology for naphthalene exposure and biodegradation. Science 249: 778-781.

Layton, A.C., C. Lajoie, J. Easter, R. Jemigan, M. Beck, and G. Sayler. 1994. Molecular diagnostics for polychorinated biphenyl degradation in contaminated soils. Ann. N.Y. Acad. Sci. 721 : 407-422.

Ogram, A. Isolation of nucleic acids from environmental sources. 1998. In Techniques in Microbial Ecology (Burlage, R.S et al., Eds), pp. 273-288. Oxford University Press, New York.

Ogram, A., and D. Bezdicek. 1994. Nucleic acid probes. In Methods of Soil Analysis, Part 2: Microbiological and Biochemical Properties (Angle, S. et al., Eds), pp. 665-687.

Soil Science Society of America, Madison, WI.

Ogram, A., and G. Sayler. 1988. The use of gene probes in the rapid analysis of natural microbial communities. J. Ind. Microbiol. 3 : 281-292.

Sanseverino, J., C. Werner, J. Fleming, B. Applegate, J. King, and G. Sayler. 1993. .

Molecular diagnostics of polycyclic aromatic hydrocarbon biodegradation in manufactured gas plants. Biodegradation 4:303-321.

Sayler, G.S., and A.C. Layton. 1990. Environmental application of nucleic acid hybridization. Ann. Rev. Microbiol. 44:625-648.

Shields, M.S., S.O. Montgomery, S.M. Cuskey, P.J. Chapman, and P.H. Pritchard. 1989. Novel pathway of toluene catabolism in the trichloroethylene-degradaind bacterium G4. Appl. Environ. Microbiol. 55:1624-1629.

Simon, M.J., T.D. Osslund, R. Saunders, B. Ensley, S. Suggs, A. Harcourt, W.C. Suen, D.L. Cruden, D.T. Gibson, and G.J. Zylstra. 1993. Sequences of genes encoding naphthalene dioxygenase in Pseudomonasputida strains G7 and NCIB 9816-4. Gene 127 : 31-37.

Stapleton, R.D., and G. Sayler. Assessment of the microbiological potential for the natural attenuation of petroleum hydrocarbons in a shallow aquifer system. Micro. Ecol. 36 : 349-361.

Tsien, H.C., and R.S. Hanson. 1992. Soluble methane monooxygenase component B gene probe for identification of methanotrophs that rapidly degrade trichloroethylene. Appl. Environ. Microbiol. 58 : 953-960.

Yen, K.M., M.R. Karl, L.M. Blatt, M.J. Simon, R.B. Winter, P.R. Fausset, H.S. Lu, A.A. Harcourt, and K..K.. Chen. 1991. Cloning and characterization of *Pseudomonas mendocina* KRI gene cluster encoding toluene-4-monooxygenase. J. Bacteriol. 173:5315-5327.

Zylstra, G.J., and D.T. Gibson. 1989. Toluene degradation by Pseudomonasputida Fl, nucleotide sequence of the tod CIC2BADE genes and their expression in *E. coli*. J. Biol Chem. 264:14940-14946.

3

Satellite Remote Sensing for Oceanographic Applications in India

H. HONNE GOWDA, V. JAYARAMAN AND V.S. HEGDE*

INTRODUCTION

India is the largest country in the Indian ocean region. The country is blessed with a long coastline of 7,500 km and a vast living and non-living resources in the seas around and in the oceanic islands. Besides, Indian ocean plays a key role in influencing weather and climate systems. The population that lives in an environment is increasingly exploring and exploiting the oceans in recent years for; marine fisheries, environmental forecasting, off-shore oil and gas, marine transport, dumping ground for disposal of wastes etc. With the increased dependence on the ocean and anthropogenic activities, have caused concern and major threat to sustaining the biological richness of oceans and coastal areas. Therefore, there is a need for a long-term management and sustainable development of ocean resources which should be based on a sound understanding of the ocean dynamics and a proper evaluation of the exploitable potential benefit of marine resources. Though the study of the Indian ocean began quite long back, our understanding of the ocean is far from

* Indian Space Research Organisation New BEL Road, Bangalore-560 094.

compete largely due to the availability of only limited observations essentially from ships/buoys. A satellite platform, with its synoptic view and capability to give spatially continuous data on a repetitive basis have recently demonstrated the potential for retrieval of many ocean geophysical parameters and understanding their inter-linkages more scientifically and harnessing of coastal and ocean resources through proper evaluation.

SENSORS FOR OCEAN APPLICATIONS

Detailed study of oceanic process are best done with satellite remote sensing sensors optimized for oceanographic applications, while coastal zone studies can be done utilizing satellites designed for land applications. Major satellite sensors available for coastal and oceanographic applications include (i) Ocean Colour Monitor (OCM), (ii) Thermal Infrared Radiometer (TIR), (iii) Scatterometer (SCAT), (iv) Synthetic Aperture Radar (SAR), (v) Altimeter (ALT), (vi) Microwave Radiometers, (vii) Imaging Spectrometers, and (viii) High Resolution imagers. These instruments provide a wealth of information on a diverse range of geophysical parameters and phenomena, such as surface or near-surface colour, sea surface temperature, wave fields, surface roughness, large scale surface topography, wind fields, etc. Measurements of these features and their variations over space and time can be interpreted to provide information relating to biological productivity, fish location, currents, sea state, surface winds, water composition and quality, sedimentation patterns, pollution and other phenomena. Integration of the complementary data from the different instruments is often used to increase the information content of the data. The high resolution imagers on the other hand will provide information on the areal extent, condition and boundary of coastal wetlands and other coastal features, which are indicative for identification of areas for preservation, conservation and development activities.

SATELLITE APPLICATIONS IN INDIAN OCEAN

Satellite remote sensing was gainfully utilised in the country for many ocean applications including coastal zone studies over the last 8 years under a major programme known as Marine Satellite Information Services (MARSIS). As an outcome of this programme, a new thrust has been given to satellite oceanography to establish and demonstrate end-to-end capability for retrieval and operational uses in the coming years. Highlights of the satellite oceanographic studies carried out in Indian ocean are outlined below.

Remote Sensing Applications in Coastal Zone Studies

Coastal zones are one of the most dynamic areas of our planet, being the meeting place of the land, the sea and the air. Owing to the existence of diverse and productive habitats important for human settlements, development and local subsistence, there has always been hectic human activities coupled with many developments. Of late, the developments are taking place in such a manner that the coastal areas are endangered with problems such as erosion, siltation, over-population, salt water intrusion, flooding, pollution, devastation of natural habitats etc. Rational development of coastal areas, which form the habitat of over 25% of the country's population, living within 60 km of the shore line, can only be achieved by understanding the various interactive processes that are operative in the coastal environment. The major issues which require immediate attention to prevent degradation of coastal areas include (i) monitoring long-term trends of dynamic changes, (ii) planning and implementation of coastal protection, (iii) formulating proper criteria for the location of industries, aquaculture and recreational activities, (iv) monitoring and conservation of critical environmental features, (v) assessing the impact of wetland conservation, reclamation of land from the sea, sand mining, dredging and recreational activities on coastal ecology, (vi) optimal management of renewable and non-renewable marine resources, (vii) controlling pollution of coastal and estuarine waters, and (viii) improvement of navigation systems, etc.

The major requisites for coastal zone monitoring and management are the availability of information on existing conditions and changes that may have occurred over the years. For example, the mangrove forests (mangals) in India now occupy less than 2000 sq.km of area, having suffered a loss of 85% since pre-agricultural times. World over the loss of mangrove forest is about 60 per cnt during this period. One of the possible explanations in India is due to the fact that unplanned expansion of aquaculture along the coastline has resulted in the destruction of mangroves and other areas of critical concern, besides causing floods during the monsoon. Continuous monitoring of these consequences along with other coastal changes calls for integrated studies which should be based on availability of information with respect to space and time on regular basis for understanding of the interactive process of coastal zone and a proper evaluation of the exploitable potential benefit of the resources on a sustainable basis. The United Nation's Conference on Environment and Development at Rio de Janeiro in June 1992 has also emphasised the importance of integrated management and sustainable development of coastal areas, exploitation of marine resources on a sustainable basis, protection of marine environment and study of the effect of the marine environment on climate.

Several national, sub-regional, regional and global initiatives launched in this regard have revealed, that the limitations posed by the conventional means of information gathering with respect to space and time, have been among the major constrains in achieving these goals. On the other hand, several scientific studies carried out using satellite data for addressing various issues related to coastal zone have proved to be extremely useful in mapping, detecting, quantifying and monitoring of these features. This is mainly because satellite remote sensing has the advantage of providing multispectral, synoptic information over large areas including inaccessible regions on a repetitive basis, which are the essential requirements for studying the coastal zone on an integrated basis. Taking the benefits from this, several scientific studies have been carried out under MARSIS programme using satellite data, mainly obtained from Indian Remote Sensing Satellites (IRS-1A/1B). These proved to be of immense use in identification and monitoring of tidal wetlands, mangrove forests, salt marshes, tidal flats, coastal land forms, assessment of potential aquaculture sites, shoreline changes, dynamics of erosion/accretion, coastal landuse/cover and inshore aspects like suspended sediment dynamics, coastal currents, near-shore bathymetry, etc. (Working Group Report, 1996). The important coastal zone studies carried out so far in India using satellite data are given in Table 3.1.

The advantages of using Geographic Information System (GIS) for integration of various thematic information derived from satellite data with other collateral data such as socio-economic and cultural data has established its immense capability in arriving at an integrated coastal zone management practices. Such an approach helps in initiating an integrated system for controlling development and other human activities that affect the condition of economic resources and the quality of environment in coastal zones. Efforts are on going to assess the areas suitable for brackishwater aquaculture and development of integrated coastal zone model through GSI coupled with remote sensing for selected regions of the coast in the country towards better planning and management of coastal zones.

The recent availability of high resolution data from second generation Indian Remote Sensing Satellite, IRS-1C & ID, which have improved spatial resolution, extended spectral bands, stereo-viewing and faster re-visit capability will further facilitate the coastal zone studies in the coming days. The major application programmes envisaged in the coming years using the data from these high resolution satellites will be to study different components of coastal processes namely prediction of shoreline changes on 1:12,500 scale every year, studies on suspended sediment and estuarine dynamics. It is also proposed to take up a modelling activity for understanding the coastal dynamics, particularly in the areas which are of highly vulnerable in nature along both east and west coast of India. Considering many significant changes occurring in the island ecosystem during the last few decades, a programme on ecosystem management to develop and evolve island ecosystem management plan has also been drawn-up.

Remote Sensing Applications in Marine Fisheries

Recognising that ocean features such as water colour, turbidity, sea state, flotsam and jetsam, wave size and direction, wind patterns and temperature distribution over the sea surface have a direct influence on the fishing strategies, efforts have been made to utilise satellite based remote sensing approach for identification and exploitation of potential fishing grounds to maximise the catch. Techniques have been developed to retrieve environmental parameters which are observable through data available from various space-borne sensors. Remote sensing satellites with their capability to cover large areas on a repetitive basis are particularly useful for nations like India having a long coastline and extensive exclusive economic zones, in mapping of valuable fishery resources and judiciously harvesting them to gain substantial economic benefit on a sustainable basis.

Identification of potential fisheries involves an understanding of the oceanic processes and interaction of the hydrobiological parameters. Fluctuation in any of the environmental conditions over the oceans affects the distribution, abundance and availability.of fisheries. While it is not possible to measure the entire spectrum of information needed to asses such changes in the marine environment, basic information on important environmental factors and processes can be retrieved by using ocean surface measurements from satellite sensors for locating marine fisheries. Towards this, techniques have been developed and operationalised for ocean colour sensing (measurement of chlorophyll pigments) and Sea Surface Temperature (SST) retrieval using data from various sensors, for identification of Potential Fishery Zones (PFZ).

Presently, in the absence of operational Ocean Colour data, efforts are being made to provide operational satellite derived fishery forecast based only on sea surface temperature gradients derived from NOAA/AVHRR data. The important thermal gradients selected for providing likely availability of fishes include thermal fronts, eddies, meandering patterns, upwelling zones, current boundaries, etc. The validation results have indicated substantial increase in the catch per unit effort in the PFZ areas as compared to non-PFZ areas.

Considering the limitations in the present fishery forecast, which is based only on SST gradients, efforts are on-going to provide integrated fishery forecasting with the scenario of availability of new sensors optimised for deriving various other ocean features in the coming years. Currently, efforts are on-going for validation of the IRS-P3 MOS (Modular Optoelectronic Scanner) sensor performance and for development of technique for ocean colour retrieval, jointly with DLR, Germany and Department of Space. Although, IRS-P3 MOS data, which was launched in March 1996, may not be useful for operational applications in our efforts to improve PFZ advisories due to its 24 day repetivity, it is providing an opportunity to our scientific community to develop techniques and algorithms for operational retrieval of chlorophyll from the planned Ocean Colour Monitor (OCM) on-board IRS-P4 (Oceansat-1). The envisaged integrated fishery forecast is expected to provide the information on the likely availability of fishes more accurately than the presently based forecast on sea surface temperature alone.

Applications of Remote Sensing in studies related to Ocean Dynamics:

Oceans and atmosphere form two important components of the climate system. The time scales-of oceanic phenomena range from a few hours and these are influenced mainly by the ocean's upper layers. The satellite systems operating at polar and geostationary orbits, are the most suitable for recording the surface and upper layer features of the global oceans on time scales of several times a day, to once in two weeks. Complementary conventional systems, e.g., ships and buoys, will provide detailed temporal and in-depth coverage at a selected few points, whereas satellite observations provide a synoptic view of an area. These combined data sets can be utilised effectively in understanding how the ocean shapes global conditions and also proper evaluation of the potential benefits of the ocean resources. Considering this, several ocean remote sensing projects were carried out in the country under MARSIS programme for retrieval of various ocean dynamic parameters using meteorological and microwave satellites data (NOAA, ERS-1 & 2, Topex-Poseidon, Geosat, etc.).

Ocean Surface Wind Stress

Wind stress is the major driving force for ocean currents and waves, and knowledge of the temporal and spatial variability of wind stress is essential in order to construct accurate predictive models of ocean behaviour and sea state. Scatterometers are used to measure near-surface winds over the oceans. Although radar altimeters and microwave radiometers have been used to measure wind speed, scatterometers are unique in that they allow measurements of vector winds (both direction and speed). The radar scatterometer on board Seasat obtained the first global measurements of winds over the ocean. In three days, the scatterometer obtained about 35,000 wind measurements providing global coverage. This amounts to more than 100 times the number of wind observations available from ships and buoys during the same period. In India, mean monthly maps of wind vector fields over 10 × 10 grids over Indian Oceanic regions for August 1992 to July 1993 have been generated and published in the form of Atlas (ISRO-SAC-SP-76-95). The maps are useful for air-sea interaction, ocean circulation studies, besides other oceanographic applications such as, storm surges, upwelling, fisheries, shipping and biogeochemical and transport studies.

Methodologies have also been developed to retrieve information on mixed layer depth, heat budget, latent heat flux and oceanic eddies utilizing satellite derived wind vector information along with other conventional data. These oceanic processes will help to understand the tropical atmospheric circulation, monsoonal forcing on oceanographic phenomena, air-sea interaction, biological activity, etc.

Ocean Waves

The surface waves are generated by surface winds and are responsible for the downward mixing of heat thereby increasing the thickness of the heat storage layer. The wave data obtained from in situ measurements and from buoys are quite meager and insufficient for any climatic study. The satellite derived information on wave spectrum (direction, speed and periods) through wave scatterometry, altimeter and SAR have been found to be feasible. The altimeter is capable of giving wave heights of the order of 20 cm.

In addition to wind scatterometers, altimeters provide useful information about sea state by measuring significant wave height and providing global data sets. The vast number of altimeter measurements over a period of time can be averaged and represented in the form of maps. Over the Indian ocean, numerical model for the forecast of off-shore surface waves upto 4 days in advance has been developed and is ready for semi-operational use (Working Group Report, 1996). This information is of great value from economic as well as safety point of view for off-shore industries, shipping companies and fishery community in planning their activities well in advance.

Ocean Circulation/Currents

Ocean circulation is controlled by winds, density structure, bottom topography and coastal boundaries. The ocean currents affect climate, biogeochemical cycles and the marine food web. Ocean currents can be detected from their influence on ocean topography. The microwave altimeter carried by SEASAT during 1978 was capable of giving data on the elevation of sea surface with an accuracy of 10 cm allowing determination of topography and roughness of sea surface. Another important satellite sensor is the microwave scatterometer designed to measure wind speed to an accuracy of 2 m/s and wind direction of 20 deg. Heat transport measurements also provide an indirect means to derive circulaüon patterns. This is possible since the main heat transport mechanism in the ocean the quasi-geographic large scale current system, which extend to a depth of upto 1000 m. Infrared radiometry is capable of giving data on the thermal front, cold wall of streams etc., which are indirectly correlated with the strength of the streams/currents.

The usefulness of satellite altimeter for measuring sea level and the ocean topography has been demonstrated by a series of altimeters of increasing accuracy and precision flown on Skylab, Geos-3, Seasat and Geosat. Data from SEASAT have allowed oceanographers to map the surface topography of the ocean and the oceanic geoid in unprecedented detail. The launch of Topex/ Poseidon mission in 1992 is providing substantial contributions to our understanding of global ocean dynamics by making precise and accurate observations of the oceanic topography using radar altimeters on a well traced satellite. The European Remote Sensing Satellites (ERS 1 & 2) which were launched in 1991 and 1995, both have altimeters and scatterometer to measure the wind speed and direction, sea level and ocean topography. Using these data, demonstration of 2-D sea surface topography and sea level variability has been carried out over the Arabian Sea and Bay of Bengal (Working Group Report, 1996).

Sea Surface Temperature

Sea Surface Temperature (SST) forms an important parameter of the oceans. Knowledge of SST changes has been recognised as an important component of climate prediction models. Solar inputs, ocean mixed layer processes and wind speeds at ocean surface have direct bearing on SST variation. SST observations in different spatial and temporal scales is essential for the understanding of a variety of processes relating to the interaction between oceans and atmosphere. For examples, small changes in the ocean surface temperature in the tropical region are found to influence the weather in the mid-latitudes. The oceans exert considerable influence on the Indian monsoon and its dynamics which has been very well correlated with thermal regimes in the ocean waters. The impact of monsoonal flow on Indian oceanic waters causing extensive upwelling regions especially in the south west coasts has been well documented. Climate studies have shown that the potential areas of cyclogenesis and formation of weather systems is closely linked to the temperature patterns of the upper oceanic layers. Information on SST are also crucial for understanding the ocean dynamics related to mixed layer depth and circulation. The SST charts showing prevailing and changing ocean environment is found to be very useful towards identifying potential fishery zones. Currently, SST is being retrieved operationally using NOAA-AVHRR data with an accuracy of ±0.7 deg k (NRSA).

Sea Mounts Detection

Indian seas are very poorly surveyed for deep ocean bathymetry and topography due to large lead time required through conventional methods, besides being laborious and expensive tasks. However, charting of sea mounts are of significant importance as they disturb the flow of the ocean currents and large amplitude internal waves around them. Also, identification of sea mounts are crucial for sea mining, submarine navigation and deep ocean circulation studies. Sea mounts in the deep ocean are the ideal fishing grounds. Satellite altimeters has been found to be very advantageous in charting the seamounts. An advanced technique using satellite altimeter has been developed for detecting

Table 3.1 : Status of Remote Sensing based Coastal Zone Studies

Project	*Achievements*
Wetland/Landfrom	Maps showing mudflats, beach, mangroves and coastal zone on 1:250,000 and 1:50,000 scale for the entire country's coast line.
Shoreline Changes	Maps showing erosional/depositional areas on 1:250,000 and 1:50,000 scales for the entire country's coastline.
Coral Reef	Maps showing type of reefs and associated features on 1:50,000 scales of the major four coral gulf reef areas viz., Gulf of Kachchh, Lakshadweep, Bay & Gulf of Mannar, A & N.
Suspended Sediment	Qualitative sediment mapping of pre and post monsoon period for Kerala and Tamil Nadu Coast.
Coastal Regulation Zone (CRZ)	Maps showing wetland areas between high and low water lines and landuse features within 500 m. strip from high water lines on 1:25,000 scale.
Coastal Zone Information System (CZIS)	Pilot scale study mainly based on remote sensing inputs is under development for providing decision support for coastal zone planners.

charted sea-mounts in the Indian Ocean (Gairola, 1995). An atlas of newly discovered and verified seamounts has been prepared. Further deduction of seamounts planned to be taken up in the future plan using ERS and Topex-Poseidon satellites data.

Marine Pollution

Pollution of the marine environment is now generally recognised as a serious global problem, with the oceans becoming a dumping ground for waste products containing hydrocarbons, heavy metals, pesticides, heated waste water and several other pollutants from various industries. Data obtained by remote sensing using satellite have been found to be advantageous in ocean pollution monitoring. Since oil normally exhibits a lower temperature than water, thermal infrared scanners are useful in the surveillance of oil spills. Passive microwave radiometers are also useful for oil surveillance since oil film modify the surface roughness of water and alter the microwave signature. A more sophisticated and considerably more costly remote sensing of soil slicks is one which utilises Side Looking Airborne Radar (SLAR) flown in jet-aircraft. It has a day and night capability and can be taken in any kind of weather conditions. Studies have also shown that ocean waste and water quality can be assessed using remotely sensed data. However, it is only in rare cases that positive identification of a pollutant is possible by remote sensing, but it is an indispensable tool in detecting the area at which the polluted discharge reaches the body of water, where the discharge goes, and the general characteristics of its dispersion pattern.

Based on the experience and expertise gained in retrieval of many ocean dynamic parameters, a major effort has been initiated in the country aimed at an end-to-end demonstration and setting up of a long term facility to analyse and use satellite data for oceanographic, marine atmosphere and coastal environment research. Under this programme, a major thrust has been to improve the retrieval accuracy of ocean parameters, assimilation of oceanographic parameters into ocean weather services and meteorological applications, besides techniques development for many newer applications. Such an effort is ultimately aimed at providing services to the down stream users through Centre for Ocean Information Services (OIS). Satellite observation of winds, currents, surface temperature and ocean colour are also useful to support various on-going international programmes on global environmental applications such as World Ocean Circulation Experiment (WOCE), Tropical Ocean and Global Atmosphere (TOGA), Global Ocean Flux Study (GOFS), Global Ocean Observation System (GOOS), Committee on Earth Observation Satellites (CEOS) and Indian Ocean Experiment. Table 3.2 give details on status of presently achievable accuracy as well as the desirable long-term accuracy planned to be achievable in the near future.

DEVELOPMENT OF INDIGENOUS SATELLITES FOR OCEAN APPLICATIONS

With the encouraging results from the MARSIS and realising the importance of availability of satellite data on a continuous basis for practical applications, efforts are on-going to develop indigenous satellites for ocean applications. A brief descriptions of these satellites which are specifically tuned to meet the operational requirements of various oceanographic applications are outlined below :

OCEANSAT-1

The first in a series of Ocean satellites planned by India towards meeting the urgent requirements of the user community is IRS-P4 mission, known as Oceansat-1. It is planned to be launched by end-1998, coinciding with the International Ocean Year. This will carry two on-board payloads viz., an Ocean Colour Monitor (OCM) and a Multifrequency Scanning Microwave Radiometer (MSMR). These payloads are expected to meet some of the critical data requirements of oceanography and ocean meteorology and will then be a forerunner to a fully operational Satellite Programme.

The OCM will have 8 narrow spectral bands operating in visible and near-infrared (NIR) bands (402-885 nm) with a spatial resolution of 360 m and swath of 1420 kms. IRS-P4 OCM will provide highest spatial resolution compared to any other contemporary satellites in the international arena during this time frame. The MSMR with its all weather capability is configured to have measurements at eight channels (6.6 GHz V&H, 10.65 GHz V&H, 18 Ghz V&H and 21 GHz V&H) with an overall swath of 1380 kms. The spatial resolution will be 120, 75, 45 and 40 kms for the frequency bands of 6.6, 10.65, 18 and 21 GHz respectively. MSMR will also be in a way a unique sensor as no other passive microwave radiometer is operating in the civilian domain today and will be useful for study of both physical and meteorological parameters. The OCM will provide unique regional and global observation for measuring biological constituents such as chlorophyll pigments and assessment of primary productivity,

Table 3.2 : Status in Retrieval of Geophysical Parameters and Future Directions

Parameter	Range	Accuracy Present	Accuracy Desirable Long Term	Instrument
Surface Wind Speed	4 - 24 m/s	± 2 m/s or ± 10%	± 1 m/s	Scatterometer, Altimeter & Radiometer
Surface Wind Direction	360°	± 20°	± 10°	Scatterometer
Wave Height	1 - 20 m	± 0.5 m	± 0.3 m or ± 10 %	Altimeter
Wave Direction	360°	± 15°	± 10%	SAR
Wave Length	50 - 1000 m	± 20%	± 10 %	SAR
SST	-2 -35° C	± 0.5° K	± 0.3° K	Radiometer
Chlorophyll	0.1-1.9 mg/cm^2	± 25%	± 10 %	OCM
Sea Surface Topography	1 - 2 m	± 3-5 cm	± 1 - 2 cm	Altimeter
Surface Currents	0 - 200 cm/s	± 1 cm/s		Altimeter
Surface Salinity	0 - 40 ppt	—	± 0.1 ppt	Radiometer
Mixed Layer Depth	0 - 200 m	—	± 10 m	Altimeter, Scatterometer
Coastal Bathymetry	0 - 50 m	± 1 m	± 0.3 m	SAR

while MSMR will be useful for measuring sea surface temperature (with poor resolution of 120 kms), atmospheric water vapour, liquid water content and sea surface winds. Both of them will provide the data once in two days, which will be useful for operational ocean applications. The details on the specifications of OCM and MSMR are given in Tables 3.3 and 3.4.

OCEANSAT-2

The Oceansat-2 is planned for launch during the time frame of 2001, which will carry a complete set of oceanographic payloads viz., Ocean Colour Monitor, Scatterometer, Altimeter, Thermal infrared radiometer and Passive Microwave Radiometer. This mission is expected to provide a full range of oceanographic observations such as wind field, wave spectrum, sea surface topography, ocean surface currents, internal waves, sea surface temperature, cholorphyll pigments etc., which are useful for providing operational oceanographic related services to the users. The details of Oceansat-2 payloads are given below:

- *Ocean Colour Monitor (OCM) :* Technical specifications are same as OCM of Oceansat-1.
- *Thermal Infrared Monitor (TIR) :* The thermal infrared scanner provides an observation capability in three spectral bands covering water vapour obsorption band (5.7-7.1 m) and split thermal infrared wavelengths (10.5-11.5 m and 11.5- 12.5 m). TIR will cover a large swath of 1500 kms and a spatial resolution of about a kilometer.
- *Wind Scatterometer :* A 3 beam wind scatterometer has been proposed at Kuband (13.97 GHz) for monitoring surface wind speed and directions over oceans. Such a configuration will reduce the four fold ambiguity in wind direction to a maximum of two fold. The expected accuracy of wind retrieval will be 2m/sec (or 10%) in magnitude and 20 deg. in direction.
- *Radar Altimeter:* A radar altimeter in Ku-band (13.6 GHz) with nadir pointing has been planned to make precise measurements of echoes from ocean surface. It will have sophisticated onboard signal processor to estimate the shape and timing of received echo. The altitude precision will be of the order of 10 cms and

measure significant wave heights in the range of 1-20 m, which will have accuracy in the range of 50 cm (or 10%) in magnitude and 20 deg. in direction.

- *Passive Microwave Radiometer (PMR):* The main objective of this sensor will be to measure the atmospheric integrated water content (both vapour and liquid) in order to compute the most problematic part of trospospheric path delay radar altimeter's signals. PMR will operate in 3 channels respectively in 18,21 and 37 Ghz.

SATELLITES FOR COASTAL APPLICATIONS

Besides the two indigenous development of Ocean satellites, efforts are also ongoing to launch 2 more advanced satellites for land applications, which are useful for mapping and monitoring of various coastal zone features with several newer applications. They are:

CARTOSAT-1 : This satellite is expected to have improved technology in terms of sensor systems as well as application potentials. It will carry only one payload-PAN with 2.5 m spatial resolution and 30 km swath. It will have a fore-aft stereo capability. The order of 2.5 m spatial resolution data will cater to the specific needs of cartographers and terrain modelling. It would be possible to map elevation differences of better than 5 m. It is expected to further facilitate the coastal application studies with mapping capability of cadastral applications. The CARTOSAT-1 is slated for end-1999 launch.

RESOURCESAT-1

The Resourcesat-1 with improved state-of-the-art payloads optimised mainly for agricultural applications is planned to be launched by end-2000. This will carry 3 band multispectral LISS-IV camera as well as improved version of LISS-III and WiFS cameras to provide enhanced applications for agriculture such as multiple crop and species level discrimination and vegetation dynamics studies. The LISS-IV camera will have a spatial resolution of better than 6 m and a swath of about 25 km with across-track steerability for selected area monitoring; it is also expected to provide improved capability for coastal zone applications, particularly in assessing the conditions of fragile mangrove ecosystems.

AVAILABILITY OF INTERNATIONAL SATELLITES

A look at the international scenario reveals the fact that a number of experimental/operational satellites have been launched with sensors designed for ocean applications during the last two decades. The first operational oceanographic satellite was the Seasat in 1978 and it functioned for three months only. During this brief period of operation, it provided a volume of data over oceans hitherto not available and demonstrated the feasibility of measuring global surface winds, wave heights, sea surface topography and ocean currents. Followed by this, several specific ocean satellites viz., NIMBUS-7, NOAA, ERS-1 & 2, Topex/ Poseidon, JERS, IRS-P3, ADEOS, Seastar and TRMM have been launched by space faring nations. The ADEOS launched by Japan in Aug. 1996, although had many important oceanographic payloads, is ceased to function recently.

Many countries are preparing for launch of more ambitious ocean observation satellites which are of partly experimental, partly semi-operation to operation instruments, to monitor the ocean on a global scale during the next 15 -20 years. These include NOAA series, NPOESS series, ENVISAT, EOS colour, ADEOS-II & III, EOS AM & PM series, METOP -1 & 2, EOS Alt 1, 2 & 3, etc., which not only offers the scientific community better opportunities to understand the oceanic processes and its environment, but also provides a challenge to explore those areas which are not yet been attempted but will be within our capability with the arrival of those satellites. Besides, several high resolution satellites such as CRSS, Quick bird, ORBVIEW and IKONOS are planned. These satellites with capability to provide a spatial resolution of around 1 m at Panchromatic band and 4-10 m at multispectral bands, are expected to further the coastal zone studies. With the scenario of availability of data from the state-of-the-art technological.

Table 3.3 : Specifications of Ocean Colour Monitor (OCM)

S.N.	*Parameters*	*Specifications*
1.	Instantaneous geometric field of view	360m (at nadir, 720 km altitude)
2.	Swath	> 1420 km
3.	Spectral bands	B1 402-422 nm B2 433-453 nm B3 480-500 nm B4 500-520 nm B5 545-565 nm B6 660-680 nm B7 747-785 nm B8 845-885 nm
4.	Camera MTF	>0.2 (at nyquist frequency)
5.	Quantization	12 bits
6.	SNR	>512 (at maximum saturation radiance)

Table 3.4 : Major Specifications of Multi-frequency Scanning Microwave Radiometer (MSMR)

Frequency (GHz)	6.6	10.6	18	21
Polarisation	——V&H——			
Orbit inclination (0)	——98.25——			
Altitude (km)	——720——			
Antenna dia (cm)	——86——			
Incidence angle (°)	42.13			
Spatial Resolution (km)	120	75	45	40
Swath (km)	——1360——			
Temp. Resolution (°K)	——1.0——			
Dynamic Temp. range (°K)	——10-330——			
Data rate (kbps)	——5——			
DC Power (W)	——76——			
Weight	——65——			

CONCLUSION

The understanding of the ocean as how each of its components contribute to its overall functioning is important in providing the ideas for long-term management and sustainable development of ocean resources. This calls for detailed studies of the ocean dynamics and a proper evaluation of the exploitable potential benefits of marine resources, including the coastal zones. Towards this, satellite observations undertaken over the last one decade in the country have demonstrated to provide successful results, which are beginning to be used on an operational mode in the coming years. The availability of number of satellites along with our own indigenous ocean satellites in the coming years will radically change the ocean observations and further facilitate in increasing self-reliance on the oceanographic and coastal data collection, its analysis for prediction of ocean state, marine resources evaluation and coastal zone management.

ACKNOWLEDGEMENTS

MARSIS programme is an outcome of the efforts of large scientific community of both ISRO/DOS and user organisations with the funding support from Department of Space. We gratefully acknowledge the efforts put in by these colleagues in realising the objectives of MARSIS programme. Thanks are also due to Ms. Malani for secretarial assistance and Ms. M. Sameena for other scientific help extended while preparing the manuscript.

REFERENCES

Gairola, R. M., Sujit Basu and Pandey, P.C. (1995) Remote detection of seamounts in the Seas around India by Geosat Altimeter Data. ISRO-SAC-TR-103-95.

Gohil, B. S. and Pandey, P. C. (1995) An atlas of surface wind vectors in the Seas around India during August 1992 - July 1993 from ERS-1 Scatterometer. ISRO-SAC-SP-76-95, p. 77.

NRSA (1997) Measurement of sea surface temperature. MARSIS project, Oceanography Division, NRSA.

Working Group Report (1996) Coastal zone studies. Recommendation to Standing Committee on Oceanography.

Working Group Report (1996) Satellite parameter retrieval for Oceanography. Recommendation to Standing Committee on Oceanography, p. 79.

4

Solid Waste Disposal Site-Selection Using Remote Sensing Technology

B.A. KANCHAN GARG, K.S. ANANTHA MURTHY AND V.R. HEGDE*

INTRODUCTION

Industrialisation, urbanisation and consequent increase in population has led to an imbalance in the use of natural resources thus damaging the ecosystem. Perhaps it has been rightly said that "people are pollution" since increase in population would lead to resource crunch that would in turn lead to many problems like improper landuse, inadequate water supply and sanitary facilities and waste management particularly in urban centres. Available data indicates that many of the urban centres do not have proper management of their drainage and waste dumps (Harish Govind, 1995; Rajiv and Ambuj, 2000).

One such urban centre is Bellary City which lies between 14°13' and 15°50' N latitude and 75°40' and 77°11'E longitude (Fig. 4.1) covering an area of about 5035.06 ha. The city with a population of 2,45,758 (Census 1991) has been classified as Class-I city and is attracting many major industries particularly steel

* Indian Resources Informartion Management Technology, Bangalore-560 040.

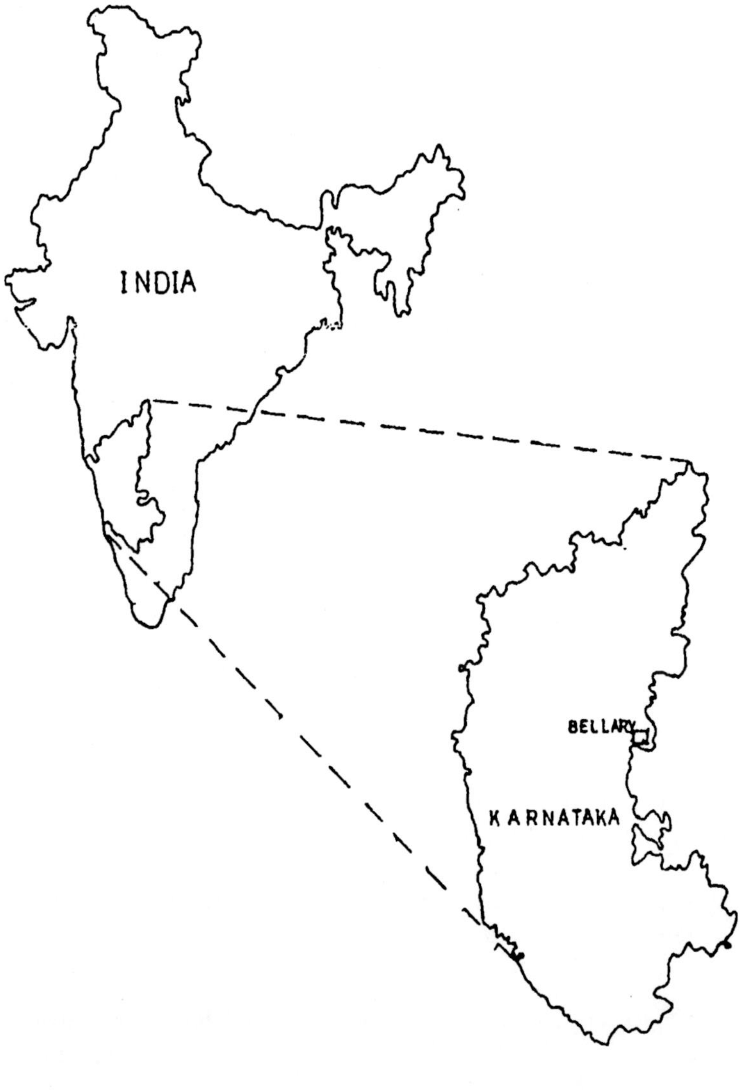

Fig. 4.1 : Location Map of Bellary City.

industries. Bellary city which is showing signs of accelerated growth since 1971 has not received any attention. Hence, the present study has been undertaken.

METHODOLOGY

An integrated approach has been adopted in the study. Population projection has been done using Arithmetical, Geometrical and Incremental increase methods (Rangawala, 1992). Visual interpretation techniques have been utilised to generate thematic maps from satellite data (IRS-IC (PAN); Scale: 1:12,500) and also to locate solid waste disposal site. The data of solid waste generated in the city has been collected from the Bellary City Municipal Council.

DEMOGRAPHY OF BELLARY CITY

Bellary city is a renowned centre for trade, commerce and industry. The city which had a population of just 34,953 in 1911 became an urban agglomeration with a population of 1,25,753 in 1971 and at present it is the seventh largest city in Karnataka State. From Table 4.1 (Fig. 4.2), it is evident that the city had a downward trend (39.99%) in population growth during 1901 and 1911. But it has been steadily increasing since 1921 onwards. During 1971-81, the increase has been more steep (61.03% decadal variation).

This increase is mainly due to improvement in trade and commerce, development of industries and other service facilities which have resulted in occupation, increased job opportunities and consequently accelerated migration.

The average growth rate of population is approximately 3 per cent and the population projected for Bellary city following Arithmetical increase. Geometrical increase and Incremental increase methods for the year 2000 A.D. and 2011 A.D. has been calculated and the average of these has been used in the study (Table 4.2, Fig.4.2).

The average growth rate of the city is slightly higher than the national growth rate of 2.5%. The density of population as per the Census is 235 persons per km^2. Which is lower than the all India density of 273 (excluding Assam and Jammu and Kashmir).

Table 4.1 : Population Figure of Bellary City from 1901 to 1991.

Sl. No.	*Censu Year*	*Population**	*Increase in Population For each Decade*	*Percentage Decadal Variation*	*Rate of Growth of Population*
1.	1901	58247	–	–	–
2.	1911	34956	–23291	–39.99	–3.99
3.	1921	39842	+4886	+13.95	1.39
4.	1931	47573	+7731	+19.41	1.94
5.	1941	56148	+8575	+18.02	1.80
6.	1951	70372	+14174	+25.25	2.52
7.	1961	85673	+15351	+21.83	2.18
8.	1971	125183	+39510	+46.12	4.61
9.	1981	201579	+76396	+61.03	6.10
10.	1991	245755	+44179	+21.92	2.19

* *Source:* Comprehensive development plan of BUDA.

Table 4.2 : Projected Population of Bellary City.

Method	*Population*	
	2001 AD	*2011 AD*
Arithmetical Increase Method	299119	352480
Geometric Increase Method	351433	502549
Incremental Increase Method	301453	359483
Average	317335	404837

Satellite data shows that the city is growing in both eastern and western parts.

The development pattern of the city indicates that the residential areas like Brucepet, Bangalore Road, Cowl Bazaar are thickly populated (Figs. 4.3 & 4.4). New residential layouts are coming up in the vacant lands around the older residential areas. Development of new layouts are taking place along the Hospet Road, Kappagal Road and Sangankal Road. There is no organised Central Business Area for the entire city and the business activity

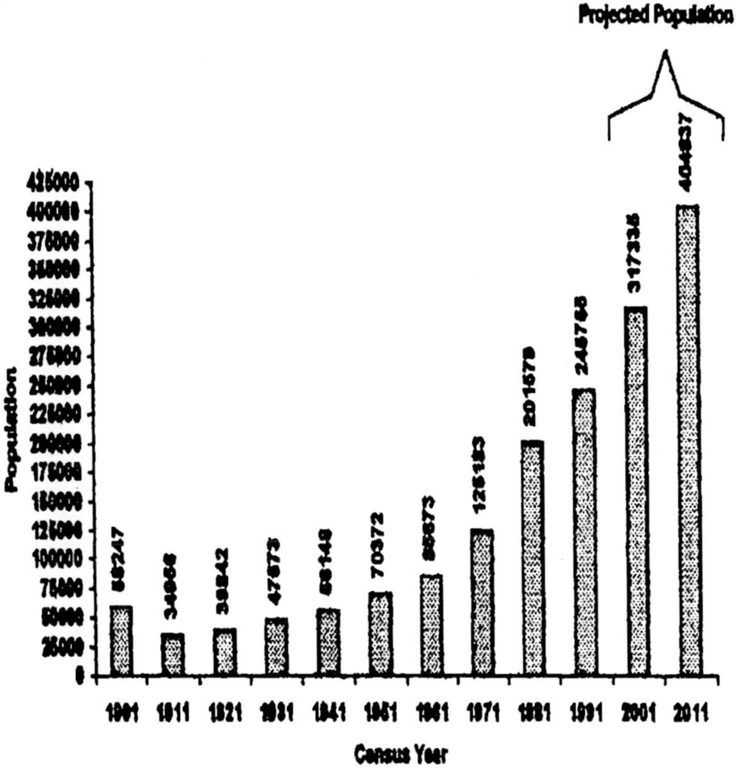

Fig. 4.2: Population Growth trend of Bellary City

is observed mainly around the bus stand, Bangalore Road, Brahmin street and Kalamma street within the Town survey limits of Bellary city.

LANDUSE PATTERN OF BELLARY CITY

The city situated in the semi-arid region, spreads over an area of 5035.06 ha. Based on the visual interpretation of IRS-IC (PAN) data (1:12,500 scale) and superimposement of the Town Survey limit from Comprehensive Development Plan (CDP) of Bellary Urban Development Authority (BUDA)(Fig. 4.3) on the interpreted map, the landuse of the city has been classified into 3 main categories.

- built-up land
- agricultural land and
- land under miscellaneous use (others)

The area under each category (present landuse) has been calculated (Table 4.3). The proposed landuse pattern of Bellary city as given by BUDA has been reclassified and drawn to 1:12,500 scale (Fig. 4.4, Table 4.4) and the area has been calculated. Comparison between the existing and proposed landuse indicates that there is an increase in the existing built-up land. The percentage of land in "others" category included in the CDP of BUDA is 13.50% which is less than that required during the town planning (15%).

The existing land utilisation rate is 6.33 ha. per 1000 population. The total land requirement would be 2008.73 ha. and 2562.16 ha. for the projected population of 317335 and 404837 by 2001 and 2011 A.D. respectively. Total agricultural land in the CDP has been reduced.

SOLID WASTE MANAGEMENT OF BELLARY CITY

Along with proper management of landuse, solid waste management also requires equal attention. The quantity of solid waste produced depends on various factors such as standard of living, occupation, habits of the contributing population, climate, industrial production and the market for waste materials (Bhide, 1972; Baldisimo, 1988). The city corporation had acquired 40 acres of land for the disposal of solid waste through composting and around 23 trenches of dimensions 16' X 32' X 8' were dug. Solid waste in the City was being disposed through composting till 1993. However, decrease in the market value for the compost has forced the Bellary City Municipal Council to dispose the solid waste being generated through open dumping and open incineration. The residents also dispose the solid waste wherever the open spaces are available. The quantity of solid waste collected by the Bellary City Municipal Council is around 18 to 20 tons/day. The present disposal site is located in the southeastern edge of the city near Rupanagudi (Fig. 4.3). The open dumping and open incineration of the waste is being carried out on the same

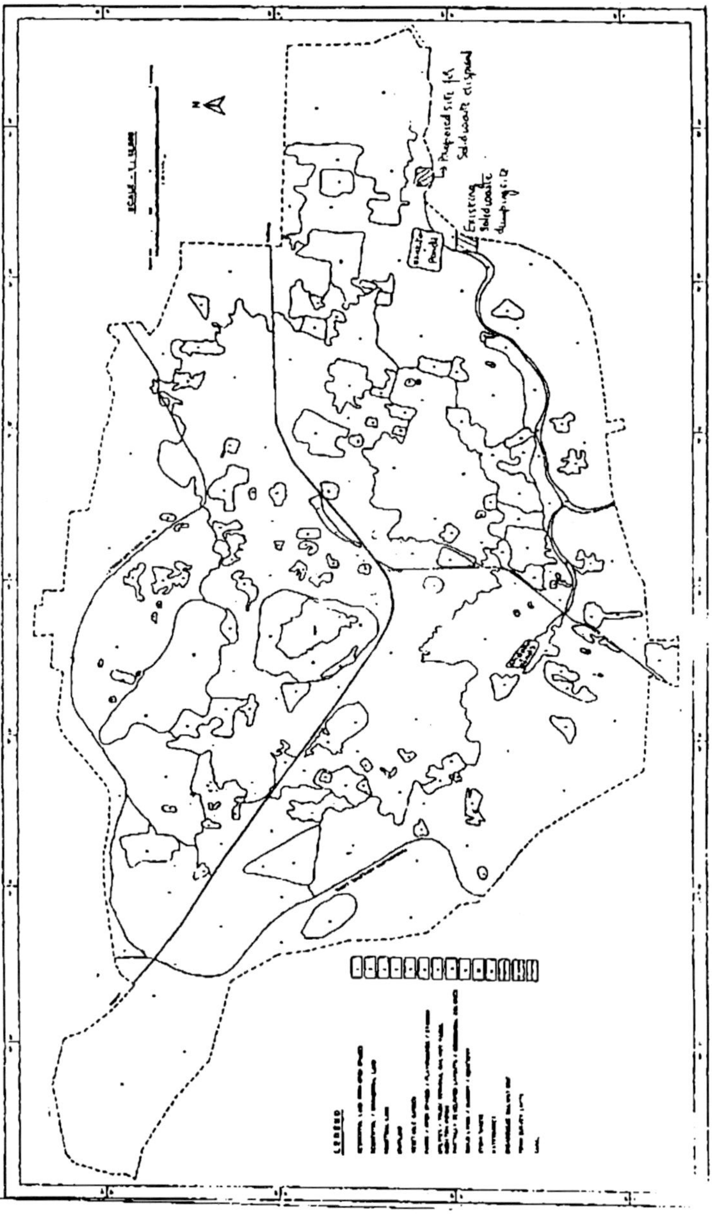

Fig. 4.3 : Existing Landuse Map of Bellary City

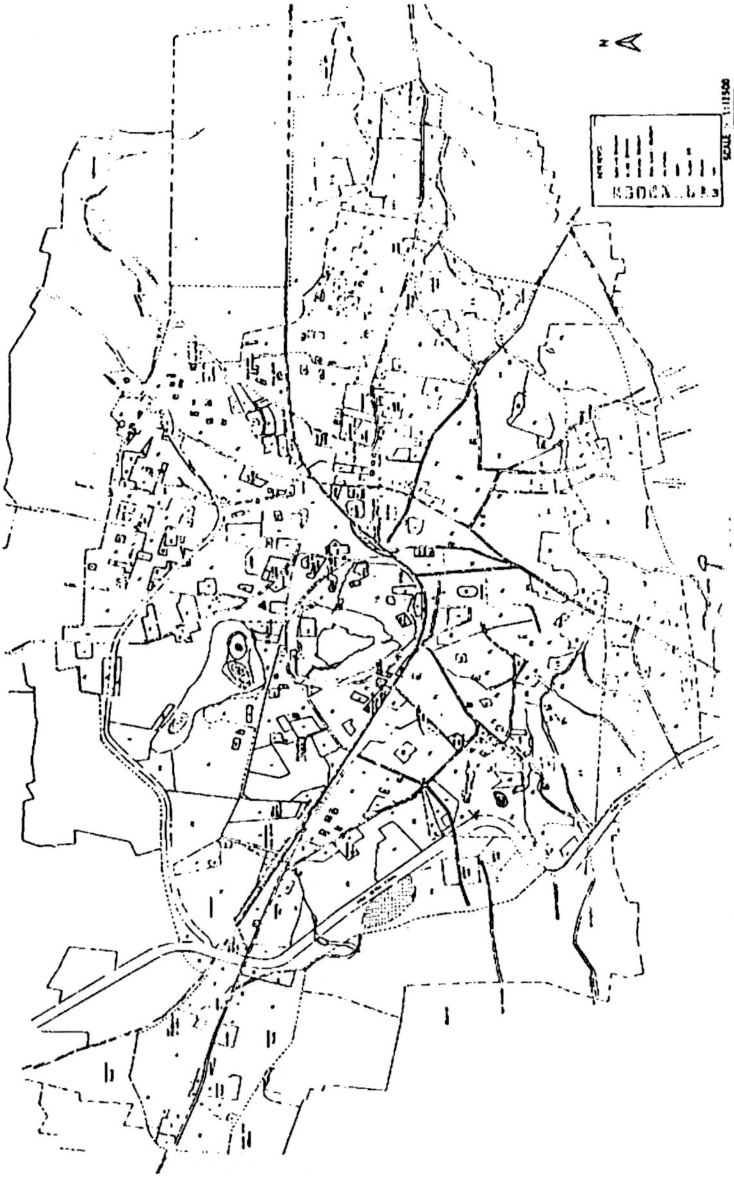

Fig. 4.4 : Comprehensive Development Plan of Bellary City (BUDA)

Table 4.3 : Proposed Landuse Analysis of Bellary City.

Sl. No.	*Landuse*	*Area in Hectares*	*Area in Per cent*
1.	Residential	1695.02	43.06
2.	Commercial	115.36	2.93
3.	Industrial	356.78	9.06
4.	Public and Semi-public	551.78	14.01
5.	Parks and open spaces	531.76	13.5
6.	Transport and Communication	668.59	17.44
7.	Vacant	-	-
	Developed areas	3936.79	100.00
8.	Water sheet	242.58	
9.	Agriculture	532.16	
	Total	4711.53	

* *Source:* Comprehensive Development Plan of BUDA.

composting site. These two methods of disposal of waste practiced by the Municipality has become a breeding ground for the mosquitoes, flies. This method of disposal is hazardous to health and leads to unaesthetic conditions in the surroundings. Moreover open incineration of solid waste leads to air pollution due to release of gases like carbon dioxide, sulphur dioxide, oxides of nitrogen and many other gases generated in the process which have ill effects on health of the citizens and the environment.

POTENTIAL USES OF RECYCLED SOLID WASTE

To overcome the difficultly in managing the solid waste, following strategies have been suggested.

1. The solid waste generated must be separated out into fractions such as biodegradable, non-biodegradable, combustible, metal scraps, plastics and glass at the source of generation.
2. The biodegradable solid waste has to be composted in the land available for composting (near Rupanagudi) and can be sold to the farmers as manure at a lower rate.
3. Closed incineration of non-biodegradable solid waste can be carried out and the heat generated during the process can be utilized for various other beneficial purposes.

Table 4.4 : Landuse Analysis of Bellary City (within the town survey limits)

Sl. No.	*Landuse*	*Area in Hectares*	*Area in Per cent*
1.	***Built up lands***		
	1.1 Residential with open spaces	856.9	17.02
	1.2 Residential/commercial	377.93	7.51
	1.3 Industrial	321.2	6.38
	Sub Total	1556.03	
2.	***Agricultural land***		
	2.1 Crop land	1809.21	35.93
	2.2 Vegetable garden	30.5	0.61
	Sub Total	1839.71	
3.	***Others***		
	3.1 Parks/open spaces/play ground/stadium	176.74	3.51
	3.2 Utilities/truck terminus/ railway yards/oxidation ponds	38.05	0.76
	3.3 Partially developed layouts/residential colonies	87.74	1.74
	3.4 Scrub lands/quarry and crematory	1103.61	21.92
	3.5 Story waste	205.45	4.08
	3.6 Water bodies	27.73	0.55
	Sub Total	1639.32	
	Grand Total	5035.06	

4. The combustible garbage can be used to produce fuel pellets since these pellets have a calorific value of around 3200 Cal/Kg and can be increased to 4000 Cal/Kg with the addition of binders and additives (Dinesh Mehta, 1995). This is one of the possibilities of using combustible garbage of Bellary city that can be looked into.
5. Metal scraps, plastics, papers and glasses can be sent to the respective industries since the substance which is a waste for one industry can be a raw material to another.
6. Venugopal (1997) revealed that NEERI and Kochi Zonal Laboratories have devised a new method for the manufacture of paper by using coconut husk. This could be adopted in Bellary City for the large quantities of coconut husk produced by the city for manufacturing pulp and paper on a small scale.

7. Conversion of solid waste to manure (biofertilizer) by using earth worms (Rajiv and Ambuj, 2000).
8. New disposal site has been selected using remote sensing technology. IRS-IC (PAN) data procured on 14th May, 1996 has been used to study the landuse/land cover of the city. Based on the field and laboratory studies, a new site has been identified on the south of Ananthapur Road (which is mostly on one side after a certain distance) (Fig. 4.3).
   (a) Disposal site should not be near the residential areas and
   (b) Since open incineration is carried out at the dumping sites, the wind direction must be considered as the gases liberated due to the decomposition of organic wastes (the monthly maximum temperature of the city is 34.03°C) and combustion/incineration of solid waste will be carried by the wind into the city thus affecting the health of the citizens. The wind direction in Bellary City varies in each season. It is from NW to SE during summer; NE to SE in winter and in rainy season, it is from West-East and SW-NW.

CONCLUSION

From the present study, it can be concluded that Bellary City does not have an effective waste management system. It can also be concluded that IRS-1C imageries (spatial resolution of 23.5m) are very powerful tools in the study of waste management. The effective management can be carried out by considering the concept of 5 R's - Remember, Reduce, Recycle, Reuse and Rethink.

ACKNOWLEDGEMENTS

The authors are highly grateful to Prof. S.G. Phene, Co-ordinator, Dept. of Environmental Science, Bangalore University, Bangalore for his encouragement throughout the course of this work.

REFERENCES

Baldimiso, J.M. (ed.) (1988). "Scavenging of Municipal Solid Waste in Bangkok, Jakarta and Manila". *Environmental Sanitation Review, December*, No. 26, Asian Institute of Technology, Bangkok.

Bhide, A.D. (1972). "Quality and Quantity of Refuse from Calcutta City". *Indian Jr. of Environmental Health*, Vol. 14 (1), pp. 81-82.

Dinesh Mehta (1995). "Urban Waste Management: Future Potents". *The Hindu Survey of the Environment*, pp. 65-69.

Harish Govind (1995). "Tiruvananthapuram: No longer Kerala's cleanest", *The Hindu Survey of The Environment*, pp 77-76.

Rangawala, S.C. (1992). *Water Supply and Sanitary Engineering (Environmental Engineering)*, 14th ed. Chorotar Publishing House, Anand, India.

Rajiv K. Sinha and Ambuj K. Sinha (2000). *Waste Management*, INA Shree Publishers, Jaipur, India.

Venugopal, P (1997). "Products from Waste". *Science Express*, Feb. 11, p. 4.

5

Carcinogens and DNA Damage

SULATHA DWARAKANATH*

INTRODUCTION

DNA damage can occur by oxidation, deamination, radiation, chemicals and others. In this article we will focus on the chemical damages caused by Polycyclic Aromatic Hydrocarbons (PAHs). PAHs are an ubiquitous group of environmental carcinogens, which are highly mutagenic and carcinogenic. They are widespread in the environment as by products of the incomplete combustion of fossil fuels and hence are present in automobile exhaust. They are also present in tobacco smoke, dyes and barbecued meat. These chemicals are first biochemically activated before getting attached to the cellular DNA in biological systems. Once the PAH binds to DNA (known as adducts), the damage may be corrected by repair mechanisms. The repair may be error-free or error-prone. If it is error-free no permanent damage is done but if it is error-prone, then a change in the genetic code will occur at the lesion site. Alternatively, if there is no repair, then mutations may occur during replication. The mutation caused and also whether the DNA is repaired is believed to be dictated by the effect the carcinogen has on the DNA structure. The effect of the slightly varied structure of stereoisomers seems to be very striking.

* Bayer Corporation Tarrytown, NY 10591.

Introduction to PAHs

Decades of works have shown that exposure to PAHs can cause genetic mutations and cancers. Hence, they constitute a serious health hazard to the human population and it is important to deduce the mechanism by which they cause mutations and ultimately tumors (Albert et al., 1977). B[a]P, B[c]P and 5-MeC (Fig. 5.1) have been extensively studied. B[a]P and 5-MeC have a "bay" region and B[c]P has a "fjord" region (Fig. 5.1). Formation of the epoxide in the bay region after activation leads to the production of a more highly reactive compound than when the epoxide is formed at other carbons. (Conney, 1982; Singer et al., 1983; Hecht et al., 1985; Jerina et al., 1986).

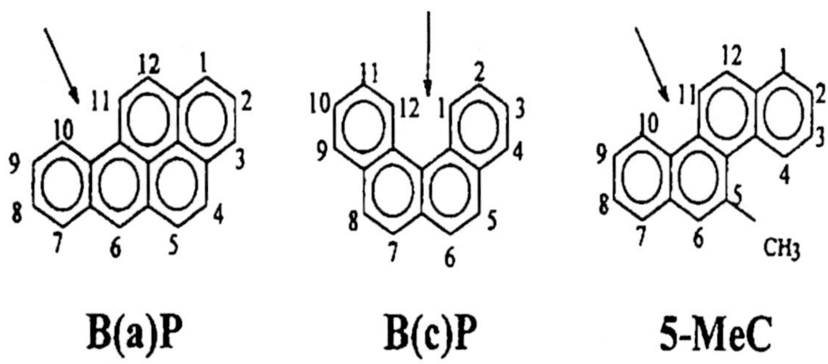

Fig. 5.1: Structures of the PAHs under investigation. The numbering of the carbon atom is shown. The arrows indicate the bay region in the case of B(a)P and 5-MeC and fjord region in the case of B(c)P.

Metabolic Activation of PAHs:

PAHs are chemically inert and are known as pre-carcinogens. The following steps are involved in the metabolic formation of the diol epoxides [reviewed in Miller (1978), Conney (1982) and Singer and Grunberger (1983)]:

(1) The angular benzo ring in the bay region is oxidized. PAH is oxidized by cytochrome P450 system using molecular oxygen,

in conjunction with cytochrome *c* reductase, that leads to the formation of a pair of enantiomeric arene oxides.

(2) Epoxide hydrolase hydrolyses the epoxide ring to form dihydrodiols.

(3) The dihyrodiols are converted to two diastereoisomeric forms i.e., *syn* and *anti*-dihyrodiol epoxides by a monooxygenase.

In the case of BP diol epoxides, the two diastereomers 7*r*,8*t*-dihydroxy-*t*9, 10-epoxy- and c9,10-epoxy-7, 8, 9,10-tetrahyrobenzo[a] pyrene (*anti*- and *syn*- BPDE, respectively) are optically resolved into (+)- and (-)- enantiomers. The absolute configuration of the two enantiomers of anti-BPDE are (+)-7(R), 8(S), 9(S), 10 (R) and (-)- 7(S), 8(R), 9(R), 10(S) [(+)- and (-)- *anti*-BPDE respectively] (Structures are shown in Fig. 5.2). In our studies, we have focused on the adducts derived from the *anti*-diol epoxide of BP, BPh and 5-MeC to guanine. The two diastereomers of BPh diol epoxide are anti- and syn-3, 4-dihydroxy-l, 2-epoxy-l, 2, 3, 4-tetrahydrobenzo[c] phenanthrene (*anti*- and *syn*- BPhDE respectively). The structures of the anti-BPhDE enantiomers (+)- 1(S), 2(R), 3(R), 4(S) and (-)-1(R), 2(S), 3(S), 4(R) are also shown in Fig. 5.2. Fig. 5.2 also depicts the enantiomers l(R), 2(S)-dihydroxy-3(S), 4(R)-epoxy-l, 2, 3, 4-tetrahydro-5- methylchrysene [(+)-*anti*-5-MeCDE)] and 1(S), 2(R)-dihydroxy-3(R), 4(S)-epoxy-l, 2, 3, 4-tetrahydro-5-methylchrysene (-)-*anti*-5-MeCDE).

These metabolically activated PAHs (*anti*- and *syn*-forms) react most commonly with the N^2 position of guanine or the N^6 position of adenine in native DNA (Fig. 5.3). The reaction mechanism is believed to involve the formation of a carbonation intermediate, which then reacts with N^2 or N^6 of guanine or adenine. The opening of this ring could be *trans* or *cis*. Therefore, each PAH diol epoxide enantiomer produces two adducts which have different absolute configuration at the exocyclic amino group-PAH linkage site. The *trans-anti* mirror image pairs of BPDE, 5-MeCDE and BPhDE are shown in Fig. 5.4. Cheng et al., (1989) have shown that in the case of the (+)-*anti*-BPDE enantiomer, >90% of (+)-trans-anti-BPDE-N2-dG adduct is formed, while in the case of (-)-anti-BPDE, about 63% of (-)-trans–BPDE, but greater proportion of -*cis-anti*-BPDE- N^2-dG (22%) and 15% (-)-*trans-anti*-BPDE-N^2-dA adducts were formed. Melikian et al., (1988) and Reardon et al.,

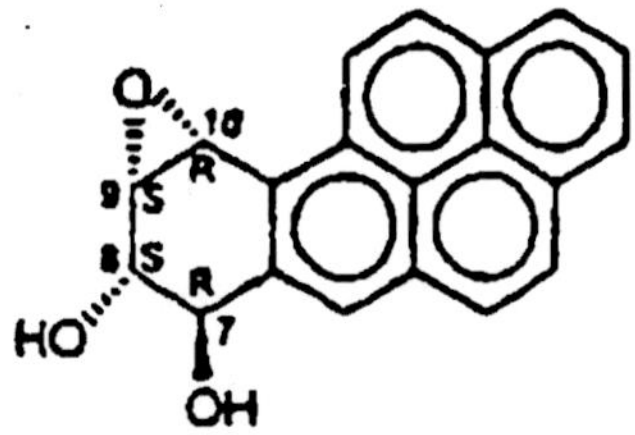

(+)-*anti*-BPDE

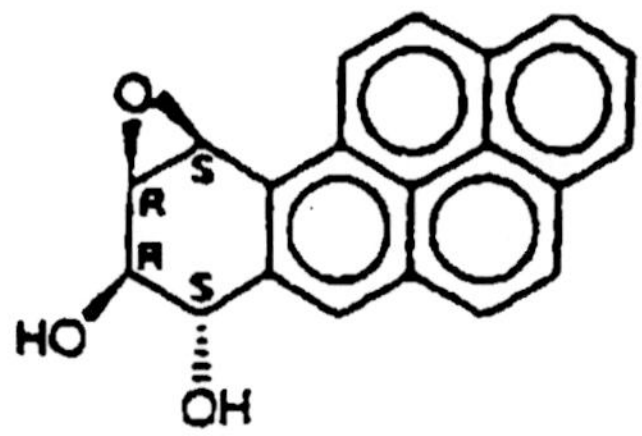

(-)-*anti*-BPDE

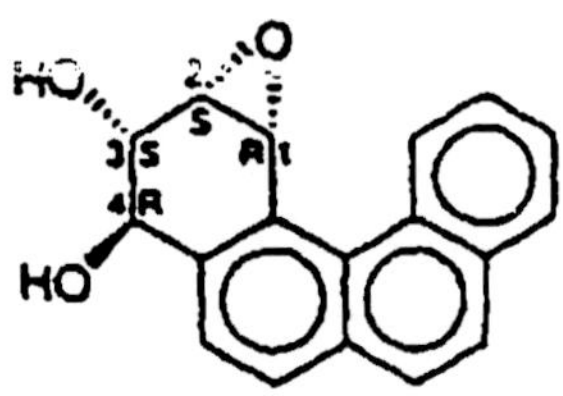

(-)-*anti*-BPhDE

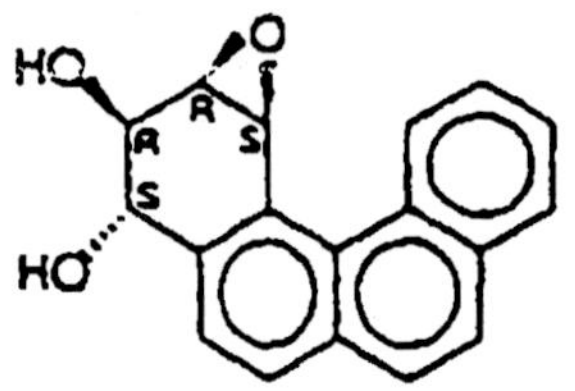

(+)-*anti*-BPhDE

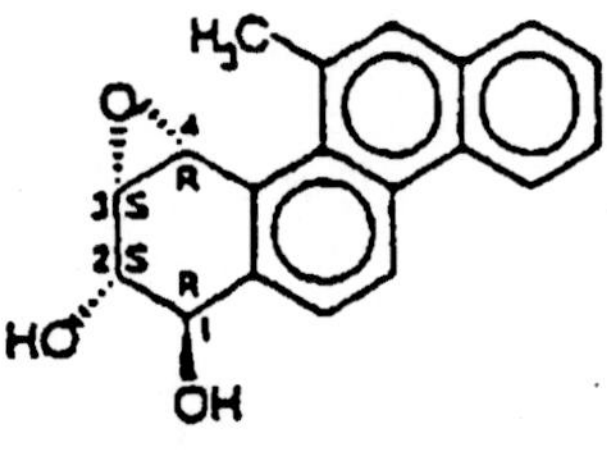

(+)-*anti*-5-MeCDE

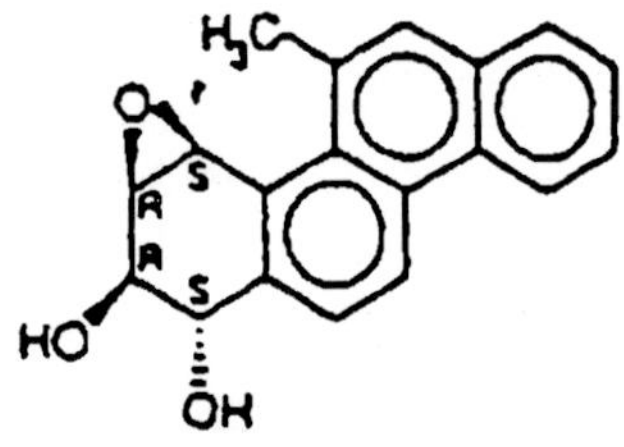

(-)-*anti*-5-MeCDE

Fig. 5.2 : Some examples ofEnantiomers of the PAH diol epoxides. Only the *anti* forms are shown here.

(1987) have shown that (+)- and (-)-*anti*-5-MeCDE also predominantly bind to the guanine exocyclic amino group by trans addition; smaller amounts of adenine adducts were also formed (factors of ~10 and ~4 of (+) and (-)-*anti*-5MeCDE, respectively) (Mellikian et al., 1988). BPhDE binds extensively to adenine in contrast to BPDE and 5-MeCDE (Dipple et al., 1989). Even though only small amounts of the guanine (+) and (-)-*anti*-BPhDE-N2-dG adducts are formed, they could be quite important because they are mutagenic (Wood et al., 1984; Bigger et al., 1989, 1992). In this article the focus will be 5-Methyl chrysene diol epoxide as

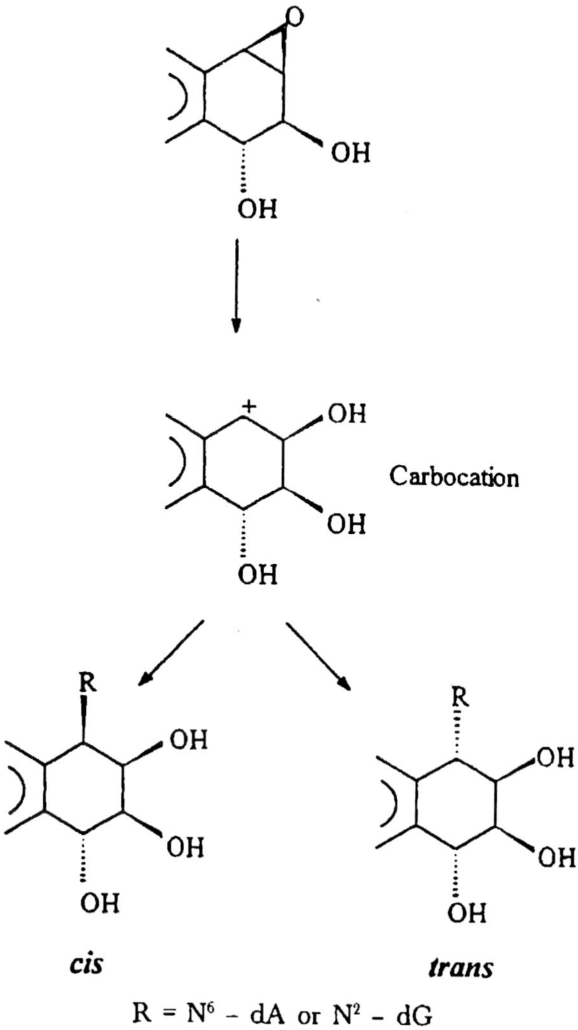

Fig. 5.3 : Opening of the epoxide ring on reaction with DNA

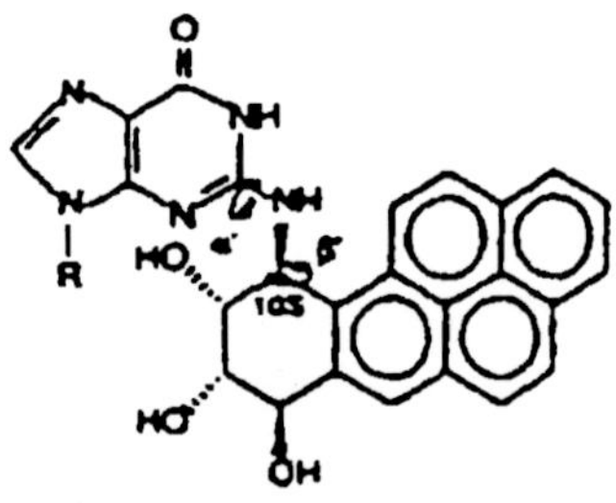

10*S* (+)-*trans-anti*-[BP]-N^2-dG

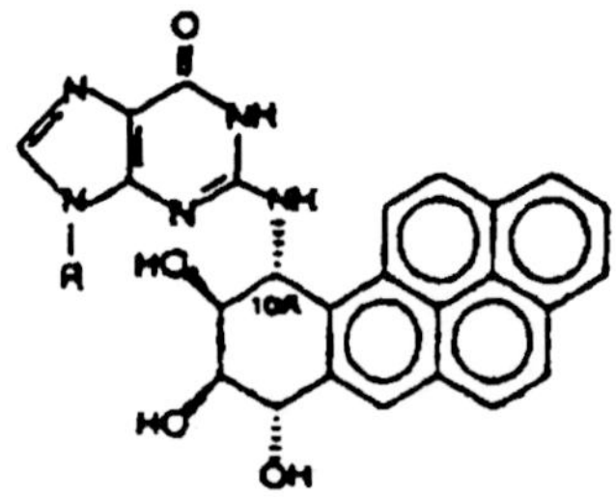

10*R* (-)-*trans-anti*-[BP]-N^2-dG

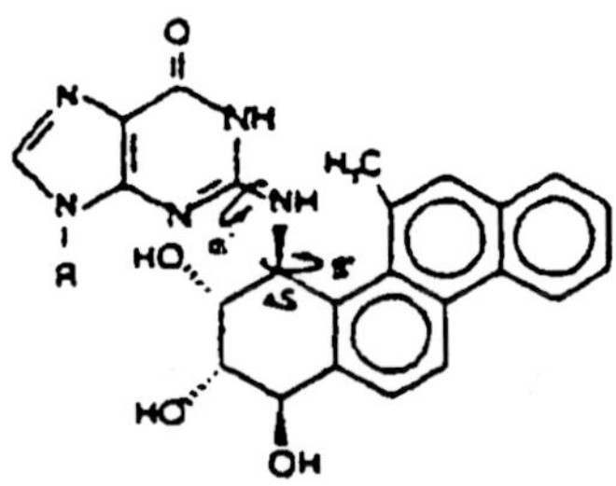

4*S* (+)-*trans-anti*-[MC]-N^2-dG

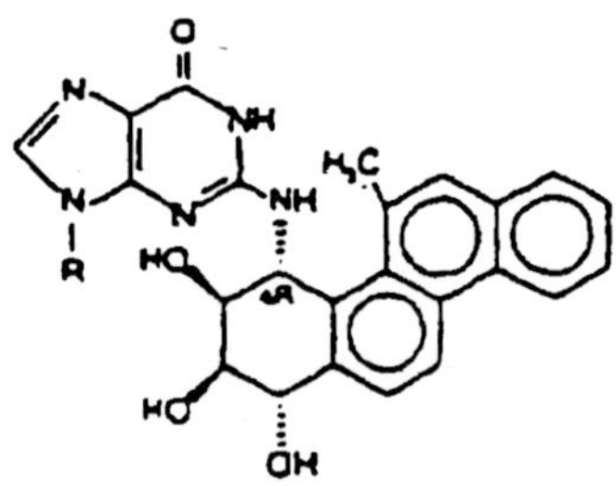

4*R* (-)-*trans-anti*-[MC]-N^2-dG

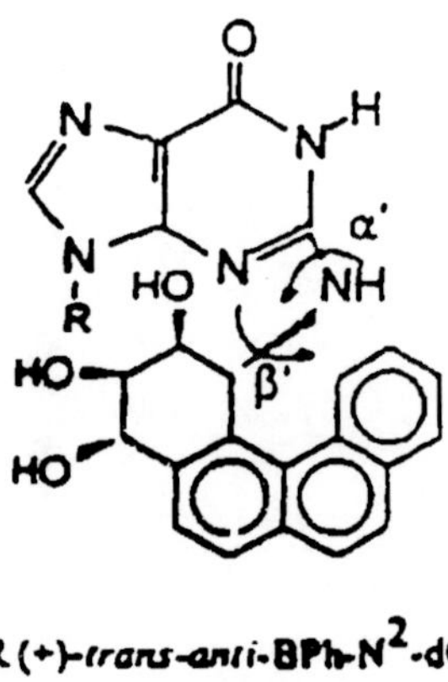

1 R (+)-*trans-anti*-BPh-N^2-dG

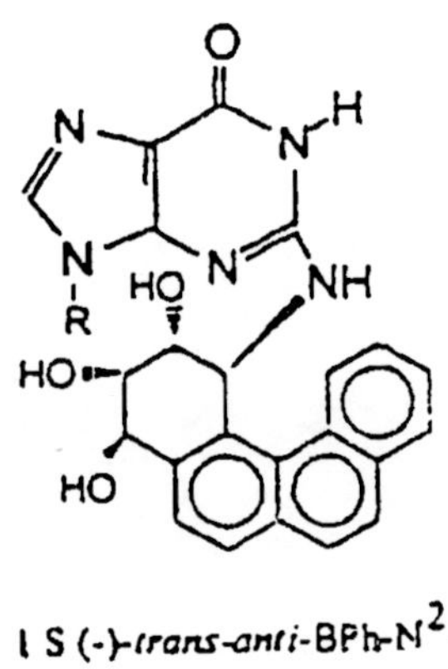

1 S (-)-*trans-anti*-BPh-N^2-dG

Fig. 5.4 : N^2-dG adducts of diol epoxides of interest. Only the *trans-anti* forms are shown.

an example and some of the work done on this PAH will be elucidated.

Background of 5-MeCDE and Significance

Chrysene is a weak carcinogen with two equivalent bay regions. Like other PAHs, chrysene is metabolically activated to highly reactive, mutagenic and tumorigenic diol epoxide metabolites (Norqvist et al., 1981; Hodgson et al., 1983). The presence of a methyl group at its 5-position (5-methychrysene or 5-MeC), strongly increases its tumor-initiating activity in mouse skin and new born mice lung and liver (Hecht et al., 1985, 1986; Amin et al., 1991; Bigger et al., 1990). Among the different diol epoxide stereoisomers produced by the oxidative metabolism of 5-MeC, only the stereoisomer (+)-5-MeCDE is strongly tumorigenic in mouse skin (Hecht et al., 1987) and mutagenic in *S. typhimurium* (Melikian et al., 1988). The molecular basis of the enhancement of the biological activity of chrysene by methyl substitution at the 5-position is poorly understood. This enhancement may be linked to a structural distortion of the bay region that results from the presence of the methyl group at the 5-position (Zacharias et al., 1984) since the isomeric 6-methyl derivative is biologically inactive (Melikian et al., 1988).

The covalent binding of diol epoxide molecules to DNA is believed to be the critical event in mutagenesis and in the initiation of the complex series of events in tumorigenesis. 5-MeCDE binds predominantly to the exocyclic amino group of guanine in DNA (Reardon et al., 1987; Melikian et al., 1984). The synthesis of oligonucleotides containing a single PAH diol epoxide-modified base at a defined position offers the possibility of carrying out detailed structural studies by NMR techniques and to correlate the structural characteristics of these adducts with their biochemical and biological activities.

Extensive studies have been carried out with (+)- and (-)-BPDE; it is well established that the (+)-enantiomer is highly tumorigenic, while the (-)-enantiomer is not (Buening et al., 1978; Slaga et al., 1979). The tumorigenicities of (+)- and (-)-enantiomers of 5-MeCDE show a similar dependence on stereochemistry (Hecht et al., 1987). The question whether DNA adducts derived from the binding of tumorigenic and mutagenic PAH diol epoxide stereoisomers differ in some systematic manner from those derived from their biologically less active enantiomer, is of significant.

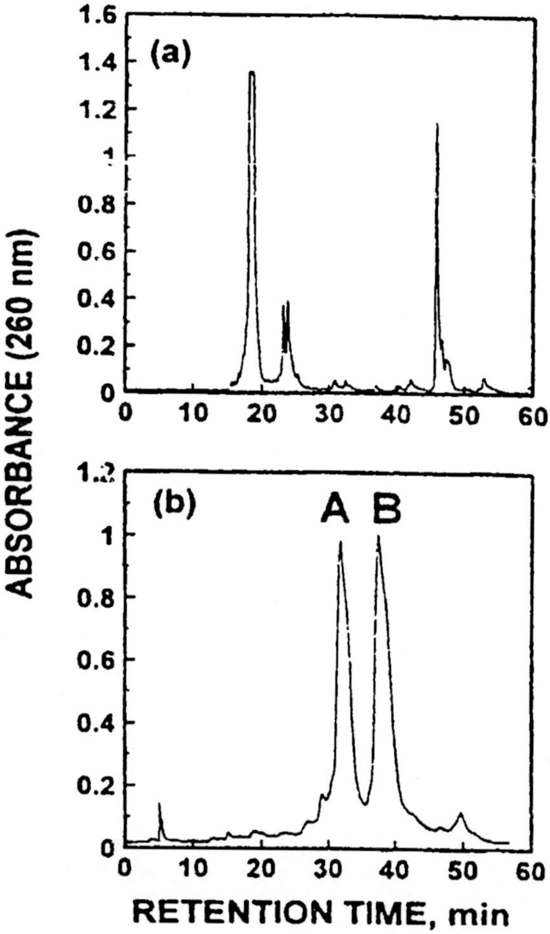

Fig. 5.5 : HPLC elution profile of the of the reaction mixture separation. (a) First separation of the reaction mixture (b) Second separation of adducts.

Both BPDE (Meehan et al., 1979; Cheng et al., 1989) and 5-MeCDE (Melikian et al., 1988) enantiomers bind predominantly to guanine residues by a trans-addition mechanism. The characteristics of site-specific BPDE-modified oligonucleotides (Cosman et al., 1992; de los Santos et al., 1992; Mao et al., 1993; Xu et al., 1995) have been studied earlier. The NMR structures of (+)- and (-)-5 MeCDE-N^2-dG-modifiedd oligonucleotide (Cosman et al., 1995; Cosman et al., 1997) and that of (+)- and (-)- BPDE-N^2-dG- modified oligonucleotides with the same sequence have been described (Cosman et al., 1992; de los Santos et al., 1992). The NMR solution structure of the (+)- 5-MeCDE adduct shows the pyrenyl moiety to be oriented towards the 5' side and that of the (-)-5-MeCDE adduct towards the 3' side of the modified strand. The orientational preferences are the same for the stereochemically similar (+)-and (-)-BPDE adducts in similar sequences.

The stereoselective resistance to enzymatic digestion of adducts derived from the binding of PAH diol epoxides (Mao et al., 1993; Stezowski et al., 1987; Dipple et al., 1987; Cheh et al., 1990; Osbome et al., 1993; Chary et al., 1996) and N-acetoxy-N-2-acetylaminofluorene (Fuchs et al., 1983) to DNA has been documented. It was shown previously by Mao et al., (1993) that modified oilgonucleotides derived by trans-addition of (+)- and (-)-BPDE to N^2-dG in oligonucleotides exhibit different degrees of resistance to digestion by exonucleases; this effect is related to different orientations of the bulky pyrenyl residues relative to the modified guanine residue and the 5'- and 3'- ends of the modified strand. The different orientation of the two adducts could produce different biological responses by enzymes such as DNA and RNA polymerases, repair enzymes and enzymes involved in recombination. The effect of different orientation in PAH diol epoxide -DNA adduct on DNA synthesis in vitro [Hruszkyewicz et. al., (1992); Shibutani et al., (1993); Chary et al., (1996); Choi et al., (1996); Li. B (1995) and in *vivo* Jelinsky et al., (1995); Moriya et al., (1997)]. There are different approaches to establishing the structure of DNA adducts. An extensively used method for double stranded DNA adduct is the NMR approach [Geacintov et al., (1997)]. We choose to use The goal of this work was to establish the orientation of the (+)- and (-)-5-MeCDE in a single stranded DNA adduct. SVPD digests only single stranded DNA from the

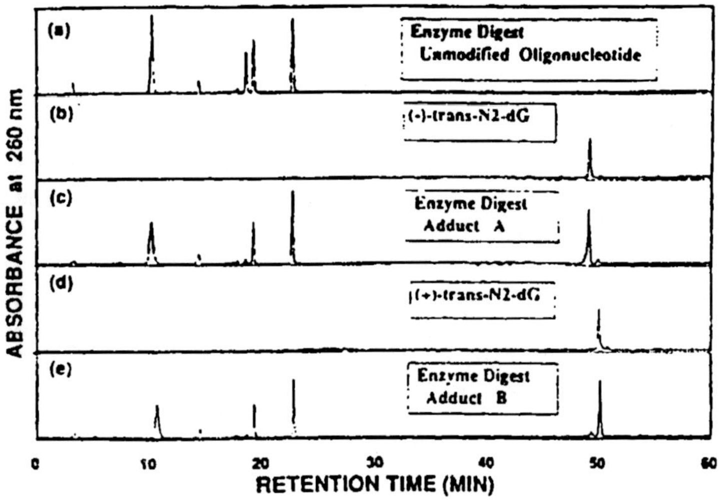

Fig. 5.6 : HPLC traces of the enzyme digested unmodified and 5-MeCDE- modified 11-mer oligonucleotide adducts compared to mononucleoside standards.

3' side. Using this property of the enzyme we characterized the stereoselective resistance to digestion by SVPD of adducts covalently modified with (+)- and (-)-5-MeCDE at the exocyclic amino group of the single guanine residue in the 11-mer d(CCATCGCTACC).

MATERIALS AND METHODS

Synthesis of site specific and stereospecific 11 mer adducts of anti-5-MeCDE.

The DNA sequence used for this purpose is as follows:

5'-CCATCGCTACC-3'

This sequence has only one guanine in the middle; thus it is easy to separate the guanine adducts formed from the reaction of 5-MeCDE with this oligonucleotide.

About 40 mg quantities of the 11-mer oligonucleotide were reacted in aqueous solution (~1.3 mg/ml) with racemic 5-MeCDE. The diol epoxide in tetrahydrofuran (1% triethylamine) solution was added step-wise over a two week period until a molar ratio of [5-MeCDE]/[strand] of 4:1 was reached. The reactions were carried out in total darkness at room temperature. The reacted oligonucleotides were separated by HPLC procedures employing a two-step approach using the Hitachi Instruments, Inc., Danbury, CT) interfaced with an IBM personal computer, and a 9.4×250 mm, 5 μm Shandon ODS- Hypersil preparative column (Keystone Scientific, Bellefonte, PA):

Step 1: Reverse phase HPLC with a 0 - 90% methanol/20 mM sodium phosphate buffer (pH = 7.0) in 60 min with a flow rate of 3.0 ml/min. This results in a separation of the modified from the unmodified oligonucleotides. The two major peaks between 23-25 min were collected separately and subject to Step 2.

Step 2: The two peaks were separated from one another using 16-18% methanol/20 mM sodium phosphate buffer gradient in 10 min followed by an 18-23% methanol/20 mM sodium phosphate gradient in 50 min with a flow rate of 3.0 ml/min.

The separated peaks were characterized by the following methods:

1. The individual oligonucleotide adducts were digested with snake venom phosphodiesterase to the 5-MeCDE-dG mononucleoside level.
2. The digestion mixtures were subjected to reverse phase HPLC analysis. In each of the two cases, the peaks due to unmodified G were missing, but new, later eluting peaks were observed which co-elute either with a (+)-*trans*- or a (-)-*trans*-5-MeCDE-N^2-dG adduct standard.
3. The UV absorption and CD characteristics of the 5-MeCDE adducts are in good accord with those of the (+)-*trans*- and (-)-*trans*-5 MeCDE-N^2-dG adduct standards.

Investigation of the efficiency of digestion of the adducts by SVPD assay.

The covalently modified N^2-dG adducts of (+) and (-)-5-MeCDE 11 mer (CCATCGCTACC) were subjected to snake venom phoshodiesterase (SVPD) digestion. SVPD digests single stranded DNA molecules from the 3' end. For BPDE N^2-dG 18 mer adducts it has been shown that the relative efficiencies of digestion differ markedly, depending on the adduct stereochemistry, in agreement with differing orientations relative to the 3' or 5' start direction (Mao et al., 1993). Different orientations of the two adducts could also produce different biological responses by enzymes such as DNA and RNA polymerases, repair enzymes and enzymes involved in recombination.

The (+) and (-)-*trans* -*anti*-5-MeCDE adducts and unmodified 11 mer were subjected to SVPD digestion at room temperature. 0.16 units of SVPD, 200ng of unlabeled sample and 60 ul of reaction buffer (0.11 M tris and 15mM $MgCl_2$, pH=8.9) were added. Unlabelled oligonucleotides were added to maintain identical concentrations, to ensure reproducibility of the results in different experiments. Controls were added to 100mM EDTA before the addition of the enzyme. The reaction was terminated by the addition of equal amounts of the sample to 100mM EDTA at 0.5min, 1 min, 2 min, 5 min and 15 min time intervals. The samples were subjected to electrophoresis on a denaturing 20% polyacrylmide gel with 7M urea and the bands were visualized by autoradiography.

RESULTS AND DISCUSSION

Synthesis of site specific and stereospecific 11 mer adducts of anti-5-MeCDE.

After reaction of the diol epoxide with the 11 mer DNA, adducts were separated by HPLC in two stages and modified undecamers were separated from the unmodified oligonucleotide and 5-MeCT hydrolysis product. The HPLC profiles are shown in Figs. 5.5a and 5.5b. In the first step (Fig. 5.5a), 5-MeCT, the hydrolysed product of 5-MeCDE and unmodified oligonucleotide elute out at ~46-47 min and 19 min respectively. Each of the two major peaks eluting between 23-25 min, which contains the two major peaks were collected separately. This was subjected to second HPLC run. The elution profile shows (Fig. 5.5b) at 32 and 38 min two fractions labelled 'A' and 'B'. These 2 fractions were collected separately. It has been shown below that the fraction 'A' and 'B' are (-)- *trans*- and (+)-*trans*-5-MeCDE-modified 11-mers respectively.

The peaks after separation were digested by SVPD to the 5-MeCDE-nucleoside level. The HPLC elution profiles of the unmodified nucleotides and 5-MeCDE-nucleoside adducts are shown in Fig. 5.6. The elution profiles of the authentic (-)-*trans*-5- MeCDE-N^2-dG and (+)-trans-5 MeCDE-N_2-dG are shown in panel (b) and (c) respectively. When these profiles are compared to the SVPD digested profiles of the unmodified oligonucleotide and adducts [panel (a), (c) and (e)], we can clearly say that fraction 'A' is (-)-*trans*-5-MeCDE adduct and fraction 'B' is (+)-trans-5-MeCDE adduct. The UV absorption spectra and their CD spectra were compared with standard (+) and (-)-*trans-anti*-5MeCDE adducts and reported ones (Reardon et al., 1987) (Not shown in this publication), which also confirm that A' is (-)-trans-5-MeCDE adduct and fraction 'B' is (+)-*trans*-5-MeCDE adduct.

Investigation of the efficiencies of digestion of the adducts by SVPD assay

Based on the model proposed by Mao et al, to demonstrate the different orientation of PDE-N^2-dG adducts (Fig. 5.7), we followed the same approach to demonstrate the orientations of the chrysenyl moieties in the oligonucleotides.

It was shown previously by Mao et al., (1993) that modified oligonucleotides derived by trans-addition of (+)- and (-)-*anti*-BPDE to N^2-dG exhibit different degrees of resistance to digestion by exonuclease. Some typical enzyme digestion patterns of the (+)- and (-)- 5-anti-MeCDE 11 mer adducts after 0.5, 1, 2, 5, 15 min of digestion at 37°C were seen. The autoradiograph is not shown in this article. A series of 5-6 bands, including bands due to intact 11-mers were observed in both cases. Consistent with the observations of Mao et al., (1993), a 5 min digestion period of unmodified 11 mer gives rise to the expected 11 bands. The 5-MeCDE-modified oligonucleotides are partially resistant to enzyme digestion until 5min but complete digestion can be obtained by 15 min.

The opposite orientation of different stereoisomers is thought to be of importance in the manner in which the enzymes (DNA and RNA polymerase, repair enzymes etc.) recognize the adducts. The differences in biological properties such as mutagenicity of

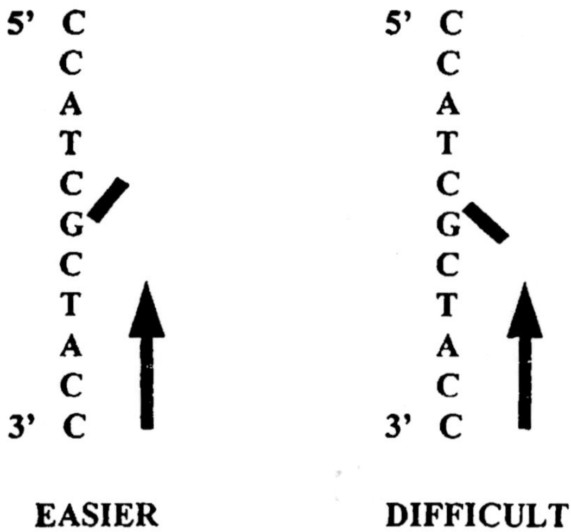

Fig. 5.7 : Model describing the relative ease of SVPD digestion of (+)-*trans* and (-)-*trans*-5-MeCDE - oligonucleotide adducts.

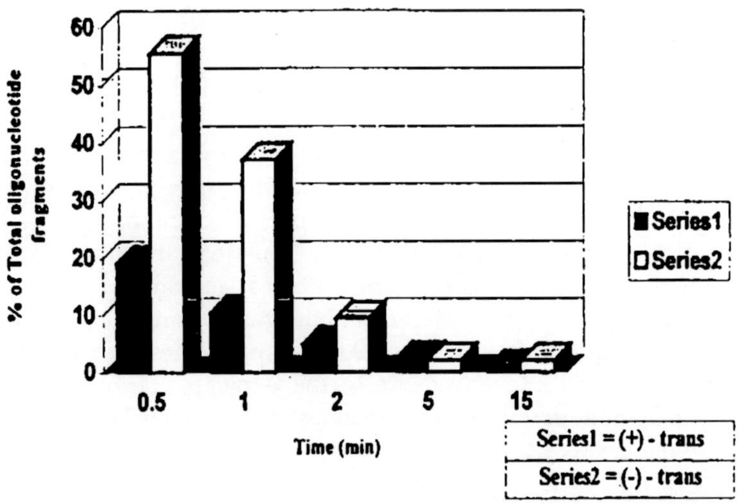

Fig. 5.8 : Rate of digestion of (+) and (-)- *trans*-SMeCDE adducts by SVPD.

these adducts may be the result of different enzymatic treatments. Upon analysis of the rate of digestion (Fig. 5.8) of the oligonucleotide fragment with the stall site at its 3'- end, it is evident that the enzyme digests past the (+)-trans lesion site more easily than past the (-)-*trans* adduct. The differences in the rates of digestion of the two adducts suggests, according to the model of Mao et al., (1993), that the chrysenyl residue in the (+)-*trans* adduct is tilted towards the 5'-direction, while in the {-)-*trans* adduct it points into the 3'-direction (Fig. 5.7). This conclusion, applicable to single-stranded (+)- and (-)-5-MeCDE-modified oligonuleotide, is supported by the NMR structural studies of Cosman et al., (1995, 1997) for the stereochemically identical 5-MeCDE-N2-dG lesions in duplex d(CCATCG*CTACC).d(GGTAGCGATGG), as indicated earlier. In the duplex, the (+)-5MeCDE residue is tilted in the 5'- direction and (-)-5-MeCDE residue is tilled towards the 3'- direction relative to the modified G residue, as observed for the same adducts by SVPD digestion in the single-stranded oligonucleotide.

A similar pattern of stereoselective resistance to enzyme digestion of oligonucleotides with (+)- and (-)-*anti*-BPDE-modified guanine residues was observed by Mao et al., (1993). Just as in the case of (+)-*trans-anti*-5-MeCDE-N^2-dG lesions, the ploycyclic aromatic residues in the (+)-*trans-anti*-BP DE-N^2-dG lesions in the single-stranded oligonucleotides were found to be tilted towards the 5 '-end of the modified strand. In the case of the (-)-trans-anti -BPDE-N^2-dG lesions, as in the case of the (-)-*trans-anti*-5-MeCDE-N^2-dG lesions, the polycyclic aromatic ring system is directed towards the 3'- end. In the case of the two *trans-anti*-BPDE -N^2-dG adducts, these adduct orientations relative to the modified guanine residues in single-stranded residues (Mao et al., 1993) are also observed in the duplexes (Cosman et al., 1992; de los Santos et al., 1992).

REFERENCES

Albert, R. E.; Bums. F. J. (1977) Carcinogenic atmospheric pollutants and the nature of low level risks. In origins of human cancer (Hiatt, H. H., Watson, J. D. and Winston, J. A., Eds.) Cold spring harbour laboratory press, New York, 289-292.

Amin, S; Mishra, B; Braley, J and Hecht, S. S. (1991). Comparative tumorigenicity in newborn mice of chrysene and 5-alkylchrysene-l, 2-diol-3, 4-epoxide. Cancer Lett. 58, 115-118.

Bigger, C. A. H.; St. John, J.; Yagi, H.; Jerina, D. M.; Dipple, A. (1992) Mutagenic specificities of four stereoisomeric benzo[c]phenanthrene dihydrodiol epoxides. Proc. Natl. Acad. Sci. U.S.A 89, 368-2372.

Bigger, C. A. H.; Strandberg, J.; Yagi, H.; Jarine, D. M.; Dipple, A. (1989) Mutagenic specificity of a potent carcinogen, benzo[c]phenanthrene (4R, 3S)-dihydrodiol (2S, IR)- epoxide, which reacts with adenine and guanine in DNA. Proc. Natl. Acad. Sci. U.S.A 86, 2291-2295.

Bigger, C. A. H; Flickinger, D.J; Strandberg, J; Pataki, J; Harvey, R.G and Dipple, A. (1990). Mutational specificity of the anti 1, 2-dihydrodiol 3, 4-epoxide of 5-methylchrysene. Carcinogenesis, 11, 2263-2265.

Buening, M. K, Wislocki. P. G, Levin. W, Yagi, H, Thakker, D. R, Akagi, H, Koreeda, M; Jerina, D. M and Conney, A. H. (1978). Tumorigenicity of the optical enantiomers of the diastereomeric benzo(a)pyrene 7, 8-diol-9-10-epoxides in new born mice: exceptional activity of (+)-7β, 8α-dihydroxy-9α, 10α-epoxy-7, 8, 9, 10-tetrahydro-benzo(a)-pyrene. Proc. Natl. Acad. Sci.USA. 75, 5358-5361.

Chary, P. and Lloyd, R. S. (1996) Impact of the stereochemistry of benzo[a]pyrene 7, 8-dihydrodiol 9, 10-epoxide-deoxyadenosine adducts on resistance to digestion by phosphodiesterases I and II and *trans*lesion synthesis with HIV-1 reverse *trans*criptase. Chem. Res. Toxicol. 9, 409-417.

Cheh, A. M; Yagi, H and Jerina, D. M. (1990). Stereoselective release of polycyclic aromatic hydrocarbon-deoxyadenosine adducts from DNA by the ^{32}P-postlabelling and deoxyribonucleasei/snake venom phosphodiesterase digestion methods. Chem. Res. Toxicol. 3, 545-550.

Cheng, S. C.; Hilton, B. D.; Roman, J. M.; Dipple, A. (1989). DNA adducts from carcinogenic and noncarcinogenic enantiomers of benzo[a]pyrene dihydrodiol epoxide. Chem. Res. Toxicol. 2, 334-340.

Choi, D.-J., Marino-Alessandri, D.J., Geacintov, Nicholas E., and Scicchtano, D.A. (1994) Site-specific benzo[a]pyrene diol epoxide-DNA adducts inhibit *trans*cription elongation by 'bacteriophage T7 RNA polymerase. *Biochemistry* 33, 780-787.

Conney, A. H. (1982) Induction of microsomal enzymes by foreign chemicals and carcinogenesis by polycyclic aromatic hydrocarbons: G. H.A. Clowes memorial lecture. Cancer Res. 42, 4875-4917.

Cosman, M.; de los Santos, C.; Fiala, R.; Hingerty, B. E.; Singh, S. B.; Ibanez, V.; Margulis, L. A.; Live, D.; Geacintov, N. E.; Broyde, S.; Patel, D. J. (1992) Solution conformation of the major adduct between the carcinogen (+)-anti-benzo[a]pyrene diol epoxide and DNA. Proc. Natl. Aca. Sci. USA 89, 1914-1918.

Cosman, M.; Xu, R.; Hingerty, B. E.; Amin, S., Harvey; R. G., Geacintov, N. E.; Broyde, S. and Patel, D. J. (1995) Solution conformation of the

(-)-*trans-anti*-[MC]dC adduct opposite dC in a DNA duplex: DNA bending associated with wedging of the methyl group of 5-methylchrysene to the 3'-side of the modification site. Biochemistry 34, 6247-6260.

Cosman, M.; Xu, R.; Hingerty, B. E.; Amin, S.; Harvey, R. G.; Geacintov, N. E.; Broyde, S. and Patel, D. J. (1997). Comparison of the solution structure of (+)- and (-)-*trans-anti*-5- MeCDE DNA duplex adduct. Proc. Amen. Assoc. for Cancer Res. 38, Abstract no: 2226.

de los Santos, C.; Cosman, M.; Hingerty, B. E.; Ibanez, V.; Margulis, L. A.; Geacintov, N. E.; Broyde, S.; Patel, D. J. (1992) Influence of benzo[a]pyrene diol epoxide chirality on solution conformations of DNA ccnvalent adducts: the (-)-*trans-anti*-[BP]G.C adduct structure and comparisons with the (+)-*trans-anti*-[BP]G.C enantiomer. Biochemistry 31, 5245-5252.

Dipple, A.; Pigott, M. A. (1987). Resistance of 7, 12-dimethyl-benzo(a) anthrance-deoxyadenosme adducts in DNA to hydrolysis by snake venom phosphodiesterase. Carcinogenesis, 8, 491-493.

Dipple, A.; Pigott, M. A.; Agarwal, S. K.; Yagi, H.; Sayer, J. M.; Jerina, D. M. (1989) Optically active Benzo[c]phenanthrene diol epoxides bind extensively to adenine in DNA. Nature 327, 535-536.

Fuchs, R. P. P. (1983). DNA binding spectrum of the carcinogen N-acetoxy-N-2- acetylaminofluorene significantly differs from the mutational spectrum. J. Mol. Biol. 177, 173-180.

Geacintov, N. E.; Cosman, M; Hingerty, B. E; Amin, S; Broyde, S.; Patel, D.J. (1997) NMR solution structures of Stereoisomeric covalent polycyclic aromatic carcinogen-DNA adducts: Principles, patterns, and diversity. Chem. Res. Toxicol. 10, 2, 111-146.

Hodgson, R. M.; Gary, P. D.; Grover, P. L.; Sims, P. (1983) Metabolic activation of chrysene by hamster embryo cells: evidence for the formation of a 'bay-region' diol epoxide-N2-Guanine adduct in RNA. Carcinogenesis 4, 1153-1158.

Hecht, S. S.; Amin, S.; Huie, K.; Melikian, A. A.; Harvey, R. G. (1987) Enhancing effect of a bay region methyl group on tumorigenicity in newborn mice and mouse skin of enantiomeric bay region diol epoxides formed stereoselectively form methylchrysenes in mouse epidermis. Cancer Res. 47 : 5310-5315.

Hecht, S. S.; Radock, L.; Amin, S.; Huie, K.; Melikian, A. A.; Hoffmann, D.; Pataki, J.; Harvey, R. G. (1985) Tumorigenicity of 5-methychrysene dihydrodiols and diolepoxides in newborn mice and on mouse skin. Cancer Res. 45, 1449-1452.

Hruszkewycz, A.M., Canella, K.A., Peltonen, K. Kotrappa, L. and Dipple, A. (1992) DNA polymerase action on benzo[a]py-rene-DNA adducts. Carcinogenesis 13, 2347-2352.

Jelinsky, S.A., Liu, T., Geacintov, N.E., and Loechler, E.L. (1995) The major N^2-adduct of the (-)-*anti*-benzo [a]pyrene diol epoxide is capable of inducing G -> A and G -> C, in addition to G -> T mutations. Biochemistry 34, 13545-13553.

Jerina, D. M.; Sayer, J. M.; Agarwal, S. K.; Yagi, H.; Levin, W.; Wood, A. W.; Conney A. H.; Preuss-Schwartz, D.; Baird, W. M.; Pigott, M. A.; Dipple, A. (1986) Reactivity and tumorigenicity of bay-region diol epoxide derived from polycyclic aromatic hydrocarbons. Adv. Exp. Med. Biol. 197, 11-30.

Li, B. (1995) Studies of Polynuclear Aromatic Diol Epoxide-DNA Adduct Structural-Biological Activity Relationships, Ph.D. Dissertation, New York University.

Mao, B.; Li, B.; Amin, S.; Cosman, M.; Geacintov, N. E. (1993) Opposite stereoselective resistance to digestion by phosphodiesterase I and II of benzo[a]pyrene diol epoxide-modified oligonucleotide adducts. Biochemistry 32, 11785-11793.

Meehan, T and Straub, K. (1979). Double stranded DNA stereoselectively binds benzo(a)pyrene diol epoxides. Nature, 277, 410-412.

Melikian, A. A.; Amin, S.; Huie, K.; Hecht, S. S.; Harvey, R. G. (1988) Reactivity with DNA bases and mutagenicity toward Salmonella typhimurium. Cancer Res. 48, 1781-1787.

Melikian, A. A.; Amin, S.; Hect, S. S.; Hoffman, D.; Pataki, J. and Harvey, R. G. (1984) Identification of the major adducts formed by reaction of 5-methychrysene anti-dihydrdiol-epoxides with DNA. Cancer Res. 44, 2524-2529.

Miller, E. C. (1978) Some current perspectives on chemical carcinogenesis in humans and experimental animals: Presidential address. Cancer Res. 38, 1479-1496.

Moriya, M., Spiegel, S., Femandes, A., Amin, S., Liu, T.-M., Geacintov, N.E., and Grollman, A.P. (1996) *Trans*lesional synthesis past benzo[a]pyrene diolepoxide-2'- deoxyguanosine DNA adducts: marked effects of host cell, sequence context, and chirality. Biochemistry, 51, 16646-16651.

Norqvist, M.; Thakker, D. R.; Vyas, K. P.; Levin, W.; Ryan, D. E.; Thomas, P. E.; Conney, A. H.; Jerina, D.M. (1981) Metabolism of chrysene and phenanthrene to bay-region diol epoxides by rat liver enzyme. Mol. Pharmacol. 19, 168-78.

Osbome, M; (1993). Investigation of the sequence specificity of carcinogen binding to DNA using exonucleases. Polycyclic Aromat. Compd, 3 (suppl.), 863-869.

Reardon, D. B.; Prakash, A. S.; Hilton, B. D.; Roman, J. M.; Pataki, J.; Harvey, R. G.; Dipple, A. (1987) Characterization of 5-methychrysene-l, 2-dihydrodiol-3, 4-epoxide-DNA adducts. Carcinogenesis 8, 1317-1322.

Salaga, T. J; Bracken, W. J; Gleason, G; Levin, W; Yagi, H; Jerina, D. M and Conney, A.H. (1979). Marked differences in the skin tumor-initiating activities of the optical enantiomers of the diastereomeric benzo(a)pyrene 7, 8-diol-9, 10-epoxides. Cancer Res. 39,67-71.

Shibutani, S., Margulis, L.A., Geacintov, N.E. and Grollman, A.P. (1993) *Trans*lesional synthesis on a DNA template containing a single stereoisomer of dG-(+)- or dG-(-)-*anti*-BPDE (7, 8-dihydroxy-*anti*-9, 10-epoxy-7, 8, 9, 10-tetrahydrobenzo[a]pyrene). *Biochemistry* 32, 7531-7541.

Singer, S.; Grunberger, D. (1983) Molecular biology of mutagens and carcinogens. Plenum Press. New York.

Stezowski, J. J; Loos-Guba, G; Schonwalder, K. H; Straub, A and Glusker, J. P. (1987). Preparation and characterization in solution of oligonucletides alkylated by activated polycyclic aromatic hydrocarbons. J.Biomol. Struct. Dynm. 5, 615-637.

Wood, A. W.; Chang, R. L.; Levin, W.; Thakker, D. R.; Yagi, H.; Sayer, J. M.; Jerina, D.M.; Conney, A. H. (1984) Mutagenicity of the enantiomers of the diasteomeric bay-region benzo(c)phenanthrene 3, 4-diol-1, 2-epoxides in bacterial and mammalian cells. Cancer Res. 44, 2320-2324.

Zacharias DE, Kashino S, Glusker JP, Harvey RG, Amin S, Hecht SS. (1984) The bay-region geometry of some 5-methylchrysenes: steric effects in 5, 6- and 5, 12-dimethylchrysenes. Carcinogenesis 5(11), 1421-30.

6

Biofilming as cause of Degradation of Bacteriological Quality of Piped Treated Water

SRINIVASA, P.C.[1], MAHESH, S.,[1] HALAPPA GOWDA, T.P.[2], VARALAKSHMI, H.D.[2], RAVISHANKAR, S.T.[2], KRISHNA NAND[2], VARADARAJ, M.C.[2] AND KESHAVA, N.[2]

INTRODUCTION

Biofilms comprise of matrix-enclosed bacterial population adherent to each other and to the surface of pipes. Biofilms are responsible for degradation of the quality of water in distribution system. The taste, odour and colour of water is affected leading to dissatisfaction of the consumers. Food processing and fermentation industries are also affected by quality of water. Biofilm formation leads to flow modification in linear systems and plugging of porous media in treatment plants.

The importance of microbial biofilm is relatively unknown because of several reasons:

1. Dept of Environmental Engineering, SJC Engineering, Mysore.
2. Dept of food Microbiology, Central Food Technological Research Institute, Mysore.

1. Microbiological quality itself is not given the importance in quality assessment as much as physical and chemical parameters.
2. WHO requires monitoring of microbial quality of source water once in a week. However, microbial analysis even once a year is very rarely done.
3. The intermittent variation of microbial quality would be apparent only if regular monitoring of the same were under taken.
4. The importance of microbial biofilming has been understood only in recent years with implication to human health.

Bacteriological deterioration of quality in water has a long history. In 1912, Gartner observed multiplication of bacteria on the bitumen coating of water reservoir. Similarly bacterial growth has been traced to organic materials like leather seals of water pumps, creosoted wood, hemp/cotton sealants as well as on PVC water pipes (Schoenen and Scholer,1985). Bacterial biofilm led to colonization of iron pipes with anaerobic sulphate reducing bacteria consequent corrosion; (Costerton et al., 1987).

The presence of invertebrates was also responsible for persistence of several pathogens in water; hence no animalcules algae or diatoms are tolerated since they could be carriers of disease-causing bacteria (WHO 1998). Persistence of bacteria in water due to attachment to plankton was comparatively less tolerant than biofilm to disinfection by chlorine.

The fluctuation of bacterial count in treated water makes an interesting study. Samples of tap water were analysed during period of 10 years for aerobic plate count at 37°C (APC 37) in Table 6.1. A wide fluctuation in APC37 count ranged between 3-25500 per ml of water.

Coliform bacteria are generally regarded as indicators of hygiene and efficacy of treatment. Coliform distribution in tap water (Table 6.2) ranged between 0.9-20,000 in 57 samples of tap water.

The presence of coliforms in water indicates either inadequate chlorination or post-treatment contamination; but in the case of well- maintained water distribution system, the presence of coliforms indicates some other cause. Fluctuation in the count of coliform with sporadic increase in number could also be from their proliferation in plumbing materials. (Burman and Colaboume 1976-1979 in Schoenen and Scholer, 1985) or in storage tanks Seidler et al., 1977).

EXPERIMENTAL

Biofilming as a mechanism for survival of indicator bacteria as well as pathogens in water was modeled with an experimental set up of PVC tubing of 20 mm dia and 5m long embedded with microslides of size 25×75×2 mm. One end of the prototype was connected to water tap (input end) and the other end was connected to a drain (output end). Water flow was maintained at a constant rate of 2 lit/min. The changes in deposition of bacteria on microslides over a period of 94 days is given in Table 6.3.

The bacterial biofilm was released from one side of the microslide over an area of 25 × 25 mm by swabbing with presterilized cotton swab and suspended in sterile phosphate buffer (pH 7.0). The serial dilutions of the bacterial suspension made in sterile saline were transferred into pour plate with plate count agar. Incubation at 37°C was for upto 72 h with daily record of the colony counts.

The gradual rise of bacterial number occurred with the period of observation which was also reflected in a rise in APC of output water. A gradual rise was evident in APC 37 count over a period of 94 days (Table 6.4). Biofilming in slow sand bed filter (Table 6.5) shows similar situation. Unless proper sanitization protocol is included, the sand filter itself would contribute to enhanced bacterial count in treated water (Ravishankar, 1993).

DISCUSSION

PVC, mild steel, cements are some of the materials used in construction of storage tanks and pipelines. Coupons made from these materials dipped into water containing biofilming bacterial suspension have been studied for biofilming characteristic of different bacteria. Structured mechanisms have been reported with interplay of polymeric substance excreted by several bacteria

Table 6.1 : APC 37 of tap water at campus at study area.

Sl. No.		*Ravishankar (1993)*	*Varalakshmi (1995)*	*Srinivasa (1998)*
1.	No. of samples	19	38	12
2.	Range	3-4736	400-25500	20-2300
3.	Mean	197	3254	800.8
4.	Standard deviation	1191	4689	1428
5.	Coefficient of Variance	197	144	79.48

Table 6.2 : Coliform count in tap water at campus at study area.

Sl. No.		*Raju (1990)*	*Ravishankar (1993)*	*Varalakshmi (1995)*
1.	No. of samples	26	19	38
2.	Range	0.9-20,000	0-4500	0-70
3.	Mean	26.4	352.5	6.28
4.	Standard deviation	8.905	181	13.05
5.	Coefficient of Variance	35.77	25.27	2.7

Table 6.3 : Growth variation in APC 37 of biofilm smear on glass slides from conduit system

Sl.No.	*Age (days)*	*CFU / cm^2 (in million)*
1.	11	1.3
2.	14	107
3.	19	400
4.	28	450
5.	40	850
6.	48	800*
7.	67	640*
8.	82	48*
9.	94	0.53*

* Algal growth reduces bacterial count

Table 6.4 : Variation in APC 37 count of input and output water in experimental biofilming conduit system.

Sl.No.	*Days*	*Input*	*Output*	*Ratio*
1.	1	20	150	7.5
2.	2	800	1600	2
3.	3	1200	4800	4
4.	5	500	4900	9.8
5.	7	200	2000	10
6.	9	30	220	7.3
7.	14	120	750	6.3
8.	18	140	250	1.8
9.	33	1000	3000	3
10.	40	2300	5200	2.2
11.	60	2100	3600	1.7
12.	70	1200	3100	2.6

(Costerton et al, 1987; LeChevallier et al, 1981; Ridgway and Olson 1981). Bulk water sampling does not necessarily expose the presence of adherent bacteria (Herson et al, 1981). It was further noticed that biofilming bacteria are less susceptible for disinfectant treatment (Nichols 1989; Gilbert 1987:1990). Engg et al, (1991) established that a range of antimicrobial agents at higher concentrations were required for controlling biofilming bacteria.

Suspended impurities like Granular activated carbon (GAC) from treatment area also contributed to biofilming mechanism. Lechevallier, (1981) found that monochloramine was superior to chlorine for disinfection. Examination of potential niches of water treatment in storage and distribution system needs to be monitored for eliminating control persistence of undesirable bacterial growth. Combination treatment of monochloramine and hypochlorite has been recommended, the first for penetratability and the second for bactericidal efficacy.

Table 6.5 : Bacteriological count of untreated and treated water using sand-charcoal-sand filter without exposure to sunlight.

Period	*sample*	*Aerobic plate count per ml*		*coutn per 100 ml*	
		37°C	*30°C*	*Coliforms*	*E.coli*
0	control	25	17	12	-
	expt	12	9	-	-
2	control	28	22	9	-
	expt	74	55	12	-
4	control	31	15	10	-
	expt	145	102	28	-
6	control	19	16	24	-
	expt	195	150	41	-
8	control	30	21	25	-
	expt	305	264	72	12
10	control	26	15	14	-
	expt	455	375	105	-
12	control	35	23	28	-
	expt	620	490	124	-
14	control	27	20	18	-
	expt	865	675	172	45
16	control	23	18	27	-
	expt	1025	840	190	52
18	control	27	19	22	10
	expt	1290	1120	5	75

Several pathogenic bacteria have been incrimented with water. Contaminated water acts as a significant vehicle for transmission of diseases through foods. Persistence and resistance of these bacteria to chlorine is shown in Table 6.6. If we have more information on the mechanism of biofilming in these bacteria preventive measures for controlling them could be developed. The status of non-culturable viable pathogens is also significant in this contact. The influence of biofilms on deterioration of treated water supplies needs wider study.

Table 6.6 : Water borne pathogens and their significance in water supplies-Bacteria (WHO, 1998)

Pathogens	*Health Significance*	*Main route of exposure a*	*Persistence in water supplies b*	*Resistance to chlorine C*	*Relative infective dose d*	*Important animal reservoir*
Campylobacterjejuni; C.coli	High	0	Moderate	Low	Moderate	Yes
Pathogenic E.coli	High	0	Moderate	Low	High	Yes
Salmonella typhi	High	0	Moderate	Low	High	No
Other Salmonella	High	0	Long	Low	High	Yes
Shigella spp	High	0	Short	Low	Moderate	No
Yersinia entero colitica	High	0	Long	Low	High	Yes
Vibro cholerae	High	0	Short	Low	High	No
Legionella spp	Moderate	C,IN	May multiply	Moderate	High	No
Pseudomonas aeruginosa	Moderate	I	May multiply	Moderate	High	No

Table 6.6 : *Contd.*

Pathogens	*Health Signific-ance*	*Main route of exposure a*	*Persistence in water supplies b*	*Resistance to chlorine C*	*Relative inf-ective dose d*	*Important animal reservoir*
Aeromonas spp	Moderate	O, C	May Multiply	High		No
Mycobacterium atypical	Moderate	I, C	May multiply	High		No

a
O=Oral ingestion; I=Inhalation in aerosol; C=Contact with skin; IN=Ingestion in immune-suppressed patients.
b
Detection period for infective stage in water at 20° C: Short = 1 week; Moderate= 1 week to 1 month; Long =More than one month.
c
When the infective dose is freely suspended in water treated at conventional of dose chlorine, Resistance moderate=Pathogen may not be completely destroyed; Resistance low=pathogen completely destroyed; Resistance high=Pathogen persists.
d
Dose required to cause infection in 50% of healthy adult volunteers.

RECOMMENDATIONS

The biofilming of water-related bacteria is important phenomenon for elucidation of bacterial persistence in the environment. Various factors responsible for adherent growth of bacteria need to be controlled. These include.

1. Flow rate, suspended impurities, organic nutrients, surface characteristics and plumbing material.
2. Control of adherent growth in treatment plants, storage tanks and distribution pipes require constant monitoring of bacterial types.
3. Sanitization protocol should include periodic flushing with efficacious combination of treatment chemicals.
4. Frequent measurement for disinfectant residual, turbidity and coliform occurrence.
5. Treatment effectivess for controlling microbial quality of water.
6. Analysis of heterotrophic bacterial population in water distribution system provides information on static water quality in reservoir and nutrient accumulation in pipe sediments that removed by pipe flushing

REFERENCES

1. Costerton, JW. Cheng, KJ. Geesey, GG. Ladd T. Nickle JC. Dasgupta M. Maerie, TJ. 1987. Bacterial biofilm in nature and disease. Annual Rev Microbiology. 41 : 435-464.
2. Engg, RHK. Padberg, FT. Smith, SM. Tan, EN. Cherubin, CE. 1991. Bacterial effect of antibiotics on slowly growing and non-growing bacteria. Antimicrob, Agents Chemother. 35 : 1824-28.
3. Gartner. Aug. Die Hygiene des Wassers. Friedr. Vieweg and Sohn braunscheig 1915.
4. Gartner, W. Uber Bakterienwachstum in Wasserreservoiren mit Innenschutzanstrichen. Journal fur Gasbeleuchtung und wasserversorgung 55 (1912) 907-908.
5. Gilbert, P. Collier PJ, Brown MRW. 1990 Influence of growth rate on susceptibility to antimicrobial agents: biofilms, cell cycle, dormancy, and stringent response. Antimicrob. Agents Chemother. 34 : 1856-68.
6. Herson, DS. Marshall, DR. Baker, KH and Victoreen, HT 1991 Association of microorganisms with surfaces in distribution systems. Journal AWWA. July 103-106.

7. LeChevallier, MW Babcock, TM and Lee, RG 1987 Examination and characterization of distribution system biofilms. Applied and Environmental Microbiology. 53 : 2714-2724.

8. Nichols, WW. 1989. Susceptibility of biofilms to toxic compounds. Structure and function of biofilms. New York. Wiley

9. Raju, NR. Studies on commercial domestic water filters and low cost filters. M. tech dissertation.

10. Ravishankar, ST. 1993 investigation on treatment of water using Sand-Charcoal-Sand filter columns with solar disinfection and UVWT system. M.Tech. dissertation.

11. Ridgway, HF. and Olson, BH 1981 Scanning electron microscopy evidence for bacterial colonization of drinking water distribution system. Applied and Environmental Microbiology. 41 : 274-278.

12. Schoenen, D and Scholer, HF. Drinking water materials field observation and methods of investigation. 1985 Ellis horwood Limited. ENGLAND.

13. Seidler, RJ., Morrow, JE., and Bagley, ST. 1977 Klebsielleae in Drinking water emanating from Redwood Tanks. Applied and Environmental Microbiology. 33 : 893-900.

14. Varalakshmi, HD. 1993. Investigation on bacterial tolerance to sunlight and ultraviolet rays in commercial UVWT system and Sand-Charcoal-Sand filter column. M. Tech dissertation. Mysore univ.

15. WHO, 1998. Guidelines for Drinking Water Quality Vol 2. Health criteria and Other Supporting Information 2nd ed. Who, Geneva pp. 10-11.

7

Evaluation of Biotreatment Options for Soil and Groundwater Contamination in Asia

JOSEPH PRINGLE*

INTRODUCTION

Since 1960, the pace of economic growth in Asia has been staggering. Although this has led to substantial gains in human welfare throughout the region, these gains have not come without costs. Rapid industrial expansion, urban population growth, and rising consumption have placed severe strains on the environment. Initially, Asian governments viewed environmental degradation as a necessary by-product of development. Over the past decade, however, the high costs of deteriorating environmental quality have become clear.

In response, governments in the region have begun to answer to the challenge. Most have recognized that environmental deterioration not only affects the quality of life of their citizens but also hinders prospects for further economic growth. To date, environmental regulators in the region have been preoccupied with tackling their most pressing tasks, such as drinking water and wastewater treatment, solid waste management, and air pollution

* National Pollution Prevention Round Table, Washington D.C. U.S.A.

control. This is understandable. Like many developing and rapidly industrializing countries, the health risks of unsafe drinking water and air pollution are greater than the risks of other environmental problems, such as hazardous waste contamination of soil and groundwater. The risks of adverse health effects due to chronic, low-level exposure to soil and groundwater contamination do exist, but these effects are difficult to demonstrate. Because of this, remediation of contaminated soil and groundwater is often viewed as a luxury that only wealthier nations can afford.

With time, this should change. Several factors may prompt industrializing countries in the region to address soil and groundwater contamination at an earlier stage of development than the OECD countries. First, the level of human exposure to industrial pollution tends to be very high in these countries. Rapid industrial expansion and poor planning has resulted in close intermingling of industry and residential areas. Surging urban population growth will exacerbate this problem. Second, governments are beginning to respond to rising public pressure to address deteriorating environmental quality. Third, the swift pace of economic growth that has contributed to Asia's environmental problems has also given them the financial resources necessary to reverse environmental deterioration.

OECD countries such as the United States, Germany, and Japan have amassed considerable expertise that can be tapped by those following in their footsteps. A better understanding of the human health risks posed by industrial pollution is inspiring more efficient regulatory regimes and more appropriate allocations of pollution control resources. Also, intense competition among many firms in the environmental industry is spawning more cost-effective pollution-control technologies. As these technologies and techniques enter the marketplace, the cost of achieving a given level of environmental quality is falling.

This is especially true of remediation technologies. Dozens of innovative technologies and techniques for remediating contaminated soil and groundwater have been developed over the past decade, delivering much more cost-effective treatment than traditional methods. Of these, biological remediation technologies hold particular promise. When applicable, they tend to be especially cost-effective, simple to use, and easy to maintain. Given the demands on scarce pollution control resources and the lack of environmental engineers in Asia, such technologies are desperately needed.

In order to speed the transfer of these technologies to the region, certain barriers must be eliminated. Most importantly, indigenous demand for remediation technologies and services must develop in these countries. Currently, the market for these goods and services in Asia, excluding Japan, is very small compared to those of OECD countries. This is due to the fact that regulatory activity and enforcement, which are almost solely responsible for driving these markets in OECD countries, are not well developed. In the few Asian countries that do have legislation addressing soil and groundwater pollution, monitoring and enforcement is lacking.

Another barrier to the transfer of biological remediation technologies stems from their relative youth. Buyers of remediation technologies often prefer traditional, proven technologies over less-expensive technologies that have yet to build a track record of successes. A few cases of inappropriate application have contributed to a perception that biological remediation technologies are risky.

Barriers to technology transfer also exist from the supply side. Suppliers of biological remediation technologies, which are predominantly from the U.S., are often content with large domestic markets and sometimes lack the resources and connections necessary to access foreign markets. Developing personal relationships with potential customers is very important to entering these markets. Often, American firms mistakenly think that offering a superior product is sufficient, that the product will sell itself. Furthermore, because regulation and enforcement are the most important drivers of markets for remediation technologies, regulatory uncertainty and corruption in the region makes the size and structure of potential markets difficult to gauge.

In order to overcome these barriers, providing adequate information about cost-effective remediation technologies to policy-makers and potential buyers in the region is critical. This will serve both to reduce reluctance on the part of politicians to address the issue of soil and groundwater contamination and to encourage firms to take preemptive steps to reduce environmental liabilities and avoid unfavorable media attention.

SOIL AND GROUNDWATER POLLUTION IN THE REGION

Though not on the top of the list of priorities for environmental regulators, the region faces a massive problem with soil and groundwater pollution resulting from accidental spillage of useful hazardous materials, and improper treatment, storage and disposal of hazardous wastes. Unregulated growth in pollution intensive sectors such as chemicals, petroleum products, basic metals, transport equipment and electronics, has resulted in rapid increase in the volume of hazardous materials used and produced in the region. Agricultural sources of soil and groundwater contaminants have grown at the same time, as farmers in the region increasingly depend on chemical inputs such as pesticides, herbicides, and fertilizers.

The volume of hazardous waste production has overwhelmed the limited capacity that exists in the region for proper treatment and disposal. As a result, most wastes are haphazardly stored on-site or are illegally dumped. Also, environmental regulators simply do not have the resources necessary for monitoring the growing number of hazardous waste generators.

Currently, the scope of this problem in the region is unknown. A few projects are underway to begin monitoring soil and groundwater pollution, but these will not see any results for some time. Despite this, most experts agree that when they do look, these studies will find a massive problem.

Although they are favorite targets of environmental activists in the region, larger enterprises are hardly the only culprits when it comes to soil and groundwater contamination. Small and medium sized enterprises probably contribute most to the problem. Small and medium sized firms constitute the bulk of industry in the region. In Taiwan for example, roughly 97 per cent of the 90,000 registered factories are classified as small to medium sized (i.e. fewer than 100 workers). Additionally, industrial authorities estimate there are at least 90,000 illegal (i.e. unregistered) factories on the island, the vast majority of which are small in size (Barnes, 1996).

Small and medium sized firms are more likely to improperly store, treat, and dispose of hazardous materials. This is so for several reasons. In addition to being largely ignored by overworked environmental regulators, many of these firms lack the resources to treat and dispose of their wastes properly. Moreover, proprietors of smaller firms are often ignorant about adverse consequences of their activities. Also, smaller firms generate smaller volumes of wastes, making it easier to dispose of it illegally without being detected; and they are not nearly as worried about unfavorable media exposure. Lastly, illegal operations risk detection by delivering wastes to government run treatment and disposal facilities.

HEALTH RISKS OF SOIL AND GROUNDWATER CONTAMINATION

As a result, populations in Asia's urban centers are exposed to thousands of different anthropogenic compounds that reside in the air, water, and soil. These agents can enter the body via inhalation, ingestion, and absorption through the skin and they are responsible for a wide range of adverse health effects such as reproductive toxicity, neurotoxicity, developmental toxicity (teratogenesis), immunotoxicity, and carcinogenesis.

While there is widespread agreement that environmental exposure to anthropogenic compounds causes adverse health effects in humans, evaluating the health effects due to exposure to soil and groundwater contamination is difficult for several reasons. Typically, humans are exposed to mixtures of many different contaminants at relatively low concentrations, over long periods of time. Investigating the adverse health effects due to mixtures of contaminants is complex. Also, adverse health effects that result from these exposures may be inconspicuous, and they may take years to manifest themselves. This makes causal relationships between exposures to individual compounds and health risks very hard to demonstrate.

In the U.S., however, the Agency for Toxic Substances and Disease Registry (ATSDR) was created in 1983 to assess the health risks of hazardous waste sites. Recently, ATSDR released a comprehensive review of toxicological and epidemiological studies researching the effects of human exposure to Superfund sites. Although its findings are far from conclusive, they at least provide policy-makers with some information to work with.

The agency reports that the studies indicate that "proximity to hazardous waste sites seems to be associated with a small to moderate increased risk of some types of birth defects and cancers." (ATSDR, 1995). Non-cancer health effects identified by the report include: birth defects, low birth-weight, speech impairments, neurological disorders, increased respiratory problems, diabetes, and urinary tract disorders

The report also cites evidence that suggests increased rates of leukemia associated with exposures to hazardous waste sites. For example, in Toms River, New Jersey, home of two Superfund sites, ATSDR found rates of leukemia among children to be eight times higher than expected, and rates of brain and central nervous system cancers to be ten times higher than normal. Overall, however, the link between hazardous waste sites and cancer is unclear at this point, and is the subject of ongoing research, since many compounds present at hazardous waste sites are known or suspected human carcinogens. Given these health findings in the aggregate, ATSDR concludes that "uncontrolled hazardous waste sites and unplanned releases of hazardous that constitute emergency events are a major environmental threat to human health." (ATSDR, 1995).

REGULATORY DRIVERS OF SOIL AND GROUNDWATER REMEDIATION IN ASIA

Markets for environmental remediation technologies and services are driven in large part by government regulation. Thus, until Asian governments respond in a meaningful way to public demands for action on this issue, these markets will remain small. Even Japan, which is far ahead of its neighbors, lags far behind the U.S. and Western Europe in this regard. Industry sources in Japan estimate that there are roughly 5,000 to 10,000 sites of contaminated soil and groundwater requiring remediation (National Trade Data Bank, 1995). Most expect that national and local regulations relevant to soil and groundwater contamination will be strengthened soon, but remedial activities will not begin in earnest until that happens (National Trade Data Bank, 1995). Some countries in the region, such as Thailand and Malaysia, have regulations that aim to punish polluters and require remediation of contaminated sites, but monitoring and enforcement are lacking.

Inadequate monitoring and enforcement plagues the entire region, which renders even the most stringent environmental controls less effective. Several factors are responsible for this. First, Asia's private sector enjoys a closer relationship with governments, which are loath to burden industry with high pollution control costs that may hinder competitiveness. Second, Asian cultures tend to espouse negotiation over confrontation. Hence, voluntary compliance by industry is relied upon heavily, and unannounced inspections are rare. Lastly, environmental regulators throughout the region face a severe lack of adequate resources and skilled manpower. In response, polluters have learned that the cheapest alternative is usually to gamble that they will not be caught. Even then, fines are often lower than the costs of cleanups, and underpaid enforcement officials are easily bribed.

Although the current picture is bleak, it should change with time, as public pressure intensifies and environmental regulators become more effective. Also, as regulators in the region begin to have success in reducing other environmental threats to human health, resources will be freed for confronting the problem of soil and groundwater contamination. Current projects underway to improve monitoring of groundwater quality should begin to shed light on the magnitude of the problem. Although the pace is slowing, the region should continue to enjoy high rates of economic growth, providing governments with the financial resources necessary for addressing environmental degradation. Lastly, the region can benefit from experiences of the U.S. and other OECD countries that have been addressing contaminated soil and groundwater problems for some time.

LATECOMER ADVANTAGES

In the past decade, several factors have put downward pressure on the unit costs of environmental remediation that may allow Asia to confront problems of soil and groundwater at an earlier stage of development than their predecessors of the OECD. Fierce competition in the market for environmental technologies has broadened the range of cost-effective technologies. OECD countries have accumulated significant experience in tackling pollution problems, and techniques for using existing technologies have been improved. Also, although there remains much work to be done, a good deal has been learned about health risks associated with human exposure to xenobiotics in the environment. This will enhance the ability of decision-makers to prioritize risks and allocate pollution-control resources more efficiently. With regard to soil and groundwater cleanup, this will permit more informed decisions about the appropriate cleanup criteria.

The enormous expense of Superfund, RCRA, and UST (underground storage tank) cleanups in the U.S. undoubtedly has discouraged less advanced countries from tackling these problems at home. In Asia, industry enjoys a cozier relationship with government relative to many OECD countries, and politicians, though responsive to public sentiment, have been reluctant to open what many consider to be a Pandora's Box and burden industry unnecessarily.

However, increased regulatory flexibility and use of more cost-effective remediation technologies will eventually mean more cost-effective cleanups in the U.S.. This may ease fears of politicians and industry in Asia and reduce industry resistance to legislation addressing soil and groundwater contamination. This fact, coupled with increasing public concern about exposure to industrial pollution, will probably lead countries in the region to tackle the problem. If and when they do, it is unlikely that cleanup criteria in the region will be nearly as stringent as those of the U.S. Also, it is likely that policy-makers will try to foster cooperation on the issue between industry and regulators, and avoid high costs of litigation characteristic of Superfund.

REMEDIATION TECHNOLOGIES

Although Asian countries face massive problems with soil and groundwater pollution, they enjoy the luxury of being able to choose from a large array of remediation technologies. The range of commercially available remediation technologies has increased substantially over the past decade due to fierce competition in the environmental industry. Many of these new technologies are far simpler, cheaper and more effective than traditional remediation strategies such as landfilling, incineration, and pump-and-treat.

Although there is a very wide range of different technologies on the market, these can be grouped into three basic categories: physical/chemical, thermal, and biological. Physical/chemical methods exploit physical and chemical properties of the contaminants or matrix to extract, contain, or degrade the contaminant into a less hazardous form. Thermal methods employ heat to volatilize, chemically convert, or immobilize a given contaminant. Biological methods include (1) bioremediation, which involves stimulating (or simply monitoring) microbial degradation of contaminants, and (2) phytoremediation, which exploits the abilities of certain plants to take up, store, and metabolize pollutants, and stimulate microbial activity in the root zone or rhizosphere.

Each of these strategies may be employed either in-place (*in situ*), or to excavated or extracted material (*ex situ*). Often, several different technologies are used in conjunction at one site. For instance, one technology may be used to extract a pollutant from the subsurface, while another is used to treat it above ground on or off-site.

In general, *ex situ* treatments are often faster, less risky, and allow more control over treatment processes of the contaminated material. This is because of the fact that the contaminated material can be thoroughly mixed, and treatment parameters such as temperature and pH can be more precisely controlled. In cases where rapid treatment is required, such as contamination that poses an immediate threat to human populations, *ex situ* treatment may be preferred. Furthermore, some contaminants and some matrices are not amenable to *in situ* treatment or containment, favoring removal. Lastly, *ex situ* strategies can often guarantee attainment of strict clean-up criteria.

In situ methods, on the other hand, have one main advantage: they tend to have a significant cost advantage over *ex situ* treatment. This is because no excavation or transport of the polluted matrix is necessary. Removal and transport of contaminated material, whether liquid or solid, are often the costliest components of a remediation project. This is true for several reasons. First, excavation of contaminated material is labor and energy intensive. Second, transport of hazardous materials is often subject to additional regulatory oversight. Third, the excavated material must be treated or disposed of, adding to the overall cost.

In situ methods permit minimal site disruption, and both contaminated soil and groundwater can be treated simultaneously, providing additional cost advantages. On the other hand, *in situ* treatments offer less control over treatment parameters, such as temperature, pH, nutrient and substrate availability. Hence, *in situ* methods often require more time, and their ability to attain strict cleanup standards is less uncertain.

In addition, the marginal costs of site remediation tend to rise fairly steeply. For example, the costs of removing 99 percent of contamination are usually far higher than removing 95 percent. In the U.S., a tremendous amount of money has been spent trying to return sites to near-pristine condition. It is unlikely that Asia will do the same. This is due to the severity environmental problems faced by Asian countries, and the scarcity of resources available to address them. Countries in the region would likely achieve a greater reduction in overall risk to human health by cleaning up a large of number of sites moderately well, rather than bringing fewer sites to pristine levels.

BIOLOGICAL REMEDIATION TECHNOLOGIES

Biological remediation technologies exploit the abilities of microorganisms and plants to metabolize and sequester contaminants from soil, groundwater, and sediments. The term *bioremediation* implies the use of biological systems to reduce the risks posed by soil and groundwater contaminants. It is typically associated with the use of microbially mediated processes. The term *phytoremediation* specifically refers to the use of plants.

Although biological treatment has been used for many decades in treating sewage, the idea of using biological techniques for treatment of soil and groundwater contamination is relatively new. When used in appropriate situations, these technologies can deliver cheap, effective treatment of contaminated soil, sediments, and groundwater. Although the specific characteristics of each contaminated site always determine what treatment methods are feasible and how much the cleanup eventually costs, biological methods tend to be the least costly when applicable. Furthermore, biological methods are capable of mineralizing contaminants. This is inherently more desirable than simply immobilizing contaminants in place or extracting and moving contaminated material elsewhere.

During the 1980s, the idea of harnessing the metabolic capabilities of microorganisms was hailed by many as the answer to the world's industrial pollution problems. Although a decade of research has instilled a bit realism in the proponents of bioremediation, the technology has great potential for cheap and effective treatment of soil and groundwater contamination.

Bioremediation capitalizes on the fact that microorganisms can use organic pollutants for food. Typically, they do this by using enzymes to facilitate reactions that yield both energy and molecules useful for cell growth. Bioremediation technologies seek to stimulate or enhance these biological processes by creating environmental conditions favorable for growth. This involves several possible strategies. First, environmental parameters can be improved by influencing the inorganic nutrient supply, oxygen concentration, availability of other electron acceptors, pH, and temperature (in the case of *ex situ* bioremediation). Second, inhibitory or toxic effects can be minimized, perhaps by removing inhibitory or toxic substrates or products. Third, selection processes can be harnessed to increase populations of the most efficient indigenous degraders, or sites can be inoculated with specialized non-indigenous organisms. Fourth, mass transfer kinetics can be enhanced, by increasing the bioavailability of the desired substrate with the use of surfactants.

Biological techniques have been successfully used to remediate soil and groundwater contaminated with organic compounds such as petroleum hydrocarbons, solvents, wood preservatives, agricultural chemicals, as well as inorganics such as heavy metals and nutrients. Some biological remediation technologies in particular, which are described below in further detail, are especially well suited for use in Asia's industrializing countries.

In the U.S. the use of these technologies has been hindered by stringent cleanup standards, which biological techniques sometimes fail to meet. This should be less of a factor in Asia. It is likely that countries in the region such as Thailand will have looser cleanup criteria, and both governments and the public may be more patient with the pace of cleanups.

IN SITU BIOREMEDIATION TECHNOLOGIES

Bioventing

Aerobic biodegradation of soil contamination is stimulated by delivery of oxygen to the subsurface. This is accomplished by injecting or extracting air through unsaturated soil at low flow rates, or simply boring holes in the soil in a passive system (US EPA, 1993). This technology is designed primarily to treat soil contamination by fuels, non-halogenated volatile organic compounds (VOCs) and semi-volatile organic compounds (SVOCs), pesticides, and herbicides. The process may be applied to halogenated organics, but is less effective. Bioventing typically costs around $15 per cubic yard of soil (US EPA, 1993) and uses simple, inexpensive, low-maintenance equipment that can be left unattended for long periods of time. Also, the technology tends to enjoy good public acceptance.

The technology requires the presence of indigenous organisms capable of degrading the contaminants of interest, as well as nutrients necessary for growth. Also, it is necessary that the contaminants be available to the organisms, and not tightly sorbed to soil particles. Bioventing is not as effective in treating areas where the water table is high, and soils with very low moisture content. Lastly, the technology is not applicable in sites where high concentrations of inorganic salts, heavy metals, or organic compounds are present, as these hinder microbial growth

Biostimulation

Biodegradation in the subsurface is stimulated by addition of water-based solutions carrying nutrients, electron acceptors or other amendments. These technologies are designed primarily to treat soil and groundwater contamination by fuels, non-halogenated VOCs, SVOCs, pesticides, and herbicides. These processes may be applied to halogenated organics, but are sometimes less effective. Although the costs of biostimulation technologies vary tremendously from site to site, these technologies tend to be among the cheapest alternatives when applicable.

The technology requires the presence of indigenous organisms capable of degrading the contaminants of interest. Also, it is necessary that the contaminants be available to the organisms, and not tightly sorbed to soil particles. Biostimulation is not applicable in sites where high concentrations of inorganic salts, heavy metals, or organic compounds are present, as these hinder microbial growth. Lastly, The circulation of water-based solutions through the soil may increase contaminant mobility and necessitate treatment of underlying groundwater. Preferential colonization by microbes may occur causing clogging of nutrient and water injection wells (US EPA, 1993).

Air-sparging

These technologies stimulate aerobic biodegradation of groundwater contamination by delivery of oxygen to the subsurface. This is accomplished by injecting air below the water table (US EPA, 1993). This technology is designed primarily to treat groundwater contamination by fuels, non-halogenated VOCs, SVOCs, pesticides, and herbicides. The process may be applied to halogenated organics, but is less effective. Air sparging can cost less than $1 per 1000 liters in favorable situations and tends to be among the cheapest remedial alternatives when applicable (US EPA, 1993). The technology uses simple, inexpensive, low-maintenance equipment that can be left unattended for long periods of time. Also, the technology tends to enjoy good public acceptance.

The technology requires the presence of indigenous organisms capable of degrading the contaminants of interest, as well as nutrients necessary for growth. Also, it is necessary that the contaminants be available to the organisms, and not tightly sorbed to soil particles. Air sparging is not applicable in sites where high concentrations of inorganic salts, heavy metals, or organic compounds are present, as these hinder microbial growth

Natural attenuation

Natural attenuation is defined by the U.S. EPA as the biodegradation, dispersion, dilution, sorption, volatilization, and/or chemical and biochemical stabilization of contaminants to effectively reduce contaminant toxicity, mobility, or volume to levels that are protective of human health and the ecosystem. Although calling this a technology is a bit tenuous, justifying its use involves careful modeling and evaluation to determine its feasibility. Target contaminants include fuels, non-halogenated VOCs, SVOCs, pesticides, and herbicides. The process may be applied to halogenated organics, but it requires longer treatment times. Also, the technology is applicable to especially hydrophobic contaminants such as high molecular weight PAHs that tend to sorb very tightly to soil particles and have very low rates of migration. Often, communities of adapted degraders will mineralize such contaminants quickly after they desorb from soil particles.

The costs of natural attenuation are associated with modeling contaminant migration and degradation rates to determine its feasibility, and monitoring costs to confirm adequate performance. Although these costs tend to be low compared with other remedial alternatives, the process does require a fair amount of technical expertise. Unfortunately, the public is often suspicious of natural attenuation due to the impression that nothing is being done.

EX SITU BIOREMEDIATION TECHNOLOGIES

Landfarming

This technology involves the application of contaminated material that has been excavated onto the soil surface and periodically tilled to mix and aerate the material. Sometimes, in cases of very shallow contamination, the top layer of the site may simply be tilled without requiring any excavation. Liners or other methods may be used to control leachate. This technology is designed primarily to treat soil contamination by fuels, non-halogenated VOCs, SVOCs, pesticides, and herbicides. The process may be applied to halogenated organics, but is less effective. Although the technology is very simple and inexpensive, it does require a large amount of space, and reduction in contaminant concentrations may sometimes be due to volatilization rather than biodegradation.

Controlled solid phase treatment

These processes include prepared treatment beds, biotreatment cells, and soil piles or composting. Moisture, heat, nutrients, oxygen, and pH can be controlled to enhance biodegradation (US EPA, 1997). These technologies differ from landfarming in that treatment processes are often enclosed to control off-gasses. Typically, excavated material is mixed with soil amendments and placed on a treatment area that includes leachate collection systems and some form of aeration. The costs of these techniques vary widely, but are among the least expensive when applicable.

Some prepared bed bioremediation techniques involve the continuous spray application of a nutrient solution into the soil and collection and recycle of the drainage from the soil pile. The drainage itself may be treated in a slurry-phase bioreactor before recycling. Vendors have developed proprietary nutrient and additive formulations and methods for incorporating the formulation into the soil to stimulate biodegradation (US EPA, 1997). Target contaminants include: non-halogenated VOCs, SVOCs, and fuel hydrocarbons. Halogenated VOCs, SVOCs, and pesticides also can be treated, but the process may be less effective and may be applicable only to some compounds within these contaminant groups (US EPA, 1997).

Like landfarming, these technologies require a lot of space, and excavation of contaminated material is required. One advantage, however, of contained ex situ methods is that toxic by-products or metabolites formed during the biodegradation process (e.g. vinyl chloride from TCE) are contained.

Slurry phase biological treatment

These technologies involve the treatment of excavated soil in the controlled environment of a bioreactor. Excavated soil is processed to separate stones and rubble, then mixed with water to a predetermined concentration dependent upon the concentration of the contaminants, the rate of biodegradation, and the physical nature of the soils. Usually slurries contain from 10 to 40 per cent solids (US EPA, 1997). Electron acceptors and nutrients are added to the reactor, and parameters such as pH and temperature are controlled to optimize biological processes. Also, the reactor may be inoculated with specialized organisms if a suitable population is not present. Both aerobic and anaerobic reaction environments may be used Target contaminants include: petrochemicals, solvents, pesticides, wood preservatives, explosives, petroleum hydrocarbons, and other organic chemicals (US EPA, 1997). The potential for treating chlorinated compounds such as TCE with anaerobic bioreactors is being investigated.

Bioreactors are favored over *in situ* biological techniques for heterogenous soils, low permeability soils, areas where underlying groundwater would be difficult to capture, or when faster treatment times are required. Like solid phase ex situ treatments, they have the advantage of containing toxic metabolites such as vinyl chloride. Slurry phase treatment tends to be faster, but more expensive, than controlled solid phase treatment.

Phytoremediation

Using plants in soil and groundwater remediation is a relatively new concept and the technology has yet to be extensively proven in the marketplace. Because of this, most information about phytoremediation comes mainly from field and laboratory research. However, the potential of phytoremediation for cheap, simple, and effective soil and groundwater remediation is generating considerable interest.

Phytoremediation may be used for remediation of soil and groundwater contaminated with toxic heavy metals, radionuclides, organic contaminants such as chlorinated solvents, BTEX compounds, non-aromatic petroleum hydrocarbons, nitrotoluene ammunition wastes, and excess nutrients (Schnoor et al., 1995).

Other uses of phytoremediation include landfill caps, buffer zones for agricultural runoff, and even drinking water and industrial wastewater treatment. Phytoremediation may also be used as a final polishing step, in conjunction with other treatment technologies.

While indeed promising, the applicability of phytoremediation is limited by several factors. First, it is essential that the contaminated site of interest is able to support plant growth. This requires suitable climate, soil characteristics such as pH and texture, and adequate water and nutrients. Second, because plant roots only go so deep, phytoremediation is practical only in situations where contamination is shallow (<5m), although in some situations with deeper contamination it may be used in conjunction with other technologies. Third, since the time requirements for phytoremediation are sometimes long relative to some conventional technologies such as landfilling and incineration, it is not suitable for situations requiring rapid treatment.

Plants facilitate remediation via several mechanisms: (1) Direct uptake, and incorporation of contaminants into plant biomass; (2) immobilization, or phytostabilization of contaminants in the subsurface; (3) release plant enzymes into the rhizosphere that act directly on the contaminants; and (4) stimulation of microbially mediated degradation in the rhizosphere.

PHYTOREMEDIATION OF ORGANIC CONTAMINANTS

Direct Uptake of Organic Contaminants

Some plants are adept at taking up and storing moderately hydrophobic organics (log K_{ow} @ .5-.3) such as BTEX compounds, chlorinated solvents, and short chain aliphatics (Schnoor et al., 1995). Less water soluble contaminants such as PAHs tend to sorb tightly to the roots and soil, and are not available to the plant. Highly soluble molecules tend not to sorb to the roots and resist transport through plant cell membranes. Once taken up, contaminants may be incorporated into plant biomass, translocated through the xylem and volatilized (Cunningham et al., 1996), or metabolized by detoxification mechanisms in the plant (Schnoor et al., 1995).

Due to their long life span, large, deep root systems, and the ability to support a diverse rhizosphere community, trees are well suited for phytoremediation. Since trees cycle vast amounts of water, they are capable of taking up large quantities of pollutants. For example, five-year-old hybrid poplars can cycle 100 liters of groundwater per day, and an acre planted with 2000 such trees could take up 45 kg of contaminant per year (Stomp et al., 1994). In a pilot study in Amana, Iowa, hybrid poplars were used effectively as a buffer for agricultural runoff (Patterson et al., 1993). Fast growing trees such as willow, poplar, and loblolly pine have also been successfully used for municipal wastewater treatment.

Some fast growing crop species are able to take up large quantities of organic contaminants as well. McMullin and co-workers found that the high lipid-content of carrot roots encourages uptake of significant amounts of lipophillic contaminants such as dichloro-diphenyl-trichloro-ethane. In a patented process, the carrots are used to take up lipophillic contaminants, and are then harvested, dried and incinerated (McMullin, 1995).

Phytostabilization of Organic Contaminants

Trees transpire large amounts of water and can therefore slow or even reverse downward migration of chemicals. This makes fast growing trees such as willow and poplar effective for slowing or halting contaminant plume migration and useful for capping landfills. In Beaverton, OR, 10,000 poplars were planted per hectare as a final cap on a municipal landfill (Schnoor et al., 1995). In addition to keeping the landfill free from leachate problems, nearby residents prefer the forest to an unsightly clay or plastic cap. Hybrid poplars have also been used to cap a landfill in Slovenia (Licht et al., 1994).

Degradation of Organic Contaminants in the Rhizosphere

Another important plant-mediated process that may be exploited for environmental remediation is enhanced degradation of organic contaminants in the rhizosphere, which is the zone of soil influenced by the plant roots. Long before [illegible] interested in phytoremediation, [illegible] naturally occurring toxic and inhibitory compounds. Such compounds are often synthesized and secreted by other plants in the attempt to eliminate the competition for space. Two defense mechanisms are employed by plants to protect themselves from xenobiotics in the rhizosphere: plant stimulation of microbial communities in the rhizosphere, and secretion of plant enzymes capable of degrading xenobiotics.

In addition to facilitating xenobiotic metabolism in the plant tissues, plant enzymes have been recognized as significant players in rhizosphere degradation. These have been shown to facilitate remediation of nitroaromatics, halogenated hydrocarbons such as chlorinated solvents and pesticides, and anilines (Schnoor et al., 1995) . In studies at the EPA laboratory in Athens, GA, researchers have isolated plant enzymes as the causative agent in several biodegradation studies. Dehalogenase, peroxidase, nitroreductase, laccase, and nitrilase enzyme systems have been identified (Schnoor et al., 1995). Plant derived dehalogenase enzymes shown to facilitate the mineralization of trichloroethylene (TCE) have sparked hopes that phytoremediation may offer cost-effective treatment of this ubiquitous contaminant. Additionally, Schnoor et al. were able to show that plant derived nitroreductase and laccase enzymes catalyze ring cleavage in 2,4,6-trinitrotoluene (TNT) and incorporate the metabolites into plant tissue (Schnoor et al., 1995).

Enhanced microbially-mediated biodegradation in the rhizosphere is also receiving much attention as a remediation strategy. A symbiotic relationship exists between plants and rhizobial microorganisms. Plants create conditions favorable for microbial growth in return for the protection of the metabolic capabilities of the organisms to degrade harmful compounds before they reach the plant. As a result, microbial biomass is often 5 to 20 times greater in the rhizosphere than that of the surrounding, bulk soil (Patterson et al., 1993). Rhizosphere typically are home to diverse populations of bacteria, fungi, algae, protozoa nematodes, mites, and insects (Patterson et al., 1993) .

Plants, which can be thought of as solar powered pumps, influence the rhizosphere in several ways. In addition to providing structure for microbial colonization, plants influence redox conditions by aerating the soil, and provide microorganisms with reliable sources of carbon, energy, and nutrients by sloughing off dead cells and exuding sugars, amino acids, vitamins, and other compounds. In return, rhizobial microorganisms perform a variety of beneficial functions for the plant. They can favorably alter physical soil characteristics, regulate root metabolic capacities, enlarge effective root surface area and uploading capacity, and in the case of leguminous plants, fix nitrogen. Most importantly for those interested in phytoremediation, they can protect the plant by detoxifying harmful xenobiotics (Cunningham et al., 1996).

While much of the focus on enhanced biodegradation in the rhizosphere to date has focused on agricultural chemicals, using phytoremediation to treat non-agricultural organics such as petroleum hydrocarbons, chlorinated solvents and PAHs is beginning to receive attention as well. Kansas State researcher Kathy Banks was able to demonstrate significantly greater rates of dissipation of total petroleum hydrocarbons (TPH) in vegetated plots versus unvegetated plots (*Environmental Science and Technology*, 1997). Plants used in this study included white clover, fescue, and Bermuda grass. Anderson and Walton found some types of vegetation were able to significantly accelerate mineralization of TCE, as well as incorporate appreciable amounts in plant tissues (Anderson et al., 1995). They monitored the fate of [^{14}C]TCE in whole plant systems composed of soil and vegetation from a contaminated field site. Four predominant plant species were used, a legume, *Lespedeza cuneata*; a grass, *Paspalum notatum*; a composite, *Solidago* sp.; and loblolly pine, *Pinus taeda*. They found production of radio-labeled CO_2 in soils planted with these species was significantly greater (2-3 times) than $^{14}CO_2$ production in non-vegetated soils, and 4-8 times greater than that of sterile controls. In another study, Qui and Sims found that the rate of disappearance of PAHs was almost doubled in the presence of eight prairie grasses (Qiu et al., 1994).

PHYTOREMEDIATION OF INORGANIC CONTAMINANTS

Remediation of metal-contaminated soils and groundwater is another potentially promising application of phytoremediation. Given that metals cannot be chemically transformed, and they can be toxic to microorganisms, metal contamination is not amenable to *in situ* treatment strategies such as microbially mediated remediation. Treatment of metal contamination therefore calls for either extraction or immobilization, and conventional treatment strategies, such as incineration, landfilling, leaching, and chemical fixation are often prohibitively expensive. Additionally, landfilling and incineration are often hampered by public hostility. Understandably, cost-effective treatments for metal contamination are desperately needed. This has stimulated considerable interest in using ability of some plants to accumulate metals in their tissues. Phytoremediation technologies that exploit this trait include: phytoextraction, rhizofiltration, and phytostabilization (Salt et al., 1995).

Phytoextraction

Phytoextraction makes use of metal accumulating plants to transport and concentrate metals in harvestable roots and shoots in order to remove metals from soil. Typically, multiple crops of metal-accumulators could be grown, followed by harvest and processing of the plant material, which could involve reclamation of metals or simply incineration and disposal.

The mechanisms of root uptake and transport within the plants are poorly understood. Solubilized metal ions may enter the root by two pathways: apoplastically (extracellularly), or symplastically (intracellularly) (Salt et al., 1995). Metal ions tend to enter plant cells via metal ion carriers or channels in an energy dependent, saturable process (Clarkson et al., 1989). Non-essential heavy metals may compete with essential metals for these transmembrane carriers, which may explain the ability of some non-essential metals to enter the cell against the concentration gradient (Salt et al., 1995) . Upon entering the roots, metals can either be stored or transported to the shoot. Xylem transport is thought to be responsible for transport to the shoot, although metals may disperse throughout the shoot via the phloem (Salt et al., 1995).

Several wild plants with the ability to accumulate very large concentrations of metals in their roots and shoots have been identified. Botanists first recognized the metal-accumulating ability of the genus *Thlaspi* over a century ago. Some have suggested this ability may have evolved as a defense against herbivores (Baker et al., 1994). Baker and co-workers were able to grow some individual *Thlaspi caerulescens* plants that accumulated more than 30,000 mg of Zn and over 1000 mg Pb per gram dried biomass (Baker et al., 1994). They also showed that the five isolated populations of the plant were adept at tolerating and accumulating metals not present in the parent soil, which suggests the mechanisms of tolerance and accumulation may be similar for different metals.

While wild metal-accumulators such as *T. caerulescens* have shown an impressive ability to accumulate metals, they tend to be slow growing and small in size. The use of such plants for phytoextraction may therefore require an unreasonable number of harvests to decontaminate a given site. Ideally, a plant well suited for phytoextraction should tolerate and accumulate metals, grow rapidly, and have the potential produce a high biomass. In the attempt to find such a plant, Dushenkov and co-workers have evaluated the heavy-metal accumulating abilities of several high-biomass crop species such as *Brassica juncea* (Indian mustard) (Dushenkov et al., 1995).

Although *Thlaspi caerulescens* has a higher tolerance for the heavy metals tested and demonstrated a higher ratio of metal accumulation to plant mass, *Brassica juncea* produces 20 times more biomass. They found *Brassica juncea* to be especially adept at accumulating lead. One strain was able to accumulate Pb at up to 3.5% dry weight in the shoots suggesting that a crop of such plants could extract 630 kg/ha of Pb in above ground biomass with a single harvest, more if some root material was harvested as well (Dushenkov et al., 1995).

It is important to note however, that these experiments were done hydroponically. In soil, the propensity of Pb to bind to clay soil particles and organic matter, and its inclusion in insoluble precipitates significantly reduces the bioavailability of Pb to the plant (Dushenkov et al., 1995). This is true of other metals as well. For this reason, biological mechanisms that enhance metal bioavailability are being investigated. For instance, in response to nutrient deficiencies, plants can secrete metal-chelating molecules (phytosiderophores) that chelate and solubilize soil-bound metals such as Fe, Mn, Cu, and Zn (Salt et al., 1995). Also, plants can reduce soil bound metal ions by specific plasma membrane-bound metal reductases; and they can adjust soil pH, which decreases metal adsorption, by exuding protons (Salt et al., 1995). Lastly, the presence of some microorganisms in the rhizosphere have been shown to enhance metal availability.

Soil amendments that can increase metal availability are also being studied. Chelating agents have been investigated by Blaylock, Raskin, and co-workers for their ability to prevent precipitation and sorption of metals (Dushenkov et al., 1995). *B. juncea* seedlings grown for 4 weeks in soil treated with a chelating agent accumulated 875 mg/kg DW Cd in the shoot, compared to 164 mg/kg DW Cd by seedlings grown in untreated soils. Additionally, numerous studies have demonstrated that lowering soil pH increases metal availability.

Rhizofiltration

Rhizofiltration is defined as the use plant roots absorb, precipitate, and concentrate toxic metals from water (Salt et al., 1995). In engineered systems, metal-contaminated water may be passed through a network of roots, which can then be harvested, dried, and either combusted and discarded or subjected to processes to reclaim metals from plant biomass. Given its potential for treating high volumes of water contaminated with low concentrations of metals, rhizofiltration enthusiasts assert it will be a cost-effective treatment for everything from industrial wastewater, to agricultural runoff, to contaminated surface water and groundwater.

The mechanisms for root accumulation include: 1) surface sorption, in which physical/chemical processes such as chelation and ion exchange lead to sorption by the root; 2) biological processes, including intracellular uptake, vacuolar deposition, and translocation to the shoots (Chaney et al., 1983); and 3) root mediated precipitation, which probably involves the release of root exudates (Dushenkov et al., 1995). Surface sorption tends to be the fastest of these, especially in the case of Pb, and root-mediated precipitation the slowest, although the relative importance of these different mechanisms is dependent on concentration. At low concentrations, surface sorption dominates, at higher concentrations, however, when sites for sorption are saturated, biological processes and precipitation assume more importance. These mechanisms also differ between metals and plant species.

Dushenkov and co-workers tested the ability of hydroponically grown plants to remove toxic metals from aqueous waste streams (Dushenkov et al., 1995). The ability of the high biomass crop plant *Brassica juncea* (Indian mustard) to accumulate Pb in the roots was compared to 24 other plant species such as *Helanthus annus* (sunflower), and various grasses such as *Agrostis tenius* (colonial bentgrass) and *Poa pratensis* (Kentucky bluegrass). While all the species tested demonstrated significant root accumulation of Pb, *Brassica juncea* (14% dry weight Pb in the roots) possessed the most favorable combination of metal accumulating ability and high-biomass production.

Additionally, *Brassica juncea* was shown to accumulate significant amounts of Cu^{2+}, Cd^{2+}, Cr^{6+}, Ni^{2+}, and Zn^{2+} as well. Given that the magnitude of metals accumulated by way of surface sorption is proportional to root mass, the ability of *Brassica juncea* to generate a large mass of roots quickly and economically makes it a promising candidate for rhizofiltration.

Rhizofiltration of radionuclides has also been successfully demonstrated. Raskin and Ensley successfully used floating rafts planted with *Helanthus annus* (sunflower) to remove radionuclides such as uranium, cesium, and strontium from a pond contaminated by the 1986 Chernobyl accident (Cooney, 1996). Also, they tested sunflower plants to remove radionuclides from water taken from the Department of Energy uranium-processing facility. They found that uranium concentrations declined by 95% within the first 24 hours, from 100-400 ppb to below 20 ppb, the EPA groundwater standard (Cooney, 1996). Ensley estimates the total costs of removing radionuclides with sunflower plants may range from \$2 to \$6 per thousand gallons treated. This compares very favorably with the costs of conventional technologies for removal of radionuclides.

Phytostabilization

Phytostabilization of inorganics involves the use of metal tolerant plants to reduce the mobility of metals in the subsurface.

Often, soils contaminated with toxic metals lack vegetation due to either physical disturbance or toxic effects of the contamination. Metal contamination of exposed soils is often more mobile due to leaching and transport by wind and water. Metal tolerant plants may be useful in reducing metal mobility that results from these mechanisms. This strategy has been used successfully to stabilize metalliferous mine-wastes by a group in Liverpool (Cunningham et al., 1993). Also, Salt and co-workers were able to demonstrate that seedlings of *B. juncea* were able to reduce the level of Pb leached from contaminated soils into the groundwater (Salt et al., 1995) .

Legumes are especially well suited for reestablishing vegetation and stabilizing degraded and metal contaminated soils. This is due to several factors: first, they accumulate nitrogen in a mineralizable form in symbiosis with rhizobia, which then becomes available to non-leguminous plants; second, they are able to grow in low nutrient conditions; and third, they are able to colonize barren habitats which are subject to strong winds and flooding (Jha et al., 1995). Jha and co-workers studied how this rhizobial symbiosis influences the ability of several wild legumes to revegetate an unreclaimed limestone quarry. They found that seeds encapsulated with polyacrylamide-entrapped rhizobia showed higher establishment, survival, and subsequent growth than uninnoculated seedlings.

FACTORS THAT GOVERN REMEDIATION TECHNOLOGY CHOICE

All of the above strategies have advantages and disadvantages. The chemical and physical characteristics of each site and the goals of the remediation project determine what technologies and techniques are appropriate. After feasible alternatives are identified, other factors come into play. Among the many considerations that govern remediation technology choice are: capital and operating costs, regulatory acceptance, cleanup time, commercial availability, residuals produced, effectiveness, space requirements, reliability, and public acceptance.

Physical factors

Physical factors include site geology such as soil characteristics, (relative amounts of gravel, sand, silt, clay, and organic matter) and the presence of layering. Soil characteristics profoundly influence the propensity of contaminants to sorb to soil particles. The hydrology of the site is also important. The depth of the water table, and speed and direction of groundwater flow determine the potential for contaminant migration. Other physical factors include temperature, the depth and breadth of contamination, and space available on the surface for equipment and treatment operations.

Chemical Factors

Chemical factors include the chemical properties of the contaminants, especially solubility and volatility. With regard to biological treatment, critical factors include: pH, nutrient availability, presence of electron acceptors such as O_2 and NO_3, and levels of toxicants such as heavy metals, inorganic salts, and high concentrations of some organic compounds.

Biological Factors

Includes the characteristics of indigenous microbial and plant communities, predation.

Regulatory Factors

Regulatory factors exert a profound influence on remediation technology choice. Factors such as cleanup criteria, or the degree of cleanliness required for a given site, and time pressures have a huge influence on the appropriateness of technologies for remediating that site

Cost

After identifying treatments that meet technical and regulatory requirements, costs differences become a critical factor in influencing technology choice. The typical cost components of a remediation project include: treatability studies, site preparation, permitting and regulatory, capital equipment, startup/fixed costs, labor, other operating costs, residue treatment and disposal, and monitoring.

Treatment Time

There often exists a tradeoff between time required for treatment and treatment cost. Simply excavating contaminated soil and carting it off to be incinerated or landfilled is often the quickest way to remediate a site. These strategies tend to be the most expensive as well. *In situ* methods are usually cheaper but often sacrifice speed of treatment.

Risk and Uncertainty

Understandably, managers of hazardous waste sites everywhere prefer to use technologies that have built up a track record of successes.

Other Factors

Other important factors include commercial availability and public acceptance.

Non-Remediation Applications

Many of the biotechnologies described above also have applications besides soil and groundwater remediation. Just as when they are applied to soil and groundwater cleanup, they possess the desirable attributes of being inexpensive, simple to use and maintain, and they are capable of mineralizing contaminants, rather than just transferring them to a different medium.

For example, technologies such as composting, slurry-phase biotreatment, rhizofiltration, and biofiltration technologies each show promise for cost-effective end-of-pipe treatments of industrial waste streams. Slurry-phase bioreactors and composting systems offer inexpensive treatment of some hazardous industrial wastes, both organic and inorganic. Rhizofiltration shows particular promise for cost-effective removal of metals from wastewater, and because it can effectively treat very low concentrations of metals, it may be useful for treating drinking water as well. Biofiltration, which entails filtering off-gases through a biologically active matrix, offers inexpensive treatment of airborne emissions of biodegradable VOCs and odors.

Leachate from landfills and agricultural runoff are major contributors to surface and groundwater pollution as well and solutions to these problems are also needed. Fast growing trees such as hybrid poplar make effective agricultural buffers and landfill caps. Currently there is a desperate need for simple, inexpensive end-of-pipe technologies in the region such as these and speeding their transfer to Asia's industrializing countries will help them to meet their pollution control objectives more efficiently.

THE APPLICABILITY OF BIOREMEDIATION TECHNOLOGIES IN ASIA

Several characteristics of potential markets for remediation technologies in Asia will likely make biotechnologies especially appropriate for use in the region. Some of these characteristics affect the suitability of environmental technologies in general, while some are specific to remediation technologies.

In general, environmental technology that has been developed in advanced western countries is not always well suited to Asia's industrializing countries. Environmental technologies that are developed for use in markets such as the United States, Germany, and Japan are often too complicated, expensive and difficult to maintain for use in less developed countries. This is because of the fact that environmental standards tend to be stricter in more advanced countries, and technologies designed to meet stricter criteria are typically more expensive and complex.

Ambient environmental standards tend to be less stringent in Asia, and buyers are reluctant to pay for more pollution control than is required. Moreover, technically complex technologies often require technically complex maintenance and repair. Given the lack of skilled manpower in the region, buyers of environmental technologies in Asian countries often complain that such technologies often make them dependent on foreign technicians for after-sales service.

In order to be appropriate for use in the region, environmental technologies should be inexpensive, simple to install, use, and maintain, and they should not require highly technical support (Rondinelli, 1994). Older environmental technologies from advanced countries, which are in the latter stages of the product cycle, are more appropriate for developing and industrializing countries as they are more apt to meet these criteria (Rondinelli, 1994). Biotechnologies such as bioventing, composting and biopiles, air sparging, phytoremediation, natural attenuation, and other *in situ* and *ex situ* biotechnologies meet these requirements. Furthermore, they tend to be the cheapest remedial alternative when used appropriately.

With regard to potential markets for remediation technologies in the region, there are several factors that enhance the appropriateness of biotreatment. These include: looser cleanup criteria relative to OECD countries, a greater deference for industry on the part of regulators and the public, longer time horizons in both the public and private sector, lack of trained manpower, and climate.

First, it is unlikely that the region's environmental regulators will mandate very stringent cleanup goals for soil and groundwater cleanup. This is due to a reluctance to burden industry with large expenses on environmental controls, a tendency which may be strengthened by slowing economic growth in the region. Furthermore, it is unlikely that the public will have the same level of access to information about levels of soil and groundwater contamination compared to OECD countries, and therefore public pressures to set lofty cleanup goals will not be as great. This, coupled with the fact that environmental regulators in the region tend to prefer pragmatism and negotiation over confrontation will probably give firms in Asia more leeway to choose their own remedy.

Looser cleanup standards also make sense for these countries from an efficiency standpoint. The marginal costs of cleanups tend to rise fairly steeply, especially as they begin to necessitate ex situ treatment. Requiring cleanup levels that permit in situ treatments will foster a greater level of overall risk reduction in the region. Given the massive scope of soil and groundwater contamination problems in the region, greater risk reductions can be achieved by cleaning up a lot of sites moderately well, rather than cleaning up a few sites to pristine levels. In addition, keeping the costs of cleanups low will reduce resistance on the part of the private sector to strengthened regulation and enforcement, and may encourage a greater level of voluntary cleanups.

In addition to looser cleanup criteria, cleanups in the region may be less subject to time pressures typical of the U.S.. Information about cleanups in the U.S. is often very accessible to the public, and pressures to complete cleanups swiftly are significant. Additionally, both the public and private sector in Asia tend to focus more on the long term when it comes to meeting environmental goals.

The likelihood of looser cleanup criteria and less pressure for speedy cleanups bodes especially well for the potential of biological remediation technologies in the region. In OECD countries, lofty cleanup goals and pressures to get cleanups done quickly are typically the most important hindrances to use of these technologies. In addition, these factors will enhance the suitability of some non-biological in situ treatments such as soil vapor extraction (SVE).

Another important factor that enhances the suitability of biotechnologies in the region is climate. Biological processes tend to move faster at higher temperatures. In OECD countries such as Germany, Japan, and the US, the pace of biotreatment slows dramatically during cold winters. Tropical climates foster year round growth of plants and microorganisms. This increases the rates of both *in situ* and *ex situ* biotreatments as well as the use of biological systems for end-of-pipe treatment of industrial waste streams. Also, in many areas throughout the region, the soils tend to be sandy which are more amenable to *in situ* biotreatment than clay soils.

BARRIERS TO REMEDIATION TECHNOLOGY TRANSFER

Biological remediation technologies are clearly needed in Asia, but their availability in these markets is limited at present. Although most of the countries tend to grant preferential tariff rates to environmental technologies, several important non-tariff barriers discourage potential buyers from using innovative technologies and foreign vendors from exploring these markets. These include lack of indigenous demand, lack of information about innovative technologies, and lack of support for U.S.-based exporters of environmental technologies.

Lack of Indigenous Demand

For biological remediation technologies to be successfully transferred to Asia, there must be indigenous demand for remediation technologies and services in these countries. Currently, this demand is very weak, and little remediation by is being done in the region outside of Japan and Singapore. The market for publicly funded environmental remediation activities will be of significant size in the future, but for now the majority of public resources are being devoted to water, wastewater, and municipal solid waste treatment.

In the private sector, the story is much the same. There have been a small number of voluntary site remediations done by firms in the region, which are driven by concerns about unfavorable media attention and future environmental liabilities. These cleanups tend to be done by the very largest firms with substantial resources available for pollution control. Also, many of these cleanups were done by multinational corporations, which tend to adhere to stricter environmental standards than those of the host country. Other than a small number of projects of this nature, private sector demand for remediation technologies and services is almost non-existent. This will not change until meaningful soil and groundwater legislation is passed and enforced.

Asian countries may be wary of enacting and enforcing meaningful soil and groundwater regulations because of the U.S. experience with the large costs of the Superfund program. In Taiwan for example, soil and groundwater legislation has been held up for three years despite public pressure for action on the issue. With time, this should change. Growing public pressure, more aggressive media coverage and NGO activities, and falling costs of remediation may ease the fears of politicians and industry, hastening the passage of soil and groundwater legislation. Furthermore, as environmental regulators are given more resources and better legal tools, levels of monitoring and enforcement will improve.

Lack of Information on Innovative Technology Performance

Another important barrier to the transfer of biological remediation technologies to Asia is lack of information. Innovative technologies such as bioremediation and phytoremediation are plagued by the fact that they have yet to build up a large track record of proven successes and are often viewed as risky by site managers. This has not been helped by some cases of misapplication of biological remediation technologies to sites where they were not suitable. This perception of riskiness has, in turn, prevented these technologies from building up the very track record they need, perpetuating an unfortunate cycle.

Site managers must weigh the tradeoff between uncertainty and cost. Often times, the small chance of failure outweighs the possibilities substantial cost savings offered by innovative technologies. This is especially true when the decision-maker is spending someone else's money for the cleanup, but may lose their job if the project fails.

To combat this phenomenon in the U.S., several programs have been created to speed the transfer of innovative remediation technologies from the laboratory to the marketplace. The Superfund Innovative Technology Evaluation Program (SITE) accomplishes this by facilitating cooperative agreements between the US EPA, technology developers, and site managers. Typically, industry or government provides the site, researchers test and refine their technologies, and EPA collects and publishes performance and cost data.

Currently, over 180 vendors of remediation technologies are enrolled in the program, and over 80 field demonstrations of new technologies have been completed. The SITE program has proved instrumental in helping these processes gain acceptance in the marketplace by lending credibility to vendors' claims about their products as well generating records of performance in the field.

In addition, several electronic databases have been created to aid site managers in choosing appropriate technologies. These include: the *Vendor Information System for Innovative Treatment Technologies* (VISITT), which contains information on 325 innovative technologies; *the Bioremediation in the Field Search System*, which provides information on soil and groundwater cleanup. The EPA also publishes *Innovative Treatment Technologies: Annual Status Report*, an annual report that contains site-specific information on over 400 projects.

These efforts to collect and disseminate information have met with considerable success in the U.S. market. As of 1994, nearly half (43%) of the source control treatment technologies selected at Superfund sites were not even available at the time of the 1986 Superfund reauthorization (SARA) (Kovalick and Kingscott, 1996). About 10% of technologies selected at these sites were bioremediation technologies (Kovalick and Kingscott, 1996).

Phytoremediation technologies are just now beginning to enter the marketplace; thus they suffer even more acutely from this lack of performance data. However, they will benefit from the fact that programs to speed development of innovative technologies are well established. Also, the intense need for cost-effective treatment of metal contamination will encourage many to choose these technologies in spite of a lack of documented successes.

Despite the fact that some applications of bioremediation technologies are now viewed as routine in the U.S., they have yet to accumulate a record of successes in Asia. Thus they are still plagued by an aura of unfamiliarity and uncertainty in the region. On the other hand, site managers in the region will undoubtedly draw on U.S. experience with site remediation, and make use of the wealth of information that is currently available about remediation projects in the U.S..

Supply Side Barriers

Although environmental technologies tend to enjoy preferential tariff rates throughout the region, there are important non-tariff barriers that hinder the transfer of biological remediation technologies from the supply-side. Currently, U.S. firms dominate the bioremediation industry. Due to the fact that the U.S. market for remediation technologies and services is by far the largest in the world (around $8.6 billion in 1994), many U.S. vendors are often content to operate solely within the domestic market (Environmental Business International, 1995). Indeed, the resources required for exploring and entering foreign markets are large, and the markets for soil and groundwater remediation in Asia are small and beset by uncertainties in regulation, monitoring, and enforcement. These uncertainties discourage many U.S. firms from exploring opportunities in the region.

Another important barrier that plagues U.S. firms is the lack of timely information about trade leads and business opportunities. American firms typically have not established a strong local presence that is necessary for success in the region (Rondinelli, 1994). This involves setting up branches or local offices, or using a local agent or partner. A local presence is critical for developing long-term relationships with officials at all levels of government. This is because Asians tend to emphasize personal relationships and trust over legal agreements when conducting business (Kenan Institute Asia, 1997).

Firms from other OECD countries such as Germany and Japan have proven more adept at these strategies and have established a stronger network of contacts within key government agencies and spent more time cultivating business relationships in the region. Furthermore, U.S. companies are hampered by the U.S. Foreign Corrupt Practices Act, which forbids payments to officials of foreign governments in hopes of winning government contracts. This confers an important advantage to non-U.S. vendors seeking public sector cleanup projects in the region, and public sector projects will probably constitute a large fraction of the market

Recently, efforts to assist U.S. exporters of environmental technology have been improved. For example, the U.S. Department of Commerce offers a wide range of export promotion programs and provides sources of information such as the National Trade Data Bank, which exporters can use to conduct preliminary market research and get news of potential projects. Also, the U.S.-Asia Environmental Partnership (USAEP), which is funded through the U.S. Agency for International Development, looks to match U.S. vendors of environmental technologies with potential buyers in Asia. USAEP has established regional offices throughout Asia. It is unclear as yet how pressures for budget cutting in the U.S. will affect these programs.

Inadequate protections of intellectual property rights in the region have also kept some vendors from exploring the market as well. According to Burt Ensley of Phytotech Inc., his firm will not even begin to look into regional markets for phytoremediation until these issues are addressed (Ensley, 1997). However, intellectual property rights concerns are less of a factor for the low-tech biotechnologies described in this paper. With regard to bioremediation technologies, fears about inadequate intellectual property rights protections are more relevant for processes that use genetically engineered or specially adapted plants and microorganisms, and advanced techniques for enhancing bioremediation such as the addition of proprietary amendments such as metal chelators or nutrient solutions. Technologies such as bioventing, composting, and air-sparging involve fairly straightforward techniques with little need of protection.

ENHANCING LINKAGES BETWEEN GOVERNMENT, ACADEMIC INSTITUTIONS, AND THE PRIVATE SECTOR

Improved collaboration between government, academic institutions, and the private sector is essential for reducing barriers to transfer of biological remediation technologies and improving the efficiency of soil and groundwater cleanup in the region. This is because many potential sources of information about soil and groundwater contamination and treatment are divided among them. Also, remediation of soil and groundwater is, after all, a business. It is important, therefore, to understand how vendors and buyers of remediation technologies and services make decisions, and how these decisions are influenced by government regulation, cultural factors, and trade barriers. Additionally, an atmosphere of dialogue and cooperation between industry, environmental regulators, and academic institutions will foster more efficient regulatory regimes and increased levels of self-monitoring and voluntary cleanup activities.

Enhancing the collection and dissemination of information about soil and groundwater contamination and the remediation technology performance is necessary to promote appropriate choices of remedies by site managers of both the public and private sector. Such information is critical to reduce the reluctance of site managers of both the public and private sector to choose innovative remedies. Governments are best equipped to gather and disseminate this information, but the private sector must be encouraged to share information about contamination and cleanups. Currently, there are significant disincentives for firms to share this information.

Academic institutions play a key role by conducting basic and applied research. This includes evaluating the health effects of contaminants, studying the behavior of contaminants in the subsurface, developing improved technologies and techniques to treat them, and evaluating policy tools available to regulators. Improving researchers' access to information about contamination and cleanup activities will expedite these efforts. Also, enhanced dialogue between academia, regulators, and firms may facilitate more productive allocations of public funding for research.

CONCLUSION

Asia's industrializing countries face massive problems with soil and groundwater contamination. Although countries in the region tend to be more tolerant of environmental health risks, public disapproval of environmental deterioration, and rising demand for water and land are driving increased pressures to begin cleaning it up. Because the environmental problems of these countries are so severe, and there are many demands on scarce resources for environmental protection, cost-effective remedies such as bioremediation technologies are desperately needed.

Biological remediation technologies are especially appropriate for use in the region because they tend to be very inexpensive, low tech, simple to use, and easy to maintain. Lower cleanup goals and less pressure for speedy cleanups relative to OECD countries such as the U.S. will enhance their attractiveness in these markets. However, in order to speed the adoption of these technologies to the region, barriers to their transfer must be reduced.

Research Needs

Improved understanding of the health and ecological effects of environmental agents is necessary. This will enhance the ability of environmental regulators in Asia to set priorities and allocate pollution control resources more efficiently. Given the scale of their environmental problems, this is critical.

Also, additional research must be done on the human health effects of soil and groundwater contamination. This will improve the ability of regulators to set appropriate endpoints for cleanups and achieve greater levels of risk reduction with a given level of pollution control resources. Again, most of this type of research will likely be done in OECD countries, and it should be made available to countries further down the development ladder.

Increased monitoring of soil and groundwater contamination in the region is sorely needed. This will serve two purposes. First, it will help regulators to assess the scope of the problem in the region and allocate pollution control resources more effectively. Second, it will increase the level of public awareness about the problem and encourage the private sector to clean up existing contamination, and take steps to prevent future contamination. Additionally, more is work needed prioritizing what should be monitored

At present, little or no incentive exists for firms to monitor and clean up their own sites. Effective policy tools for encouraging a greater level of self-monitoring and voluntary cleanups must be developed. Such tools would greatly ease the burden on overworked regulators.

To further reduce the costs of remediating soil and groundwater, a better understanding of the behavior of contaminants in the subsurface is needed. This will yield two important benefits. First, it will lead to improved technologies and techniques for soil and groundwater remediation. Second, it will aid in prioritizing contaminated sites, and facilitate more informed decisions about cleanup goals. With regard to biological remediation technologies, further work is necessary to better understand how organisms can facilitate soil and groundwater remediation, and how we may enhance these processes.

Lastly, additional research is necessary to improve our understanding of decision-making by both buyers and sellers of remediation technologies and services. At present, government regulation sometimes creates disincentives to use the most cost-effective remedies for soil and groundwater pollution, and trade barriers can delay their commercial availability in industrializing countries such as Thailand. A better understanding of the behavior of buyers and sellers of remediation technologies can help reduce such hindrances to the use of bioremediation technologies in Asia and elsewhere.

LIST OF ABBREVIATIONS

BMR	Bangkok Metropolitan Region
BTEX	benzene/toluene/ethylbenzene/xylene
CERCLA	Comprehensive Environmental Response, Compensation, and Liability Act
CO	carbon monoxide
DIW	Department of Industrial Works (Thailand)
EPA	Environmental Protection Agency
GDP	gross domestic product
GNP	gross national product
MOI	Ministry of Industry
MOSTE	Ministry of Science, Technology, and Environment (Thailand)
NO_x	nitrogen oxides
O_3	Ozone
OECD	Organization for Economic Cooperation and Development
PAH	polycyclic aromatic hydrocarbon
PCB	polychlorinated biphenyls
PCD	Department of Pollution Control (Thailand)
RCRA	Resource Conservation and Recovery Act
RIC	rapidly industrializing country
SO_2	sulfur dioxide
SPM	suspended particulate matter
SVE	soil vapor extraction
TCE	trichloroethylene
UST	underground storage tank
Zn	zinc
Cd	cadmium
Cu	copper
Fe	iron
TPH	total petroleum hydrocarbons

REFERENCES

Agency for Toxic Substances and Disease Registry. (1995). Annual Report to Congress. Washington DC.

Anderson, T.A., and Walton, B.T. (1995). Comparative Fate of [^{14}C]Trichloroethylene in the Root Zone of Plants from a former Solvent Disposal Site. *Environmental Toxicology and Chemistry* 14, 2041-2047.

Baker, A.J.M., Reeves, R.D. and Hajar, A.S.M. (1994). Heavy metal accumulation and tolerance in British populations of the metallophyte *Thlaspi caerulescens. The New Phytologist* 127, 61-68.

Barnes, Robert. (1996). Taiwan – Industrial Waste Treatment Equipment. US Department of Commerce, International Trade Administration. Washington DC.

Bruestle, Arthur. (1993). East Asia's Urban Environment. *Environmental Science and Technology*. 27, no. 12, 2281.

Cassarett and Doulls. (1991). *Toxicology: The Basic Science of Poisons.* McGraw-Hill. New York.

Chaney, R.L. (1983). in *Land Treatment of Hazardous Wastes*. Parr, J, Marsh, E.M. eds. Noyes Data Corp. Park Ridge. 50-76.

Clarkson, D.T. and Luttge, U. (1989). Mineral Intrusion: Divalent Cations, Transport and Compartmentalization *Prog. Botany* 51, 93-112.

Cooney, C.M. (1996). *Environmental Science and Technology* 30 no. 5, 194A.

Cunningham, S.D., Berti, W.R. (1993). Remediation of contaminated soils with green plants: An overview. *In Vitro. Cell. Dev.. Biol.* 29, 207-212.

Cunningham, S.D., Ow, D.W. (1996). Promises and Prospects of Phytoremediation. *Plant Physiology* 110, 715-719.

Dushenkov, V., Kumar P.B.A.N., Motto, H., Raskin, and I. (1995). Phytoextraction: the Use of Plants to Remove Heavy Metals from Soils. *Environmental Science and Technology* 29, 1232-1238.

Dushenkov, V., Kumar P.B.A.N., Motto, H. and Raskin, I. (1995) Rhizofiltration: The Use of Plants to Remove Heavy Metals from Aqueous Streams. *Environmental Science and Technology* 29, 1239-1245.

Engineering Science, Thai DCI Co., Ltd. and System Engineering Co., LTD. (1992). *National Hazardous Waste Management, Volume 2.* Report submitted to the Office of National Environment Board. Bangkok.

Ensley, Burt. (1997). Personal Correspondence. Monmouth Junction, NJ.

Jha, K.P., Nair, S., Gopinathan, M.C. and Babu, C.R. (1995). Suitability of rhizobia-inoculated wild legumes *Argyrolobium flaccidum, Astagalus graveolens, Indigo gangetica,* and *Lespedeza stenocarpa* in providing a vegetational cover in an unreclaimed limestone quarry," *Plant and Soil.* 177. 139-149.

Kato, Kazu. (1996). Legal Frameworks for Environmental Governance in East and Southeast Asian Countries. East-West Center, Honolulu.

Kenan Institute of Private Enterprise, Kenan Institute Asia. (1997). *Thailand Environmental Technologies Export Market Plan.* Bangkok.

Kovalick, Walter, Kingscott, John. (1996). Emerging Issues and Technology in Site Remediation. Office of Solid Waste and Emergency Response, US EPA. Washington, DC.

Licht, L.A., Marinko, A. and Macarol, B. (1994). *Proceedings of the Second International Conference on Energy and Environment.* Energy and Environment Research Center, University of North Dakota. Grand Forks, ND.

McMullin, E. (1995). An Absorbing Idea. *California Farmer*. Feb. 1995. 20-24.

Nguyen, Mai-Ann. (1995). Thailand: The Environment and the Law. Thailand Environment Institute. Bangkok.

O'Connor, David. (1994). *Managing The Environment with Rapid Industrialization: Lessons from the East Asian Experience.* OECD. Paris.

Patterson, K.G. and Schnoor, J. (1993). *Journal of Environmental Engineering.* 119 no. 5. 986-93.

Pfaender, Fred. (1996). Personal Conversation. Chapel Hill N.C.

Qiu, X., Shah S.I., Kendall E.W., Sorenson, D.L., Sims, R.C. and Engelke, M.C. (1994). Grass-Enhanced Bioremediation for Clay Soils Contaminated with Polynuclear Aromatic Hydrocarbons. *In Biormediation Through Rhizosphere Technology*. T.A. Anderson and J.R. Coats eds. ACS Symposium Series 563. American Chemical Society. Washington DC.

Rondinelli, Dennis. (1994). Thailand's Expanding Markets For Environmental Technologies and Services: Opportunities for U.S. Trade and Investment. Kenan Institute of Private Enterprise. Chapel Hill, N.C.

Rondinelli, Dennis. (1992). Business and Environmental Protection: The Emerging Investment Opportunities in Asia. *Business and the Contmeporary Worl.* 144-159.

Salt, D.E., Blaylocj, M., Kumar, N., Dushenkov, V., Ensley, B., Chet, I. and Raskin, I. (1995). Phytoremediation: A Novel Strategy for the Removal of Toxic Metals from the Environment Using Plants. *Biotechnology* 13. 468-474.

Schnoor, J.L., Licht, L.A., McCutcheon, S.C. and Wolfe, N.L, Carreira, L.H.. (1995). Phytoremediation of Organic and Nutrient Contaminants. *Environmental Science and Technology* 29. 7, 317-323A.

Stomp, A., Han, K. and Wilbert, S., Gordon, M., Cunningham, S. (1994). Genetic Strategies for Enhancing Phytoremediation," *Annals of the New York Academy of Sciences* 721, 481-491.

Thailand Development Research Institute (1995). *The Monitoring and Control of Industrial Hazardous Waste: Hazardous Waste Management in Thailand.* Bangkok.

Tilstra, Diane. (1996). Thailand – Solid Waste Disposal Equipment / Services. U.S. Department of Commerce, International Trade Administration. Washington DC.

United Nations. (1990). *World Urbanization Prospects* Geneva.

United Nations Development Programme. (1996). *UNDP Human Development Report.* Oxford University Press. New York.

Unknown Author. (1997). TPH Soil Cleanup Aided by Ground Cover. *Environmental Science and Technology* 31 no. 5, 214A.

Unknown Author. (1993). A Billion Consumers. *The Economist* Oct. 30. 1993.

US Agency for International Development. (1991). *Ranking Environmental Health Risks in Bangkok.* Washington DC.

US Department of Commerce. (1995). Japan – Remediation Svcs / Soil / Groundwater Remediation. Washington DC.

US Department Of Commerce. (1996). Thailand – Solid Waste Disposal Equipment / Services. International Trade Administration. Washington DC.

US EPA. Remediation Technologies Roundtable. (1993). *Cleaning Up the Nation's Hazardous Waste Sites: Markets and Technology Trends.* Washington DC.

US EPA. (1997). *Clu-In.* Washington DC.

US EPA, Federal Remediation technologies Roundtable. (1993). *Remediation Technology Screening Matrix and Reference Guide.* Washington DC.

World Bank (1994). *Managing Urban Environmental Quality in Asia.* Washington DC.

World Bank. (1995). *World Data 1995.* World Bank. Washington DC.

8

Contaminated Land Bioremediation : Back to Nature and the Future ?

LEE HAUGHTON AND STAN FROST*

INTRODUCTION

Soil is one of the key resources for sustainability and survival. It is has to be conserved is ways which retain its vital characteristics but so often we take it for granted. It is a complex ecosystem but our interest in soil is largely utilitarian. We manipulate and often exploit the soil to produce the plants it supports to feed ourselves. This situation occasionally leads to a degrading of the quality of the soil. Often, too, there may be willing or accidental contamination from industrial sources or degradation caused by salination and waterlogging. Land degradation is recognised as the loss of fertility or potential utility by changes in irreplaceable features or communities of organisms substituting one organic form for a lower organic form (Barrow, 1991).

Bacteria and fungi are natural recyclers capable of transforming natural and synthetic chemicals into sources of energy and raw materials for their own growth. This implies that biological processes supplement chemical or physical remediation processes. In the

* Environmental Resources University of Salford School of Environment and Life Sciences, Allerton Building, Frederick Road, Salford M6 6PU, UK.

medium to short term there might well be advantages in concentrating on and encouraging biological processes that are lower in cost and more environmentally benign (National Agricultural Library. U.S. Department of Agriculture, 2000).

Bioremediation is becoming important for the clean-up of contaminated soils around the world. The amount of bioremediation information continues to grow. This is particularly true of the internet. A search engine primed with the word bioremediation will return many thousands of hits. Sites such as *www.epa.gov/ORD/WebPubs/biorem* (Biodegradation Documents), *www.sgcleanup.com/bioremediation.html* (Soil and Groundwater Cleanup, corn) *weber.edu/Bioremediation/* (the Centre for bioremediation), and others, lead onto numerous pages and linked sites. In the UK, a 1988 survey of derelict land for the Department of the Environment suggested that there were up to 27,000 hectares of derelict land that could be contaminated (Pratt, 1993).

Bioremediation is becoming popular as a remedy for contaminated soil. In the UK, land is usually reclaimed by landfill or capping. Landfilling of contaminated soil moves the contamination from one place to another without treating it. Landfill space is increasingly in short supply and in the UK a Landfill Tax is making this form of remediation even more expensive. Landfill is less satisfactory environmentally than methods which remove contaminants from the soil or which break them down into harmless substances. More treatment technology is needed that is environmentally friendly, competitively priced, appropriate for large areas and based upon bioremediation (The UK Royal Commission on Environmental Pollution report "Sustainable use of soil" 1996).

After negotiations lasting over two years, the first mobile plant licence for the bioremediation of contaminated land was issued in 1998. Negotiations, which involved the Environment Minister, the Department of Environment Transport and the Regions, the Environment Agency and the Environmental Industries Commission, resulted in the initiation of on-site remediation. The company 'Bio-logic' owns four Mobile Plant Licences which allow them to carry out remediation projects throughout the UK. This eliminates the need for other documentation such as Waste

Management Licences. The licence leaves site with BioLogic and so avoids concerns associated with blight issues.

The Mobile Plant Licence must be sent to the Environment Agency or the Scottish Environmental Protection Agency for approval of particular site specific conditions. The earthworks contractors mobilise to excavate the contaminated soils to remove to a Waste Treatment Area surfaces capable of remaining intact when heavily trafficked in wet weather.

Once out of the ground, the soils are generally screened to remove any large lumps of demolition rubble such as concrete, brickwork, or old steel pipes and ironwork. The screened soils are laid out on the treatment area in the trapezoidal shape known as windrows or biopiles. These are usually 4-6m wide and up to 100m or more long. An average windrow will contain up to 500m^3 of soil. Treatment products are now added to the new windrows. These can take the form of solid composting substrates, fungal additives, nutrient blends in solid or liquid form and other organic soil conditioning agents. The constructed windrows are now sampled for chemical testing and regularly monitored to ensure that the treatment is maintained at an optimum level.

Machines turn and aerate the soil, providing the added oxygen to stimulate microbial activity. To protect the windrows from rainfall, the piles are covered in tarpaulins or bioremediation fleece. The latter is placed over the soils using an automated windrow covering machine which is placed on the back of a tractor. This allows the windrows to be covered and uncovered in a matter of minutes. When chemical testing shows that remediation is complete, some weeks later, the soils can be removed from the treatment area and replaced in the excavation, landscaping area or wherever the soils are to be reused (Bio-Logic Remediation Ltd, 2000). This new technology and legislation emphasises that soil is a valuable resource - it should be reused, not thrown away.

BIOREMEDIATION

Bioremediation is the use of micro-organisms in a technological process to decompose toxic pollutants into less harmful compounds

(Norris, et al., 1993). This process can be carried out *in situ* (processes occurring in un-excavated soil) or *ex situ* (excavated soil either taken off-site or treated on site) (Norris et al, 1993; Baker and Herson 1994; Rasul Chaudhry, 1994; Martin and Bardos, 1996).

The majority of bioremediation technologies were initially developed to treat petroleum hydrocarbon contamination (National Research Council, 1997), to immobilise contaminants (reactants) or to transform them to chemical products no longer hazardous to human health and the environment. Where contaminants pose no significant risk to water supply wells or surface water bodies bioremediation may be an appropriate strategy (BEST, 1999). End products this biodegradation process will include carbon dioxide, water and other compounds with little deleterious effect upon the environment (Baker & Herson, 1994).

Bioremediation is the optimisation of biodegradation and may be enhanced by:

(1) fertilising (adding nutrients such as carbon, nitrogen and phosphorous.) and/or (2) seeding with suitable microbial populations. This is enhanced or engineered bioremediation It speeds up the biodegradation process by augmenting the natural microbial communities, or by the addition of communities which may have been absent from the soil that are particularly suited to the pollution problem in hand.

Intrinsic bioremediation is the biological degradation of chemicals in the environment utilising existing microbial communities. It is not a "do nothing" approach. It involves:

(i) characterising the fate and transport of the chemicals to determine if their distribution is being effectively controlled;
(ii) evaluating the factors that will affect the long-term maintenance of the intrinsic processes; and
(iii) monitoring the intrinsic processes to ensure that requirements are being met (Beak, 2000).

Intrinsic Bioremediation is often the most cost effective method available for land contamination. Even in the most contaminated soils, indigenous microbial activity can be enough to clean the soil effectively. Microbial communities within contaminated ecosystems tend to be dominated by those organisms capable of utilising or surviving toxic contamination. These communities are typically less diverse than those in non-stressed systems (Baker & Herson, 1994).

Once the soil has been fertilised and/or seeded, control of temperature, water and oxygen content can be used to speed up the process or reduce the negative effects of factors such as air pollution (See Appendix 1). Bioventing is for contaminants that are more readily broken down aerobically than anaerobically. Oxygen is delivered to unsaturated soils by forcing air through the soil to enhance the biodegradation of organic contaminants. It maximises *in situ* biodegradation while minimising contaminant volatilisation (Norris, et al, 1993).

Engineered *in situ* bioremediation using bioventing was first applied in 1972 to clean up a spill of gasoline from a pipeline in Ambler, Pennsylvania. The process has changed radically since this first application. Water was first used to carry the oxygen around the soil, then hydrogen peroxide and finally injection of oxygen into the ground improved the bioremediation rate (National Research Council, 1997).

Bioremediation has proven to be an important soil/ groundwater remediation technology because it:

- Harnesses naturally-occurring biochemical processes;
- Destroys or immobilises contaminants rather than transfers them from one environmental media to another; and
- Conserves limited financial resources due to shortened cleanup times and can involve a lower capital expenditure compared to many other remediation technologies.

Soil remediation has superseded the established technologies of excavation followed by either incineration or landfilling (National Research Council 1997).

Other Technologies

The 1996 Review of Full Scale Treatment Technologies for the Remediation of Contaminated Soil, prepared by the Waste and Soil Treatment Division of the Centre for Research into the Built Environment for the Royal Commission on Environmental Pollution (Martin & Bardos, 1996) listed the treatment technologies available at the time and their relative cost (Table 8.1).

Table 8.1 : A comparison of soil remediation treatment costs

Treatment	*Approx. Cost/Tonne Soil*
Biological	£5-170
Chemical	£12-600
Physical	£20-170
Solidification/Stabilisation	£17-171
Thermal	£30-750

Biological treatments compare favourably with alternative methods. Treatment periods generally last from 2 to 48 months, about the same for chemical or thermal methods. Physical processes (soil washing and soil vapour extraction) is faster, rarely lasting more than one year. Solidification is almost instantaneous.

There is also the option of not cleaning the site and changing the end use when the contamination is not severe. However, local and regional planning laws may not allow this. Also, the end use of the site will not be as financially attractive (Failey & Scrivens, 1994).

Bioremediation (when used in isolation) does not require environmentally damaging processes such as chemicals or heat treatment. It has a beneficial effect upon soil structure and fertility (Martin & Bardos, 1996) but with limitations on its effectiveness:

- Susceptibility to inhibition by other toxic contaminants such as heavy metals;
- Low biodegradability of some contaminants such as chlorinated solvents;
- Possible residual contamination after treatment, such as using hydrogen peroxide as an oxygen provider;
- The potential formation of intermediate compounds which are more toxic than the original treatment, such as DDD and DDE from the breakdown of DDT(Martin & Bardos, 1996; Failey & Scrivens, 1994).

GENETIC RESEARCH

The successful use of micro-organisms in bioremediation depends upon the development of a basic understanding of the genetics of a broad spectrum of micro-organisms. Micro-organisms adapted to degrade specific pollutants have been found among populations growing naturally at polluted sites with genes that govern a wide variety of metabolic processes, including the degradation of environmental pollutants. (National Agricultural Library. U.S. Department of Agriculture, 2000).

Bacterial genetic information in the form of DNA occurs in the plasmids. Many bacterial plasmids contain the genes necessary for bioremediation which can be exchanged between bacteria. 'Blending' of bacterial traits creates organisms which biodegrade contaminants where the two 'unblended' bacteria would be unable to survive. Such natural genetic changes occur in a variety of environments (Baker & Herson 1994).

Genetic mutations can arise as a result of environmental conditions. For example, some bacteria degrade highly toxic trichloroethylene (TCE) only when the less toxic compound toluene was present. A minority of the bacteria mutate to degrade the TCE without toluene-however, those bacteria capable of TCE degradation may lack other traits, such as tolerance to salts, heavy metals and acid soils, which necessary for their use in bioremediation (National Agricultural Library. U.S. Department of Agriculture, 2000).

As more and more research is conducted into the modification of genetically created microorganisms for increased use in bioremediation governments and the public have concerns over the implications. Fears of the unforeseen effects of these 'un-natural' modifications are founded upon a lack of knowledge. One example of the problems of using genetically modified bacteria is their containment when used on-site.

Currently, *in situ* degradation processes can be measured by tracers and gas generation (National Agricultural Library. U.S. Department of Agriculture, 2000). The processes cannot be measured directly and there is uncertainty over how much of the process of *in situ* bioremediation is the biodegradation of the contaminant due to the actions of the microorganisms, or due to natural dispersion and attenuation (Norris, et al., 1993). If a genetically modified organism is introduced into and uncontrolled site, one cannot be sure that the organism is not going to be transported to nearby areas, with possible detrimental results. The effectiveness of introduced microorganisms reflects their ability to compete with indigenous species. The need to contain these organisms (to ensure that indigenous microorganisms are not superseded) means that they are generally only used in bioreactor treatments and enclosed or biological treatment lagoon reactors usually used for slurries with high water content (Baker & Herson, 1994; National Research Council, 1997).

CONCLUSION

Bioremediation has now become an established technology for a wide range of contaminants. There are certain contaminants that are beyond this technology so it would be unwise to assume that because bioremediation can provide an environmentally sound, cost-effective treatment, all attempts on all sites will be completely successful. This applies equally to all treatment methods. In spite of, or perhaps because of its success, there still needs to be a guarantee that individual treatment efforts are both appropriate and effective. It is important from a regulatory perspective that treatments which do not fulfil their objectives or which cause additional, perhaps unforeseen, environmental problems are identified early and that lessons are learned for the future. Bioremediation is successful using existing non-genetically modified techniques. Caution needs to be introduced when considering the use of genetically modified bacteria. The question of 'why introduce the genetic risk?' needs to be asked.

After a somewhat protracted series of negotiations lasting over two years, Bio-Logic became the proud owner of the UK's first Mobile Plant Licence for the bioremediation of contaminated land in 1998. Negotiations, which involved Environment Minister Michael Meacher, the DETR, the Environment Agency and the Environmental Industries Commission, resulted in the on-site remediation it is now possible to carry out today. Bio-logic now owns four Mobile Plant Licences which allow them to carry out remediation projects throughout the UK.

At each of Bio-Logic's sites, a Mobile Plant Licence is in place, eliminating the need for Waste Management Licences. As the licence leaves site with BioLogic, it also removes concerns associated with blight issues.

For each new project, a Mobile Plant Licence must be sent to the Environment Agency or the Scottish Environmental Protection Agency for approval. This will be returned to Bio-Logic with any site specific conditions in place.

By the time the Mobile Plant Licence is in place, the main earthworks contractor will be ready to mobilise to site to begin excavating the contaminated soils. Firstly the proposed Waste Treatment Area must be prepared. These can be any surface which is capable of remaining intact when heavily trafficked in wet weather. In the past, this has included old football pitches, compacted fill, concrete slabs, and tarmac surface. The excavation works can now commence.

Fig. 8.1 : Excavation of Contaminated Soil

Once out of the ground, the soils are generally screened to remove any large lumps of demolition rubble such as concrete, brickwork, or old steel pipes and ironwork. The screened soils are now laid out on the treatment area in the trapezoidal shape known as windrows or biopiles. These are usually 4-6m wide and up to 100m or more long. An average windrow will contain up to 500m^3 of soil. Treatment products are now added to the new windrows. These can take the form of solid composting substrates, fungal additives, nutrient blends in solid or liquid form and other organic soil conditioning agents. The constructed windrows are now sampled for chemical testing and regularly monitored to ensure that the treatment is maintained at an optimum level.

Fig. 8.2 : Constructing Windrows

The 'Goliath' machine then turns and aerates the soil, providing the added oxygen needed to stimulate microbial activity. To protect the windrows from rainfall, the piles are covered in tarpaulins or bioremediation fleece. The latter is placed over the soils using an automated windrow covering machine which is placed on the back of a tractor. This allows the windrows to be covered and uncovered in a matter of minutes where the tarpaulins can take up to an hour to remove and replace.

Fig. 8.3. The 'Goliath' Soil Aeration Machine

When successful results are received from the chemical testing laboratory some weeks later, the sign off sheets are presented to the client and once signed the soils can be removed from the

Fig. 8.4: Covering Windows with 'Fleece'

treatment area and replaced in the excavation, landscaping area or wherever the soils are to be reused. (*Source:* Bio-Logic Remediation Ltd, 2000).

REFERENCES

Baker, K.H. and Herson, D.S. 1994. *Bioremediation.* New York. McGraw-Hill Inc.

Beak, 2000. Beak International Incorporated / Beak Consultants Incorporated. *Frequently Asked Questions* [on-line]. Available internet: *http://www.beak.com/Services/BioRemediation/BioFAO.htm* Accessed on 1/5/00

BEST, 1999. Bioremediation, Education, Science and Technology homepage [on- line] Available internet: http://bark214-3.berkeley. udu/BEST/ 'Besbiorem.html, Accessed on 1/5/00.

Bio-logic Remediation Ltd. 2000. *Bio-logic Remediation Homepage* [on-line], Available internet: *www.biologic.com.* Accessed on 1/5/00.

Failey, R.A. and Scrivens, A.J. 1994. Contaminated Land. Assessment and Redevelopment. Cheltenham, UK. Stanley Thornes Publishers.

National Agricultural Library. U.S. Department of Agriculture, 2000. *Biotechnology For The 21st Century: NEW HORIZONS* [on-line]. Available internet: *http://www.nal.usda.gov/bic/bio21/envron.html#bio*, Accessed 1/5/00.

National Research Council, 1997. *Innovations in Ground Water and Soil Cleanup: From Concept to Commercialization.* Washington D.C. National Academy Press.

Norris, Hinchee, Brown, McCarty, Semprini, Wilson, Kampbell, Reinhard, Bouwer, Borden, Vogel, Thomas and Ward. 1993. *Handbook of Bioremediation.* Boca Raton, Florida. Lewis Publishers.

Pratt, M. 1993. *Remedial Processes for Contaminated Land.* Warwickshire, UK. Institution of Chemical Engineers.

Rasul Chaudhry, G. 1994. *Biological Degradation and Bioremediation of Toxic Chemicals.* London. Chapman and Hall Publishers.

UK Royal Commission on Environmental Pollution, 1996. *Sustainable use of soil.* London. HMSO.

9

Environmental Clean-up by Marine Bacteria

K. Latha, N. S. Kavitha and A. Hilda*

The possibilities for conjugal transfer of a well characterised degradative plasmid (SAL) from terrestrial to marine bacteria was investigated. The SAL plasmid is known to metabolise aromatic organic compounds like sodium salicylate and sodium benzoate. Conjugal transfer was comparatively high in matings conducted on nutrient plates than in broth mating. The number of transconjugants obtained increased with an increase in incubation time. Conjugal transfer varied significantly with the change in incubation temperature. Maximum transfer was observed at 10°C. The viability of the donor, *Pseudomonas putida, in* sea water gradually decreased and after 16 days of incubation reached the viable but non-culturable (VNC) state. The SAL plasmid was stably maintained in the transconjugants under stress conditions (500 mM salicylate) whereas under non-stress conditions they progressively lost the plasmid. The transconjugants expressed the SAL plasmid encoded metapathway enzymes involved in the

* CAS in Botany, University of Madras, Guindy Campus, Chennai-600 025, India.

degradation of the organic compound, sodium benzoate and salicylate. Therefore by introducing catabolic genes into the marine bacteria it is possible to enhance their degradative potential for a near clean environment.

Key words : marine bacteria; SAL plasmid; expression and stability of SAL plasmid

INTRODUCTION

The critical role of the microorganisms in the degradation of organic pollutants is well known. Although the microorganisms capable of degrading organic pollutants and their catabolic pathways have been investigated intensively, information on the microbial degradation of xenobiotics in extreme environments such as high salinity, alkalinity and acidity is still very limited (Oren, 1992).

The pollution of the marine ecosystem has tremendously increased in the recent years with the ever increasing discharge of industrial effluents, municipal sewage and oil spills due to accidents (Burton et al., 1982). Only very limited work has been done on the potential of marine bacteria to degrade toxic compounds. Besides their usefulness in terms of biotechnological

Table 9.1 : Antibiotic sensitivity

Organism	*Antibiotics (MIC - 25 $\mu g/\mu l$)*
Donor	
P. putida (SAL)	Am^+ Cm^+ Gn^- Kn^- Tet^+
Recipients	
Micrococcus sp.	Am^- Cm^+ Gn^+ Kn^+ Tet^+
M. luteus	Am^+ Cm^+ Gn^+ Kn^+ Tet^-
B. pumilus	Am^+ Cm^+ Gn^+ Kn^+ Tet^-
M. varians	Am^+ Cm^+ Gn^+ Kn^+ Tet^-
M. halobius	Am^- Cm^+ Gn^+ Kn^- Tet^+
H. salinarium	Am^+ Cm^+ Gn^+ Kn^+ Tet^+

+ Resistant ; - Sensitive ; Am Ampicillin ; Cm Chloramphenicol ; Gn Gentamicin ; Kn Kanamycin ; Tet Tetracycline.

scale up is not known. The compounds shown to be degraded at high salt concentrations by marine bacteria include chlorophenols (Boone et al., 1989), organophosphorus compounds (DeFrank and Cjeng, 1991), nitroaromatic compounds (Oren et al., 1991), and formaldehyde (Oren et al., 1992). However, introduction of specialized marine bacteria into the marine waters to deal with the degradation of specific pollution events has not been attempted so far. The genetics of most of the marine bacteria completely remains unexplored due to the lack of suitable genetic tools which has classically hampered the genetic manipulation of marine bacteria (Ventosa et al., 1998).

The studies on plasmid transfer (by conjugation) in aquatic systems have acquired greater relevance since the amount of wastewater dumped grows daily, both in urban and rural enviroriments (Astroga et al... 1992). Although conjugal transfer of plasmid mediated genes has been demonstrated in a variety of environments such as rivers (Grabow et al., 1975; Bale et al., 1988), lakes (O'Morchoe et al., 1988), ponds (Gowland and Slater, 1984), sewage (Oren et al., 1991), not much is known about the gene exchange in the marine ecosystem. Terrestrial bacteria containing catabolic plasmids have been isolated from a wide variety of environments. These plasmids harbour genes for the degradation of recalcitrant compounds. Since many of the pollutants that accumulate in the terrestrial environment also find their way to the marine environment, it would be a useful strategy if the catabolic plasmids from terrestrial bacteria are transferred to marine bacteria which will enable the marine bacteria to degrade the otherwise persistent compounds. Therefore by introducing catabolic genes into the marine bacteria it is possible to enhance their degradative potential for a near clean environment.

Once the possibility of transfer is demonstrated it is essential to study the influence of abiotic and biotic factors of the system in which the process of transfer takes place (Bale et al., 1988; Astroga et al., 1992). However, biotic and abiotic factors affecting transfer in aquatic systems are rarely investigated (O'Morchoe et al., 1988; Astroga et al., 1992). It is also generally assumed that the organism carrying a newly introduced plasmid will lose its plasmid unless the environment imposes a selective pressure for

Table 9.2 : Conjugal transfer of SAL plasmid from *P. putida* to marine isolates at 30°C during broth mating

Incubation Time (min)	*No. of Transconjugants (CFU/ml)*					
	Micrococcus sp.	*M. luteus*	*B. pumilus*	*M. varians*	*M. halobius*	*H. salinarium*
30	9.9×10^2	-	-	-	-	-
45	1.4×10^3	6.0×10^2	6.0×10^2	9.7×10^2	7.1×10^2	9.7×10^2
60	5.2×10^4	1.4×10^3	4.9×10^3	3.5×10^3	1.2×10^3	4.4×10^3
120	7.9×10^5	4.0×10^3	9.9×10^4	9.9×10^4	1.9×10^3	1.1×10^5
180	7.1×10^6	4.3×10^5	4.1×10^6	9.0×10^6	4.7×10^5	3.7×10^6
240	2.7×10^8	4.2×10^6	7.7×10^7	1.1×10^7	8.9×10^6	2.4×10^7
300	6.1×10^8	9.9×10^7	5.7×10^8	1.2×10^7	2.9×10^7	9.0×10^7

plasmid maintenance (Freter, 1984: Sobecky et al., 1992). This paper presents evidence for the conjugal transfer of a well characterised degradative plasmid (SAL) from terrestrial to marine bacteria. The influence of incubation time and temperature on conjugal transfer, the expression of the catabolic plasmid in the marine bacteria and the stability of the introduced plasmid under stress (500mM sodium salicylate) and non-stress conditions in the marine isolates have also been investigated.

MATERIALS AND METHODS

Microorganisms and Growth conditions

Pseudomonas putida DSM 4478 harbouring the 85 kb SAL plasmid obtained from the Institute of Microbial Technology, Chandigarh, India, was used as the donor. Of the six marine isolates used as recipients, two strains, namely *Micrococcus halobius* ATCC 21727 and *Halobacterium salinarium* MTCC 1595 were obtained from the Institute of Mcrobial Technology, Chandigarh, India and the remaining four were isolated from seawaters collected from Foreshore Estate, Chennai. Based on the biochemical characteristics outlined in Bergey's manual (Buchanan and Gibbons, 1974), they were identified as *Bacillus pumilus, Micrococcus luteus, Micrococcus varians* and *Micrococcus* sp.

The donor strain was grown in Luria Bertani (LB) medium (Peptone 1%, Yeast extract 0.5%, NaCl 0.5%) and the marine isolates were grown in Upper Bay Yeast extract (UBY) medium (Caldwell et al., 1989) (NaCl 0.5%,KCI 0.016%, Paptone 0.1%, Yeast extract 0.1%, Agar 1.5%, aged seawater 75 ml, distilled water 25 ml).

Antibiotic sensitivity testing

The sensitivity of bacteria to various antibiotics was determined by plating them on UBY (for the marine isolates) and LB agar (for the donor) supplemented with various concentrations of ampicillin, chloramphenicol, gentamicin, kanamycin and tetracycline. The minimal concentration inhibiting bacterial growth was determined (MIC).

Table 9.3 : Conjugal transfer of SAL plasmid from *P. putida* to marine isolates at 30°C during plate mating

Incubation Time (min)	*No. of Transconjugants (CFU/ml)*					
	Micrococcus sp.	*M. luteus*	*B. pumilus*	*M. varians*	*M. halobius*	*H. salinarium*
30	1.0×10^3	6.0×10^2	7.1×10^2	4.1×10^2	7.1×10^2	3.9×10^2
45	2.4×10^3	1.0×10^3	2.1×10^3	2.1×10^3	1.0×10^3	1.0×10^3
60	6.0×10^4	4.2×10^3	5.4×10^4	8.7×10^3	2.5×10^4	6.4×10^3
120	4.9×10^6	3.1×10^5	1.1×10^5	1.0×10^5	3.0×10^5	4.9×10^5
180	8.7×10^6	7.1×10^6	7.3×10^6	3.9×10^6	5.7×10^5	5.9×10^6
240	6.1×10^8	9.7×10^7	9.9×10^7	1.1×10^7	8.7×10^6	6.4×10^7
300	9.4×10^8	5.2×10^8	7.0×10^8	9.6×10^7	5.1×10^7	5.7×10^8

CONJUGATION EXPERIMENTS

Broth mating

The cell density of overnight cultures of donor and recipient bacteria was adjusted to an O.D of 1.0 at 600nm (approximately 10° CFU per ml). 100 µl of culture from the donor as well as from the recipient were aliquoted into a sterile test tube to which 800 µl of sterile LB broth was added and was incubated at 37°C on a rotary shaker. At regular intervals, aliquots of this suspension were serially diluted and plated on UBY medium amended with various antibiotics (MIC-25 µg/µl). The plates were incubated at 37° C for 24 h and then observed for the growth of transconjugants. They were replica plated onto UBY medium amended with antibiotics. Only those colonies that grew in all the plates were scored as transconjugants.

Plate mating

The density of the donor and recipient cultures were adjusted to 10° CFU per ml and 100 µl of donor and recipient culture were spotted on UBY agar plate. The plates were incubated at 37° C. At required time intervals the bacterial spots were moistened with phosphate buffer (0.1 M, pH 7.0) and gently scrapped with a sterile glass rod. The cell suspension was serially diluted and plated on UBY medium amended with appropriate antibiotics (MIC-25 µg/µl). The colonies so obtained were replica plated on antibiotic amended UBY medium and the number of transconjugants formed were scored.

Stability of transconjugants

The stability of the SAL plasmid in the transconjugants and also in the natural host *P. putida* DSM 4478 was studied under stress (500mM sodium salicylate) and non-stress conditions. The overnight grown tansconjugants were centrifuged and the cells were transferred to about 100 ml of sterile sea water and incubated for a period of 30 days. On every third day about 100 µl aliquots were plated on UBY agar and the resultant colonies were replica plated and scored for plasmid loss.

Table 9.4 : Influence of mating temperature on conjugal transfer of SAL plasmid from *P. putida* to marine isolates after 6h of incubation

Temperature	No. of Transconjugants (CFU/ml)					
	P. putida vs *Micrococcus sp*	*P. putida* vs *M. luteus*	*P. putida* vs *B. pumilus*	*P. putida* vs *M. varians*	*P. putida* vs *M. halobius*	*P. putida* vs *H. salinarium*
3°C	$9.7 x 10^6$	$9.7 x 10^6$	$9.4 x 10^6$	$4.7 x 10^7$	$9.4 x 10^6$	$9.8 x 10^6$
10°C	$5.1 x 10^8$	$7.3 x 10^7$	$1.9 x 10^8$	$4.6 x 10^8$	$3.1 x 10^8$	$3.2 x 10^8$
30°C	$3.1 x 10^8$	$7.4 x 10^7$	$7.2 x 10^7$	$1.1 x 10^8$	$5.7 x 10^8$	$6.2 x 10^8$
50°C	$9.2 x 10^7$	$4.9 x 10^7$	$1.0 x 10^7$	$6.9 x 10^8$	$4.7 x 10^7$	$5.2 x 10^8$
60°C	$6.9 x 10^6$	$6.3 x 10^6$	$6.9 x 10^6$	$6.9 x 10^6$	$5.7 x 10^6$	$4.9 x 10^6$

Viability of donor in sea water

Freshly grown cells of P. *putida* were centrifuged and resuspended in sterile seawater and were incubated on a rotary shaker for -a period of 30 days. The viable count of the cells was determined at every 24 hours by plating them onto LB medium.

Enzyme assays

Catechol 2, 3-oxygenase and 2-HMS hydrolyase were assayed from cell free extracts as described by Chakrabarthy (1972). Enzyme assays were performed using a Beckman DU 40 Spectrophotometer at 25°C. Total protein in the supernatants was determined by the method of Bradford (1976).

RESULTS AND DISCUSSION

The sensitivity of the bacteria to various antibiotics was tested and the minimal inhibitory concentration (MIC) was found to be 25 μg/μl for all the organisms tested. The sensitivity of the donor and recipients to various antibiotics at MIC formed the antibiotic markers for further conjugation experiments (Table 9.1).

The antibiotic markers for the donor, P. *putida* was Am^+, Tet^+, Kn^-. For the marine recipients the following antibiotic markers were used *Micrococcus* sp. : Am^- ,Kn^+; *M. luteust* Kn^+,Tet^- ; B. *pumilus* and *H,* varians : Am^-, Tet^-; *M. halobius:* Am^-, Gn^+; H. *salinarium : Gn^-, Kn^+.*

Since the donor, *P. putida* could not grow on marine agar the transconjugants were screened based on their ability to grow on UBY agar amended with specific antibiotics.

Intially, broth mating experiments were carried out to determine the initial cell density of both the donor and the recipients. Mating experiments were carried out at D/R ratios of 1:1, 2:1 and 1:2. The influence o-f donor/recipient ratios on the conjugal rate of transfer is summarised in Fig. 9.1. There was not much variation in the conjugal frequency in the range of ratios used. Of the six marine isolates used. *Micrococcus* sp. underwent conjugation from the initial period (30 min.) itself. The number of transconjugants obtained was about 10^6–10^7 at the end of the mating period (6 h). According to Gauthier et al., (1985) the

Table 9.5 : Stoichiometry of 2-HMS formation from catechol by heattreated extracts of Donor and Transconjugants grown on salicylate after 48 h of incubation*

Organism	*2-HMS formed (μM)*			
	Catechol (μM)			
	100	*200*	*300*	*400*
Donor				
P. putida (SAL)	71.5	149.3	287.1	354.2
Receipients				
Micrococcus sp	45.6	97.2	184.4	245.3
M. luteus	39.1	78.8	154.6	206.2
B. pumilus	44.2	88.1	178.2	177.9
M. varians	43.1	88.1	177.9	241.2
M. holobius	44.2	84.3	181.4	248.7
H. salinarium	44.8	89.4	182.6	253.5

* Total 2-HMS formed has been determined from absorbance at 375 nm using a molar extinction coefficient of 3.0×10^4 (7).

Table 9.6 : Specific activities of enzymes of meta pathway in cellfree extracts of Donor and Transconjugants grown on salicylate

Organism	*Specific activity (units)*	
	Catechol 2, 3 oxygenase	*2-HMS hydrolyase*
Donor		
P. putida (SAL)	1.33	1.23
Recipients		
Micrococcus sp.	0.64	0.68
M. luteus	0.63	0.65
B. pumilus	0.66	0.64
M. varians	0.68	0.69
M. halobius	0.67	0.67
H. salinarium	0.68	0.67

* Specific activities of the enzymes have been described as micromoles of product formed or substrate used per minute per milligram of protein.

majority of plasmids transferred maximally at donor to recipient ratios of 1:1, while Graves and Riggs (1980) reported that transfer frequencies were higher at ratios < 15 and > 1. However, in the present study the donor recipient ratio did not have any influence on the number of transconjugants obtained. Thus, the initial cell density of the D/R was taken as 1:1 for further conjugation experiments.

The results of broth mating experiments conducted with *Pseudomonas putida* (harbouring the SAL plasmid) and the marine isolates are summarised in Table 9.2. Maximum transconjugants was obtained with *Micrococcus* sp. which underwent conjugation right from the initial period (i.e. 30 min) of incubation. The number of transconjugants detected was between 10^6–10^7 at the end of the mating period (after 6 h) for the rest of the marine isolates. The number of transconjugants obtained was more in plate mating when compared to broth mating (Table 9.3). The conjugal transfer in plate mating ranged to a maximum number of transconjugants obtained to the order of 10^7 in most marine isolates and a minimum of 10^6 in the case of *M. varians* and *M. halobius*.

As evidenced in the present, study, gene transfer in broth was significantly less than that on agar medium which may be due to the fact that cell to cell contact is more in solid surface than in liquid medium. Similar conclusion was obtained by Genthner et al., (1988) who reported that solid environment was superior to liquid environment for plasmid transfer.

Interestingly, the transfer of SAL plasmid from *P. putida* to the marine isolates varied with incubation time (Tables 9.2 and 9.3). The number of transconjugants obtained increased with the increase in incubation time. However, there was no further increase in the number of transconjugants after 6h of incubation period. These results are in accordance with the results of Rochelle et al., (1989) who detected maximum transfer after 5 h of incubation. Furthermore, *Micrococcus* sp. showed maximum transfer than the other marine isolates.

Further, the number of transconjugants obtained was found to vary significantly with the change in incubation temperature

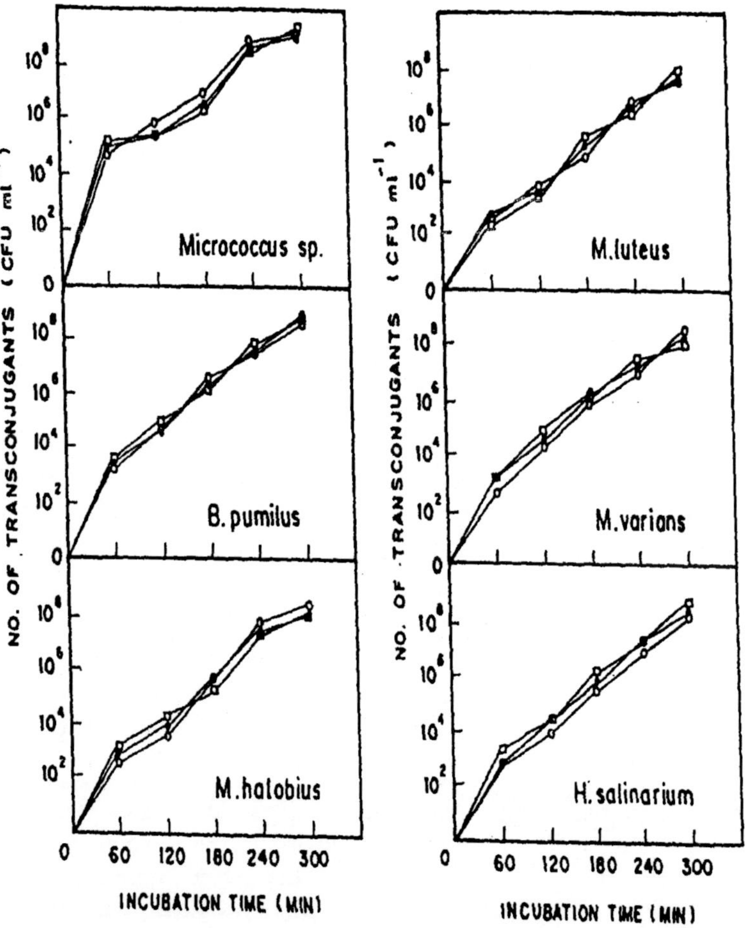

Fig. 9.1 : Influence of donor/recipient ratios on conjugal transfer of SAL plasmid from *P. putida* to marine isolates. Symbols: (●) 1 : 1 D/R ratio ; (○) 2:1 D/R ratio ; (□) 1:2 D/R ratio.

and maximum number of transconjugants were obtained at 10°C. Minimum transfer, of about 10^5 was observed at 3°C and 60°C (Table 9.4). However, in the case of *M. vcurians* and *H, salinarium* the number of transconjugants obtained was comparatively more which was around 10^7 at 50°C. The rate of conjugal transfer was more or less same in most marine isolates at the temperature range from 10-50°C. On the contrary O'Morchoe et al., (1988) showed the occurrence of maximum transfer frequencies between 20–30°C.

The viability of *P. putida* in seawater was assessed by incubating the bacterium (1×10^9CFU per ml) in marine water without any added nutrients. There was a progressive loss of colony forming ability (Fig. 9.2). The colony count dropped down to almost nil after nearly two weeks of incubation. Microscopic examination showed the presence of cells which suggest that the cells lost their colony forming activity completely after 16 days of incubation in marine water. The result is in accordance with that of Byrd and Colwell (1990) who demonstrated that when *E. coli* cells were placed in sterile artificial sea water, the cells became viable but non-culturable. It is generally agreed that when a terrestrial bacterium is released into the ocean it enters into a viable-but-non culturable (VNC) state (Roszak and Colwell, 1967) and the VNC state of a bacterium appears to be a compromise for the strategic survival of the organism when exposed to conditions less optimal for growth, such as the exposure to saline environments for non-marine bacteria (Colwell et al., 1976).

To assess the stability of the introduced plasmid, the transconjugants were studied under both, stress and non-stress conditions. In the present study, the introduced plasmid in the marine isolates was stably maintained under stress conditions even after 30 days of exposure whereas under non-stress conditions plasmid loss occurred after 24 days (Fig. 9.3). It is generally assumed that added metabolic burden due to a newly introduced plasmid would most probably result, either in complete or partial loss of the plasmid unless the environment imposes a selective pressure for stable plasmid maintenance (Freter, 1984).

The donor, *P. putida* harbouring the SAL plasmid is known to metabolise salicylate via meta pathway (Chakrabarthy, 1972).

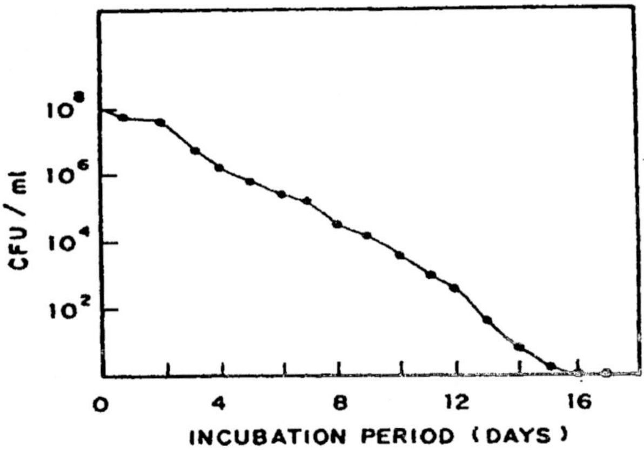

Fig. 9.2 : Viability of the donor, *P. putida* (SAL) in seawater.

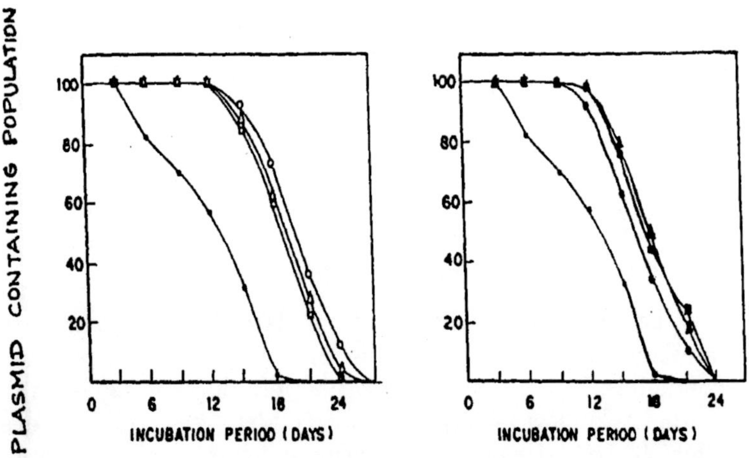

Fig. 9.3 : Stability of the plasmid in the donor and transconjugants under non-stress condition. Symbols: (•) *P. putida* (SAL) ; (○) *Micrococcus sp.*; (▲) *M. luteus;* (□) *B. pumilus;* (●) *M. varians;* (▲) *M. halobius;* (■) *H. salinarium.*

Catechol is the central metabolic intermediate which is converted to 2-HMS by catechol 2, 3-oxygenase. Therefore, in order to determine the expression of SAL plasmid in the marine recipients, cell-free extracts of transconjugants were assayed for the accumulation of 2-HMS (Table 9.5). While the donor cells yielded 71.5 μM of 2-HMS after 48h of incubation with 100 μm of salicylate, all the transconjugants yielded 2-HMS ranging from 39.1–45.6 μM. However, when the concentration of salicylate was increased to 400 μM, the stoichiometry of conversion of salicylate to 2-HMS also increased both in donor as well as in the transconjugants (Table 9.5). The expression of SAL plasmid in the marine recipients was further substantiated by the results shown in Table 9.6, where the specific activities of the enzymes, catechol 2, 3 oxygenase and 2-HMS hydrolyase in the donor were about 1.33 and 1.23 units while that of the transconjugants, the values ranged from 0.64 to 0.69 units. These data clearly confirm the functional expression of the SAL plasmid encoded genes in the marine recipients.

Although the stoichiometric conversion of salicylate to 2-HMS by the transconjugants was comparatively lower than the donor strain, the fact that the marine isolates which could not utilise salicylate have acquired the capability to grow on salicylate after conjugation. These results clearly suggest the expression of plasmid encoded genes viz. the catechol 2, 3 oxygenase and 2-HMS hydrolyase that are involved in the catabolism of salicylate.

Therefore, by the introduction of catabolic genes into marine bacteria it is possible to biodegrade organic pollutants at high salt concentrations. This study poses only as a model system whereby the introduction of catabolic plasmids to the marine bacteria will enhance their degradative potenial.

ACKNOWLEDGEMENTS

The authors express their thanks to the Director, Centre for Advanced Studies in Botany, University of Madras, Chennai, for providing necessary facilities and the CSIR for the financial assistance.

REFERENCES

Bale, M.J., Day, M.J., Fry J.C., 1988. Novel method for studying plasmid transfer in undisturbed river epilithon. Appl. Environ. Microbiol. 54, 2756-2758.

Boone, D.R., Johnson, R.L., Chen, D.C., Mathrani, I.M., Mah, R. A., 1989. Methanogenesis and reductive dechlorinations in an alkaline, hypersaline sediments and groundwater. pp. 205-215. *In* M. S. Da Costa, J. C. Durate and R. A. D. Williams (ed.). Microbiology of Extreme Environments and its Potential for Biotechnology. Elsevier Applied Science, London.

Bradford, M.M., 1976. A rapid and sensitive method for the quantification of microgram quantities of protein utilizing the principle of protein dye binding. Anal. Biochem. 72, 248-254.

Buchansan, R.E., Gibbons, N.E., 1974. Bergey's Manual of Determinative Bacteriology, 8th Edition. Williams and Wilkins. Baltimore.

Byrd, J.J., Colwell, R.R., 1990. Maintenance of plasmids pBR322 and pUC8 in nonculturable *Escherichia* coli in the marine environment. Appl. Environ. Microbiol. 56, 2104-2107.

Caldwell, B.A., Ye, C., Griffiths, R.P., Moyer, C.L., Morita, R.Y., 1989. Plasmid expression and maintenance during long-term starvation--survival of bac:teria in well water. Appl Environ. Microbiol. 55, 1860-1864.

Chakrabarthy, A.M.. 1972. Genetic basis of the biodegradation of salicylate in *Pseudomonas* sp. J. Bacteriol. 112, 815-823.

Colwell, R.R., Sayler, G.S., Nelson, J.D., Justice, A., 1976. Microbial mobilisation of mercury in aquatic environments pp. 473-481. *In* Nrigau (ed) Environmental biogeochemistry Vol 2. Michigan, Ann Arbor Sciences.

DeFrank, J., Cjeng, T.C., 1991. Purification and properties of an organophosphorus acid anhydrase from a halophilic bacterium. J. Bacteriol. 173. 1938-1943.

Fernandez-Astroga, A., Muela, A., Cisterna, R., Iriberri, J., Barcina, R., 1992. Biotic and abiotic factors affecting plasmid transfer in *Escherichia coli* strains. Appl. Environ. Microbiol. 58, 392-398.

Freter, R., 1984. Factors affecting conjugal transfer in natural bacterial communities, pp. 105-114. *In* M. J. Klug and C. A. Reddy (ed.), Current persp ctives in microbial ecology, American Society for Microbiology, Washington, D.C.

Genthner, F.J., Chatterjee, P., Barkay, T., Bourquin, A.W. 1988. Capacity of aquatic bacteria to act as recipients of plasmid DNA. Appl. Environ. Microbiol. 54, 115-117.

Gowland, P.C., Slater, J.H., 1984. Transfer and stability of drug resistance plasmids in *Escherichia coli* strains K12. Microb. Ecol. 10, 1-13.

Grabow, W.O.K., Prozesky, O.W., Burger, J.S. 1975. Behaviour in a river and dam of coliform bacteria with transferable or non-transferable drug resistance. Water Res. 9, 777-782.

O'Morchoe, S. B., Ogunseitan, O.G., Sayler, S., Miller, R.V., 1988. Conjugal transfer of R 68.45 and FP5 between *Pseudomonas aeruginosa* strains in a freshwater environment. Appl. Environ. Microbiol. 54, 1925-1929.

Oren, A., Gurevich, P., Henis,Y., 1991. Reduction of nitrosubstituted aromatic compounds by the halophilic anaerobic eubacteria *Haloanaerobium pravalenes* and *Sporohalobacter marismotui* Appl. Environ. Microbiol. 57, 3367-3370.

Oren., A., Gurevich., P., Asachi, M., Henis, Y., 1992. Microbial degradation of organic pollutants at high salt concentrations. Biodegradation 3, 387-398.

Rochelle, P. A., Fry, J.C., Day, M.J., 1989. Factors affecting conjugal transfer of plasmids encoding mercury resistance from pure cultures and mixed natural suspensions of epilithic bacteria. J. Gen. Microbiol. 135, 409-424.

Roszak, D.B., Colwell, R.R., 1987. Survival strategies of bacteria in the natural environment. Appl. Environ. Microbiol. 51, 904-907.

Saye, D.J., Miller, R.V., 1989. The aquatic environment: Consideration of Horizontal gene transmission in a diversified habitat, pp. 223-259. *In* S.B. Levy, and R.V. Miller (ed.). Gene Transfer in the Environment. McGraw-Hill Publishing Company.

Sobecky, 'A. P., Schell, M.A., Moran, M.A., Hudson, R.E., 1992. Adaptation of model genetically engineered microorganisms to Lake waters : Growth rate enhancements and plasmid loss. Appl. Environ. Microbiol. 58, 3630-3637.

Truper, H.G., Severin, J., Wohlfahrt, A., Muller, A., Galinski, E.A., 1991. Halophily, taxonomy, phylogeny and nomenclature. pp. 3-7. *In* F. Rodriguez-Valera (ed) General and Applied aspects of Halophilic Microorganisms. Plenum Press, New York.

Ventosa, A., Nieto, J.J., Oren, A., 1998. Biology of moderately halophilic aerobic bacteria. Microbiol. and Mol. Biol. Reviews. 62. 504-544.

10

Use of *Mucor piriformis* for Steroid Transformation

P. Shubha Maiya, K.M. Madyastha* and S.B. Sullia

A *Mucor* sp. isolated from the soil and identified as *Mucor piriformis* was used to study the mode of transformation of 1,4 Androstadiene 17β-ol, 3-one (I). The major metabolite isolated and identified was 14α, 17β-dihydroxy 1,4 androstadiene 3-one (IA). Among the minor metabolites, 14α-hydroxy 1,4 androstadiene 3,17-dione (IB), 7α, 17β-dihydroxy 1,4 androstadiene 3-one (IC), 7α-hydroxy 1,4 androstadiene 3,17-dione (ID) and 7α,14α,17β-trihydroxy 1,4 androstadiene 3-one (IE) have been characterised. Metabolites IB to IE appear to be hitherto unknown. Time course studies suggested that the transformation is initiated by hydroxylation at 14α position leading to the formation of a dihydroxy metabolite (IA). Incubation of IA with the organism resulted in the formation of IB suggesting the presence of 17β-hydroxy steroid dehydrogenase activity. *M.piriformis* has been shown to carry out 14α-hydroxylation in the steroid molecule containing a 1,4 diene-3-keto system.

• Department of Organic Chemistry, Bio Organic Chemistry section, Indian Institute of Science, Bangalore.

Department of Microbiology, Jnana Bharathi, Bangalore University, Bangalore.

INTRODUCTION

Microorganisms have been successfully utilised to carry out critical steps in the synthesis of a number of pharmaceutically important steroids. The remarkable aspect of microbes is their ability to carry out reactions with high catalytic efficiency and specificity. They are also eco friendly. The environmental pollution due to various chemical reagents can be restricted to a considerable extent if biocatalysts are used.

Although fungi belonging to the genus *Mucor* have been shown to carry out various transformations in C19 and C21 steroids (Charney and Herzog , 1967, Iizuka and Naito, 1981, Mahato et al. 1989, Cabral *et al.,* 1997) very little work has been carried out on the transformation of steroids using *M. piriformis. Mucorales* are known for their versatile nature in introducing hydroxyl functions at different positions in the steroid molecule (Charney and Herzog, 1967, Holland and Riemland, 1985, Madyastha and Srivatsan, 1987) . However, details of many of these transformations are covered by patents (Charney and Herzog, 1967) *Mucorales* are also known to possess 3-keto and 17-keto oxido-reductase activities (Charney and Herzog, 1967, Holland and Thomas, 1982, Holland et al., 1984).

M. piriformis was shown to carry out 6β, 9α. and 11α hydroxylations of steroids (Eroshin 1962, Murray and Peterson 1957). We have previously studied the transformation of progesterone (Madyastha and Srivatsan 1987) Androst-4-ene-3,17-dione, Testosterone (Krishnan et al., 1991) 17α-hydroxyprogesterone and 16 dehydroprogesterone (Madyastha and Joseph, 1994) dehydroepiandrosterone and pregnenolone (Madyastha and Joseph, 1995) using a versatile fungal species isolated in our laboratory, identified as *Mucor piriformis* (Madyastha and Srivatsan, 1987). This organism was shown to initiate transformation of these steroids by introducing a hydroxyl function at the 14α position, except in 17α-hydroxyprogesterone. The inability of the organism to effect hydroxylation at the 14α position in the 17α-hydroxyprogesterone could be due to the spatial proximity of the 17α group to the 14α position (Singh et al., 1967). It was observed that bulky substitution at 17α position sterically hinders hydroxylation at 14α position and directs it to 7α position. Earlier studies (Madyastha and Joseph, 1995) have also demonstrated that pregnenolone and dehydroepiandrosterone were not hydroxylated at the 14α position. This again suggests that 4-ene-3-one structural unit of A-ring is an obligatory structural requirement for the organism to carry out 14α hydroxylation. Since 14α hydroxylation was noticed with steroids having a 4-ene-3-one structural unit, it is of interest to find out whether the organism would accept a steroid molecule in which 4-ene-3-one structural unit is cross-conjugated with an additional double bond as seen in 1,4 Androstadiene 17β- ol, 3-one (I). Hence the transformation of (I) using *M. piriformis* was undertaken. Infact, transformation of this steroid has never been studied so far using *M. piriformis*.

MATERIALS AND METHODS

The steroid 1,4 Androstadiene 17β-ol, 3-one was obtained from Sigma (St. Louis, Mo., USA).

Microorganism

The organism used in the present study was isolated from soil and identified as *Mucor piriformis.* It was maintained and propagated on Potato Dextrose Agar (PDA) slants at 4°C and subcultured periodically.

Cultivation of the organism and Steroid Transformation

Flasks (500ml) containing 100ml of sterile Czapek Dox medium (Prema and Bhattacharyya, 1962) were inoculated with spore suspension (2.78 × 10^8 spore count ml) from a 5-day-old culture grown on PDA slant and incubated at 28° - 30°C on a rotary shaker (220rpm) for 24h. After this growth period , 30mg substrate (I) in acetone (1ml) was added to each flask and incubation was continued for a further period of 48h.

Extraction and Isolation of Metabolites

At the end of the incubation period, the contents of the flask were pooled and filtered, and the filtrate (broth) and the mycelium were extracted separately with chloroform. Since the mycelial extract contained only the added substrate, it was not processed further. The broth extract after evaporation to dryness was subjected to column chromatography to isolate metabolites in the pure form.

Chromatography

Metabolites were separated by column chromatography on silica gel using various percentages of isopropanol in benzene as eluents. Thin layer chromatography (TLC) was carried out on glass plates (0.75mm silica gel GF_{254} plate) using chloroform/methanol(94 : 6 v/v, system 1) chloroform/methanol (92:8 v/v, system 2) and chloroform/methanol (88:12 v/v, system 3).

High performance liquid chromatographic (HPLC) analyses were carried out on a system consisting of a Shimadzu model LC 10 AD pump, 25 μl Shimadzu HPLC syringe for injection and Waters model 450 variable UV-Vis spectrophotometer detector (sensitivity at 0.2 AUF). The analyses were performed on a micro-porasil normal phase column using chloroform/methanol (95:5 v/v or 90:10 v/v) solvent system at room temperature. The eluents were monitored by UV detector (254 nm) and the metabolites formed were quantified on the basis of the area under each peak.

Spectral studies

Infrared (IR) spectra were recorded in nujol mull on a Perkin-Elmer model 781 IR spectrophotometer ^{1}H-nuclear magnetic resonance (NMR) spectra were recorded on a JEOL 90MHz or Brucker WH-270 MHz spectrometer. Mass spectra (MS) were recorded on a JOEL-JMX-DX-303 instrument with JMA-DA-5000 data system.

Results and discussion

A batch of 30 flasks were inoculated with *M. piriformis* and the substrate added at the end of the incubation period, the contents of all the flasks were pooled and processed as described in Materials and Methods. The broth extract (≅ 780 mg) upon examination by TLC (system 1) revealed the presence of five metabolites (Rf values : 0.25, 0.56, 0.48, 0.29, 0.19) besides the substrate (I : Rf 0.67) and a few minor metabolites. Out of these one of the metabolites (Rf 0.25) was formed in large quantity. The total fraction was subjected to column chromatography on silica gel (20g) using varying percentages of isopropanol in benzene as the eluent. The unmetabolized substrate (I, Rf 0.67, solvent system 1) (≅100mg) was eluted almost exclusively with benzene alone.

The fraction (≅115mg) eluted with 1 % isopropanol in benzene showed on TLC (solvent system 1) the presence of 3 major (Rf: 0.56, 0.48 and 0.29) and few minor metabolites. These metabolites with Rf: 0.56(IB), 0.48 (IC) and 0.29 (ID) were further purified by preparative TLC (solvent system 1). The metabolite (IE) with Rf 0.19 was eluted with 1.5 per cent isopropanol in benzene. The major metabolite (IA) with Rf 0.25 was eluted with 2 per cent isopropanol in benzene. This metabolite was recrystallized using methanol.

The five metabolites (IA, IB, IC, ID and IE) (Fig. 10.1) which were separated and purified were subjected to various spectral analyses like NMR.MS and IR. Based upon these spectral analyses, the metabolites were identified as 14α, 17β-dihydroxy, 1,4-androstadiene-3-one (IA), 14α hydroxy-l, 4-androstadiene-3, 17 dione (IB), 7α, 17β-dihydroxy-l, 4-androstadiene-3-one (IC), 7α hydroxy-l, 4-androstadiene-3, 17-dione (ID), and 7α, 14α, 17β-trihydroxy-l,4-androstadiene-3-one (IE). In the present investigation, chemical structures of the metabolites were assigned mainly by comparing the ^{1}H-NMR spectra of the starting material or the major metabolite. As far as we know, some of the metabolites (IB,IC,ID and IE) appear to be hitherto unknown. *M. piriformis* offers a means of preparation of these steroids which are difficult to prepare by conventional methods. Since the organism produces IA as the major metabolite, it can be used as a reagent to prepare this novel steroid.

Cursory examination of various metabolites (IA to IE) isolated from the culture medium as well as from resting cell experiments carried out with I, indicated that the organism carries out two enzyme mediated reactions : (i) hydroxylation at 14α and 7α positions, (ii) oxidation of the 17β hydroxyl group. Time course experiments have clearly suggested that hydroxylation at 14α position is the first step involved in the transformation sequence resulting in the formation of dihydroxy compound viz, 14α, 17β-dihydroxy 1,4 androstadiene 3-one (IA) (Fig. 10.1) . Incubation

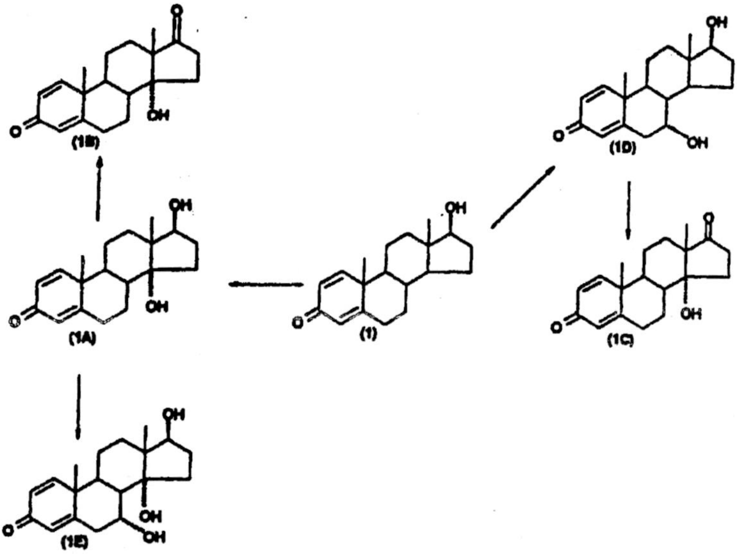

Fig. 10.1 : Pathway of the transformation of 1, 4 Androstadiene 17β-ol, 3-one by *Mucor piriformis*

of IA with resting cells yielded a metabolite identified as 14α-hydroxy-1,4 androstadiene-3,17-dione(IB). This suggests that 14α hydroxylated metabolite IA, further gets oxidised at the 17 position resulting in the formation of IB which indicates the presence of a 17β-hydroxy steroid dehydrogenase enzyme in *M. piriformis.* The other two metabolites IC and ID are formed as a result of hydroxylation at the 7α position and oxidation at the 17 position. It appears that the 17β hydroxy steroid dehydrogenase has a broad specificity as it accepts both 7α and 14α hydroxy-1,4 androstadiene-3-one as a substrates. Earlier it was observed that steroids with 4-ene-3-one structural unit and a bulky substitution at 17α position, when used as substrate, the organism carries out the initial hydroxylation at 7α position instead of at 14α position. In the present study it was noticed that the organism carries out hydroxylation at 7α position even though there is no bulky substitution at the 17α position. This unusual behaviour could be due to the presence of 1,4-diene-3-one unit in the substrate. However, it should be noted here that both 7α hydroxylation and oxidation of the 173 hydroxy group are not the major activities of *M.piriformis.* In fact the organism predominantly carries out hydroxylation at 14α position.

14α hydroxylase activity was also demonstrated at the cell-free level (Madyastha and Joseph, 1993). Preliminary studies have demonstrated that microsomes isolated from the cell-free extracts prepared from cell culture of *M. piriformis* catalysed the conversion of I into its 14α hydroxylated compound in the presence of NADPH and O_2 - Although our preliminary studies did indicate that the 14α hydroxylase is localised in the microsomes, the hydroxylase activity seen was very low. Hence more experiments have to be carried out to find out the optimal conditions for isolating cell-free extract with better hydroxylase activity.

One of the significant observations made in the present study was that both steroid-induced and glucose grown (uninduced) cells of *M.piriformis* efficiently transforms I into various metabolites (IA to IE) and the HPLC profiles of the metabolites are similar. In both the cases, IA was the major metabolite formed indicating the constitutive nature of 14α hydroxylase. However, more experiments have to be performed to confirm these findings.

Since *M.piriformis* very efficiently transforms I to its 14α hydroxy derivative, it was of interest to find out whether entrapped cells also retain the 14α hydroxylase activity. The Ca-alginate cells were prepared as reported earlier.(Qureshi and Tamhane, 1985, Takahashi et al., 1990, Mahmond and Rehm, 1986). The free spores were able to transform I to an extent of 35% at the end of 24h, whereas the immobilized spores were able to biotransform I to an extent of 36 and 78% at the end of 12 and 24h respectively. The formation of the major metabolite IA, and IB was also observed. However, the metabolites IC to IE were formed in trace amounts. Considering the preliminary studies further work can be done and the proficiency of the immobilized spores can be tapped. As they have the advantage of being reused this method can be adopted industrially for economic feasibility.

In fact, *Mucor piriformis* can be used as an efficient agent to prepare some of the novel C-19 steroids.

REFERENCES

Cabral, J.M.S., Airesbarros, M.R., Pinheiro, H. and Prazeres, D.M.E. (1997) Biotransformations in organic media by enzymes and whole cells. J. Biotechnol. 59 : 133-143.

Charney, W., Herzog, H.L.(1967) Microbial transformations of steroids : a handbook, Academic Press. New York.

Eroshin, V.K. (1962) Capacity of *Mucorales* fungi to oxidise steroid S. Mikrobiologiya 31 : 608-615.

Holland, H.L., Riemland, E.(1985) Microbial hydroxylation of steroids 10. Rearrangement during epoxidation and hydroxylation and the step wise nature of these enzymic reactions. Can J Chem 63 : 1121-1126.

Holland, H.L., Thomas, E.M. (1982) Microbial hydroxylation of steroids 8. Incubation ofCn halo and other substituted steroids with Cn hydroxylating fungi. Can J Chem 60 : 160-164.

Holland, H.L., Chenchaiah, P.C., Thomas, E.M., Mader, B., Dennis, M.J. (1984) Microbial hydroxylation of steroids 9. Epoxidation of a -3-keto steroids by *Rhizopus arrhizus* ATCC 11145, and the mechanism of the 6β hydroxylase enzyme. Can J Chem 62 : 2740-2747.

Iizuka, H., Naito, A. (1987) Microbial conversion of steroids and alkaloids. University of Tokyo Press, Tokyo.

Krishnan, R., Madyastha, K.M., Visvamitra, M.A. (1991) The crystal structure of 14α, 17β dihydroxy androst-4-ene-3-one monohydrate and 14α, 17β-dihydroxy androst-l,4-diene-3-one monohydrate steroids. 56 : 440-445.

Madyastha, K.M., Joseph,T. (1993) Studies on the 14α hydroxylation of progesterone in *Mucor piriformis*. J. Steroid Biochem. Molec. Biol. 45 : 563-569.

Madyastha, K.M., Srivatsan, J. (1987) Novel transformation of progesterone by a *Mucor* sp. Can J Microbiol 33 : 361-365.

Madyastha, K.M., Joseph, T. (1994) Transformation of 16-dehydroprogesterone and 17α-hydroxyprogesterone by *Mucor piriformis* Appi Microbiol Biotechnol 41 : 170-177.

Madyastha, K.M., Joseph, T. (1995) Transformation of dehydroepian drosterone and pregnenolone by *Mucor piriformis*. Appl Microbiol Biotechnol 44 : 339-343.

Mahato, S.B., Banerjee. S., Podder. S. (1989) Steroid transformations by microorganism III. Phytochemistry 28 : 7-40.

Mahmond, W. and Rehm, H.J. (1986) Appl Microbiol Biotechol 23 : 305.

Murray, H.C., Peterson, D.H. (1957) 8. Hydroxy-11-deoxycorticosterones. U.S. patent No.2, 800, 490.

Quereshi, N., and Tamhane, D.V. (1985) Appl Microbiol Biotechnol. 21 : 280.

Singh, K., Sehgal, S.N., Vezina, C. (1967) Transformation of steroids by *Mucor griseocyanus.* Can J Microbiol 13:1271-1281.

Takahashi, M., Ochi, H., Haneko. T., Suzuki, H. and Tanaka, H.(1990) Biotechnology Letters 12 : 280.

11

Biotechnology for Clean Environment by *Chrysosporium Keratinophilum*

KAVITHA, N.S., HILDA, A. AND LATHA, K.*

Environmental pollution and degradation of ecosystem have assumed significance owing to increase in the accumulation of wastes from industries, agriculture and poultry. Annually several thousand tons of feather wastes have been discharged into the surrounding environment. Microorganisms could minimise regulatory problems of uncontrolled accumulation of waste feathers.

Major component of feather is keratin which is an insoluble fibrous protein. Keratins are highly resistant to hydrolysis by weak acids, alkalies and also to enzymatic digestion. The durability of keratin is due to cross-binding of closely packed polypeptide chain in which cystine molecules are held together by disulphide bonds. However, the keratinophilic fungi are known to degrade this highly resistant proteins.

In the present investigation, non-pathogenic keratinophilic fungus namely *Chrysosporium* was isolated from soil samples and its biodegradability was evaluated. *Chrysosporium keratinophilum* plays a significant role in the degradation of feathers from White

* Centre for Advanced Studies in Botany, Buindy campus, Chennai-25.

Cornish (IC_2), White Rock (IR_2), Nandanam Broiler I (B_1) and Nandanam Broiler II (B_2). The biochemical changes during the degradation of feather was studied by analyzing the culture fluid. The release of nitrate, nitrite, cysteine and cystine components during the process of biodegradation was studied. There was also a change in pH of the medium towards alkalinity. The release of nitrate and nitrite increased with the increase in the incubation period when IC_2, IR_2 and B_2 were used as keratin source. The release of cysteine was maximum at the 60th day of incubation when B_1 was used. While the cystine release was maximum at 50th day when IR_2 was used as substrate.

Key words : *Chrysosporium keratinophilum;* feather degradation; biochemical release.

INTRODUCTION

Environmental pollution and degradation of ecosystem have assumed significance owing to increase in the accumulation of wastes from industries (Oblisami, 1996), agriculture and poultry. Every year more than 20,000 tons of feathers are produced as waste by poultry farming (Vogt and Stute, 1975). This uncontrolled production of large amount of feather has created massive pollution problem in the surrounding environment. Biodegradation, could minimise regulatory problems of uncontrolled accumulation of waste feathers.

The major component of feather is keratin which is an insoluble structural protein. Keratins are highly resistant to hydrolysis by weak acids, alkalies, ethanol or salt solution (Page, 1950) and also to enzymatic digestion (Weary et al., 1965). The durability of keratin is due to cross-binding of closely packed polypeptide chains in which cystine molecules are held together by disulphide bonds (S-S). However, the keratinophilic fungi have been frequently isolated from soil, where they colonize various keratinous substrates, degrade them and add the mineral content to the soil (Kunert, 1989).

Feathers of birds are most suitable substrate for the survival of many fungi in nature (Sarangi and Ghosh, 1990). Ramesh (1996) isolated a number of pathogenic and non-pathogenic fungi

STRUCTURE OF CYSTEINE (-SH-)

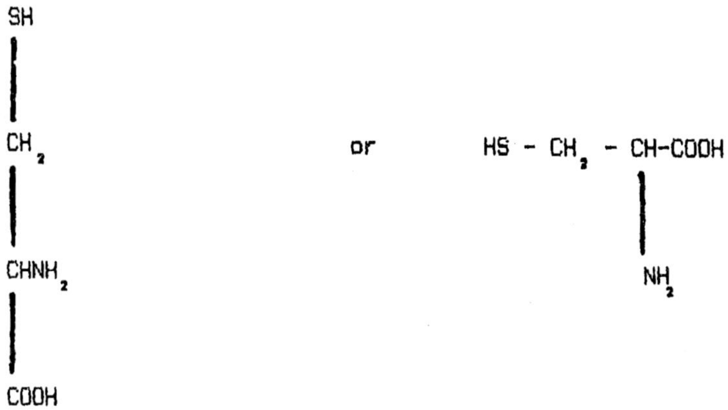

STRUCTURE OF CYSTINE (-S-)

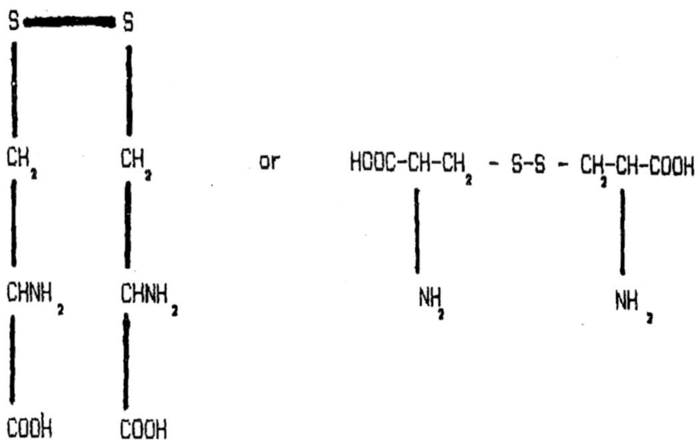

from keratin substrates and studied their intensity and the type of hair degradation. However, there is no detailed investigation on the degradation of feathers. Therefore, the present investigation was envisaged to study the biodegradation of feathers.

MATERIALS AND METHODS

Chrysosporium kerakinophilum was isolated by feather baiting technique. The fungal species were grown and maintained on Sabouraud Dextrose Agar (SDA) medium.

Dextrose	-	40 g
Peptone	-	10 g
Agar	-	25 g
Distilled water	-	1000 ml
pH	-	5.6

Feathers chosen for the present study

Poultry strain - Broiler

a. White Cornish (IC_2)
b. White Rock (IR_2)
c. Nandanam Broiler I (B_1) and
d. Nandanam Broiler II (B_2)

Different strains of fowls' feather were collected from Poultry Research Station, Madras. The organism was grown in mineral medium with 500 mg of feather as a sole source of nitrogen and carbon (Day et al., 1968) and incubated for the period of 60 days. At 10 days intervals, the biochemical changes associated with biodegradation was evaluated by analyzing the culture filtrate. The changes in pH, the release of nitrogenous compounds (nitrate & nitrite) and soluble sulphydryl compounds (cysteine & cystine) were determined.

Changes in pH

The changes towards alkalinity of the culture filtrate was measured by using pH meter (Orion pH meter) with a glass electrode.

2.3 Determination of nitrate release (NO_3)

Nitrate in the culture filtrate was estimated by UV - Spectrophotometric method of Goldsmith et al., (1973). To an aliquot of the sample (1 ml), 0.5 ml of concentrated hydrochloric acid was added and made upto 5.0 ml with distilled water. Absorbance at 210 nm was recorded after 15 min. incubation. The nitrate release was calculated using a standard graph prepared with sodium nitrate.

Determination of nitrite release (NO_2)

Nitrite release was estimated by the diazo-coupling method using sulphanilamide and NED, following the procedure of Hewitt (1975).

1. 1.0% Sulphanilamide in 1.5N HCl

2. 0.02% N– l-naphthyl ethylene diamine dihydrochloride

To a known volume of the sample (1 ml) made upto 2.0 ml with distilled water, 1.0 ml each of the two reagents were added in the order given above. Solution was mixed well and allowed to stand for 15 min. The absorbance at 540 nm was measured using distilled water with reagents as blank. Nitrite was quantified by referring to a standard graph prepared with potassium nitrite.

2.5 Determination of cysteine (S-H) (Ramakrishna et al., 1979)

To 4 ml culture filtrate, 15 ml of potassium biphalate hydrochloric: acid buffer (pH 3.3), 2 ml 0.2%. metol solution and 3 ml of 0.01M potassium dichromate solution were added successively. It was diluted to 25 ml with distilled water in a calibrated volumetric flask and the absorbance was measured at 510 nm after 160 min. of incubation against a corresponding reagent blank prepared in the same manner. The cysteine content was computed from an appropriate calibration curve.

Determination of cystine (S-S) (Ramakrishna et al., 1979)

10 ml of culture filtrate was treated with requisite quantity of 1N potassium hydroxide to maintain a pH of 9.2. After adding 1 ml of 2% sodium borohydride solution, the resulting mixture was warmed on a waterbath which was preheated to 50°C for 30 min. Then the excess sodium borohydride was destroyed by dropwise

Fig. 11.1 : Changes in pH during biodegradation of feathers

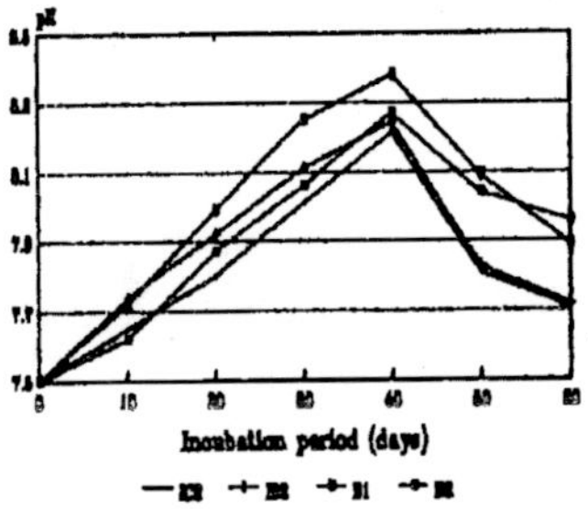

IC_2	-	WHITE CORNISH
IR_2	-	WHITE ROCK
B_1	-	NANDANAM BROILER I
B_2	-	NANDANAM BROILER II

Fig. 11.2 : Release of nitrate from feathers

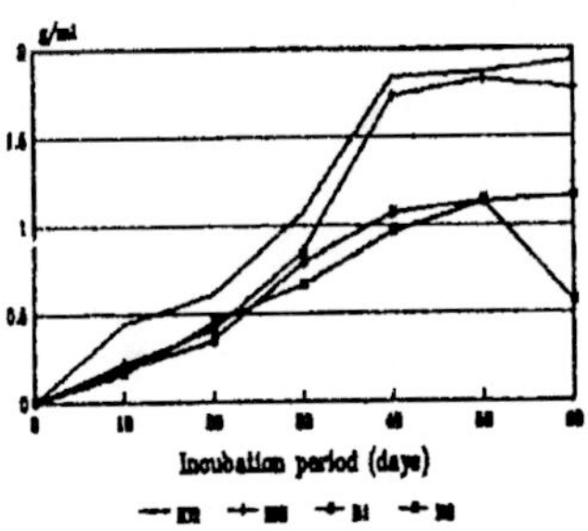

Fig. 11.3 : Release of nitrite during biodegradation

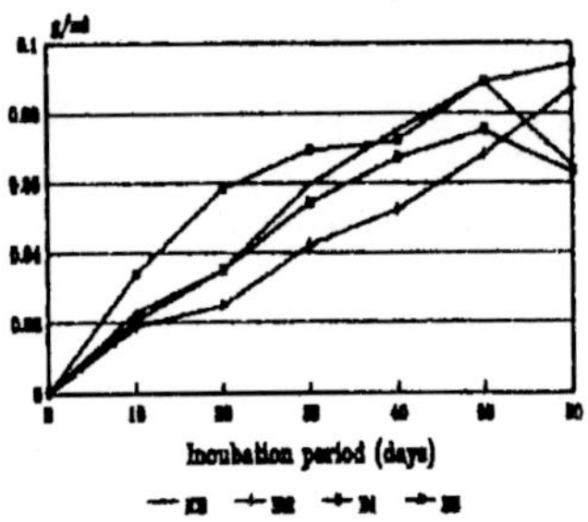

Fig. 11.4 : Release of cysteine from feathers

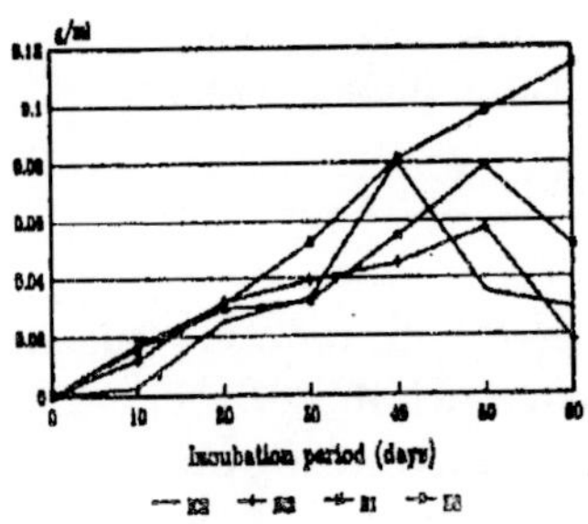

Fig. 11.5 : Release of cystine from feathers

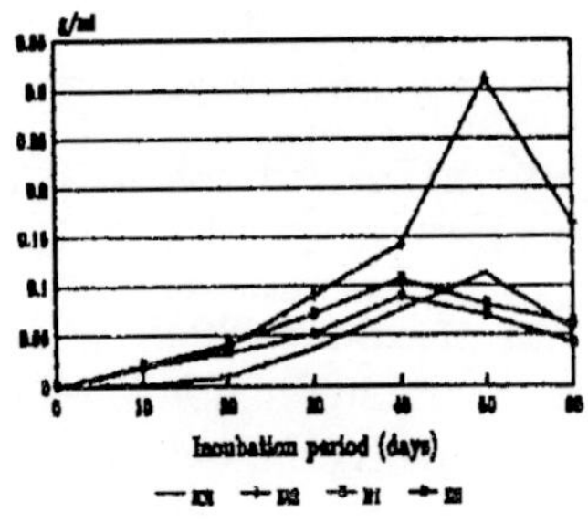

addition of 10 ml of 10% acetic acid. Finally the solution was diluted to 50 ml after the pH of the solution was brought to 5.0. 4 ml of this solution was used for the determination of cystine following the method used for cysteine.

RESULTS AND DISCUSSION

During the process of biodegradation, there was a gradual increase of pH, into the alkaline phase for all the 4 feathers till 40th day of incubation. However, the pH change was maximum when B_2 was used as keratin source (Fig. 10.1). Such an alkalinization of the medium may be due to the excretion of excess nitrogen via deamination and ammonia excretion (Kunert, 1989). Keratin degradation involves rupturing the disulphide linkages between the peptide chains of keratin molecules by some extra and intracellular enzymes collectively called keratinase (Singh et al., 1995).

The liberation of *nitrogenous compounds* gradually increased with the increase of incubation period when IC_2 (nitrate), IR_2 (nitrite) and B, (nitrate & nitrite) were used as the keratin source (Figs. 11.2 & 11.3). However, when the depletion of nitrate and nitrite from the culture media resulted in the reduction of fungal growth.

The release of *sulphydryl compounds* namely, cysteine was maximum when B_1 was used as substrate (Fig. 11.4). Whereas the release of cystine was niaximum at 50th day of incubation when IR_2 was used (Fig. 11.5). During the process of degradation the -S- sulfo groups (–S.SO_3 H) and sulfhydryl groups were formed by sulfitolysis (Kunert, 1976).

Accumulation of cystine may be a direct reduction of disulphide bridges of keratin (Noval and Nickerson, 1959). Chicken feathers, a type of eukeratin which contained histidine, lysine and arginine in a definite proportion of 1:4:12 and 3-5% sulphur nearly all of which is in the form of cystine (Singh et al., 1996). The low values of cysteine during initial growth phase also indicate that fungi at first utilised more readily digestible material (non-keratinous protein) till it was exhausted. The actual keratinolysis occurs only after the degradation of non-keratinous protein (Malviya et al., 1992).

Chrysosporium keratinophilum could degrade this highly resistant keratin of feather. With an increasing world-wide concern for the environment it is possible to use *Chrysosporium keratinophilum* for the degradation of enormous quantity of waste feathers. Biodegradation leads to recycling the wastes and thus maintaining the environmental quality of the biosphere (Atlas, 1986).

ACKNOWLEDGEMENTS

The authors express their thanks to the Director, Centre for Advanced Studies in Botany, University of Madras, Chennai, for providing the necessary facilities.

REFERENCES

Atlas, R.M. 1966. Basic: and practical microbiology, Mac Millan publication. New York.

Day, W.C., Tonic, P., Stratman, S.L., Leeman, U. and Harman, S.R. 1968. Isolation and properties of an extracellular protease of *Trichophyton granulosum.* Biochim. Biophys. Acta **167**: 597-606.

Frierdrich, A.B. and Antranikian, 6. 1996. Keratin degradation by *Fervidohacterium pennsvorans,* a nova I thermophilic anaerobic species of the order Thermotogales. Appl. Environ. Microbiol. **62(8)**: 2875-2882.

Goldsmith, J., Livoni, P., Noaberg, C.L. and Segal, I.H. 1973. Regulation of nitrate uptake in *Penicillium chrysogenum* by ammonium ion. Plant physiology **52** : 362-367.

Hewitt, E.J. 1975. Assimilatory nitrate-nitrite reduction. Annual review of plant physiology **26** : 73—100.

Kunert, J. 1976. Keratin decomposition by dermatophytes. II- presence of S- sulfocysteine and cysteic acid in soluble decomposition products. Z. Allg. Mikrobiol. **16** : 97-105.

Kunert, J. 1989. Biochemical mechanism of keratin degradation by the Actinomycete *Streptomyces frediae* and the fungus *Microsporum gypseum* : A comparison. J. Basic. Microbiol. **29** : 597-604.

Malviya, H.K., Parwekar, S., Rajak, R.C. and Hasija, S.K. 1992. Evaluation of keratinolytic potential of some fungal isolates from gelatin factory campus. Ind. J. of Experiment. Biol. **30** : 103-106.

Mercer, E.H. and Verma, B.S. 1963. Hair digested by *Trichophyton mentagrophytes.* An electron microscope examination. Arch. Dermatol. (Chicago) **87** : 357-360.

Noval, J.J. and Nickerson, W.J. 1959. Decomposition of native keratin by *Streptomyces fradiae.* J. Bacteriol. **77** : 251-263.

Oblisami, G. 1996. Microbiology in protecting and rehabilitating the environment, 37th annual conference of Association of Microbiologists of India.

Page, R.M. 1950. Observations on keratin digestion by *Microsporum gypseum.* Mycopathologia **42(5)** : 591-602.

Ramakrishna, R., Siraj, P. and Prakasa Sastry, C.S. 1979. Spectrophotometric method for the direct determination of cysteine in the presence of other naturally occurring amino acids. Current Science **48(18)** : 815-816.

Ramesh, V.M. 1996. Studies on keratinophilic fungi characterization of keratinolytic potential and fungitoxic evaluation of some plant extracts. Ph.D. thesis. University of Madras.

Sarangi, S.M. and Ghosh, G.R. 1990. Episoochory transmission of keratinophilic fungi by the Eusynancthropic common house sparrow in Orissa. Kavaka. **18** : 14-18.

Singh, C.J., Geetha Singh, B. and Sundara Singh, B. 1995. Biodegradation of certain keratin substrates invitro by some keratinophilic fungi. Advances in Plant Sciences. **8(2)** : 271-276.

Singh, C.J., Geetha Singh, B. and Sundara Singh, B. 1996. Morphological, keratin colonising and degrading characters of some keratinophilic fungi and related dermatophytes isolated from Keoladeo national park, Bharatpur.

Tschiersh, B. and Mothes, K. 1963. Amino acidai Structure and distribution (eds. Marcel Florkin and Howard S. Mason) In : Comparative Biochemistry – A comprenhensive treatise, vol. 5. pp. 1-90.

Vairinhos, F., Mallace, W. and Nicholas, D.J.D. 1989. Simultaneous assimilation and denitrification of nitrate by *Bradyrhizobium jabonicum.* J. Gen. Microbiol. **135** : 189-193.

Vogt, H. and Stute, K. 1975. Scheinbare Aminosaure Verdaulichkeit des Federmehls bei Legehennen. Arch. Beflugelkd. **39** : 51-55.

Weary, P.E., Canby, C.M. and Cowley, E.P. 1965. Keratinolytic. activity of *Microsporum canis and Microsporum gypseum,* J. Invest. Derma. **44** : 300-310.

12

Glucose Dependent Bioaccumulation of Cadmium by A Metal Tolerant Yeast

Sridevi Panchapakesan, J. S. Melo, K.C. Bhainsa and S.F. D'Souza*

A cadmium tolerant yeast strain (MIC - 80 ppm) isolated from metal bearing soil was also found to accumulate high amounts of cadmium. The accumulation of cadmium was enhanced considerably in the presence of glucose. One of the notable observation as compared to the extensively studied *Saccharomyces cerevisiae* is its ability to internalize glucose even in the presence of metal. Thus simultaneous addition of glucose and metal did not significantly affect the bioaccumulation efficiency. Presence of metabolic inhibitors like azide reduced the metal uptake ability. Metal uptake was also reduced with dead cells. The viability of the cell was found to be essential for bioaccumulation. Bioaccumulation from binary mixtures indicated that Cu^{2+}, Ni^{2+} and Ca^{2+} reduced cadmium uptake whereas monovalent cations like K^+ and Na^+ had no effect. The bioaccumulation capacity depended on the ionic strength of the external medium. This yeast can be a potential candidate for bioremediation application.

* Nuclear Agriculture and Biotechnology Division, Bhabha Atomic Research Centre, Mumbai-400085, India.

INTRODUCTION

With the rapid growth of industrialization the introduction of toxic heavy metals into natural waters has also increased at a drastic rate. One of such heavy metals which has attracted attention of environmentalists is cadmium. Cadmium is found in the effluents of electroplating, smelting, alloy manufacturing, mining, pigment, fertiliser, pesticide and rubber manufacturing industries and thereby poses a potential hazard to the life forms as well as the environment (Murphy and Siegal, 1982). It is therefore important to establish an efficient and low cost method for the abatement of this pollution.

Microorganisms are potent biosorbent materials and may provide the technology required for metal removal especially as they show metal specificity and could adsorb metal ions from dilute solutions (Brady & Duncan, 1994). The uptake of metal ions by microorganisms may be either through biosorption to cell walls, entrapment in extracellular capsules, precipitation, oxidation reduction reactions or through bioaccumulation, a metabolically/

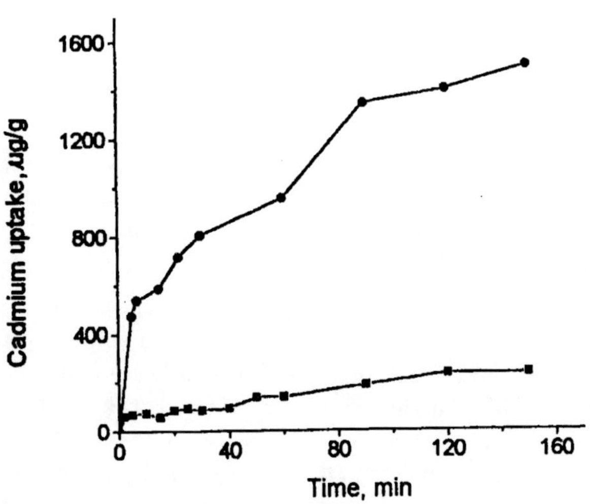

Fig. 12.1 : Kinetics of Cadmium uptake —■— without glucose —●— dead cells.

energy dependent process which depends on the viability of a cell in a waste environment (Volesky, 1995).

Most studies on the application of microorganisms for the removal of metals have been done using easily available biomass such as industrial wastes from fermentation or marine algae (Puranik et al., 1995). Isolating a metal tolerant strain having uptake potential would be of greater interest, as it would not only survive in such polluted environments but could also accumulate the metal. The present paper delineates some of the cadmium bioaccumulation characteristics of a metal tolerant yeast isolated from metal bearing soil from the metal industry zone (Thane creek, Mumbai) and the effects of various external factors on the process .

MATERIALS AND METHODS

Materials

Cadmium chloride was obtained from Qualigens. Yeast extract and malt extract were obtained from Difco laboratories (Detroit, USA).

All solutions were prepared with double glass distilled water. Glassware were prepared for use by washing with detergent, rinsing and leaving them overnight in an acid bath. The glassware were then rinsed with distilled water and heat dried. The culture was maintained on Yeast extract - Malt extract Agar (YM) (g/L) [Yeast extract : 3; malt-extract : 3; peptone : 5 and glucose:1].

Metal tolerance of the yeast isolate

The yeast isolate was studied for its ability to tolerate heavy metals. The stock metal solutions of cadmium chloride, mercuric chloride and cobalt chloride were prepared separately in sterile double distilled water. This was added to sterile YM broth so as to give the required final concentration of respective metals. Further dilutions with YM broth were carried out to obtain lower concentrations of the respective metals. An overnight broth culture of the yeast was inoculated into tubes and incubated at 37°C for 24 h. The results were recorded as presence or absence of growth on comparison with appropriate controls.

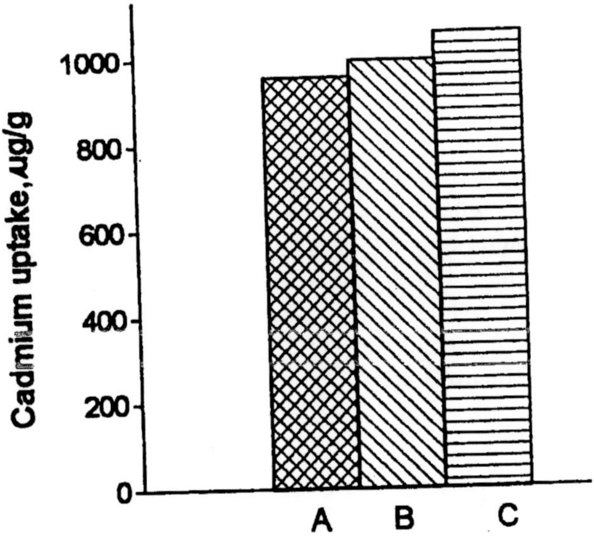

Fig. 12.2 : Effect of glucose treatment on cadmium uptake; A— control, B— no preincubation, C— washed.

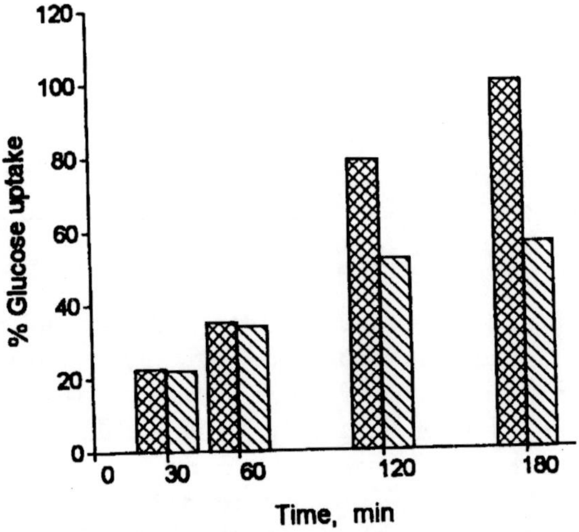

Fig. 12.3 : Effect of cadmium on glucose uptake; [cross-hatched] absence of cadmium [hatched] presence of cadmium.

Culture conditions

For metal cation accumulation experiments the organism was grown at pH 6.8 in medium (g/L) KH_2PO_4,2.72; K_2HPO,5.22; $(NH_4)_2SO_4$, 1.98; $MgSO_4.7H_2O$,0.12; $FeSO_4.7H_2O$, 0.0022; glucose, 10.0 and yeast extract, 1.0. Flasks (1L) containing 300 ml medium were incubated (125 rpm) on an orbital shaker. The cells were harvested after 48 h of growth and were washed twice with distilled water. The amount of cells used in these studies are indicated as wet weight of the packed cells.

Bioaccumulation assay

The cell suspension to be used in bioaccumulation assay was adjusted to 10 mg/ml by dilution with 5mM sorbitol pH 5.0 (adjusted by HCl). The yeast cell suspensions were agitated in Erlenmeyer's flasks on an orbital shaker (125 rpm) at 37°C. Glucose (0.1%) was added to the flasks 30 min prior to the addition of the metal stock solution and were incubated for appropriate intervals of time. The samples collected were immediately centrifuged at 10,000g for 5 min and then washed twice with 5mM sorbitol. The cells were digested with sorbitol (5ml) and 55 per cent HNO_3 (0.2 ml) on a boiling water bath for 60 min. The samples were cooled and then analysed for metal content by atomic absorption spectrophotometer (Perkin - Elmer Model 380).

Effect of Glucose treatment on cadmium uptake

Glucose (0.1%) and metal stock solution (40 mg/ml) were added together to the cell suspension and were studied for uptake. Another set of cells were pre-incubated with glucose (0.1%) for 30 min, washed twice, suspended in metal stock solution (40 mg/ml) and studied for metal uptake.

Glucose uptake by the cells in presence and absence of cadmium

The cell suspensions were prepared as per the requirement for the bioaccumulation assay and the glucose uptake by the cells was studied at periodic intervals. The samples collected were analyzed for residual glucose content in the supernatants by

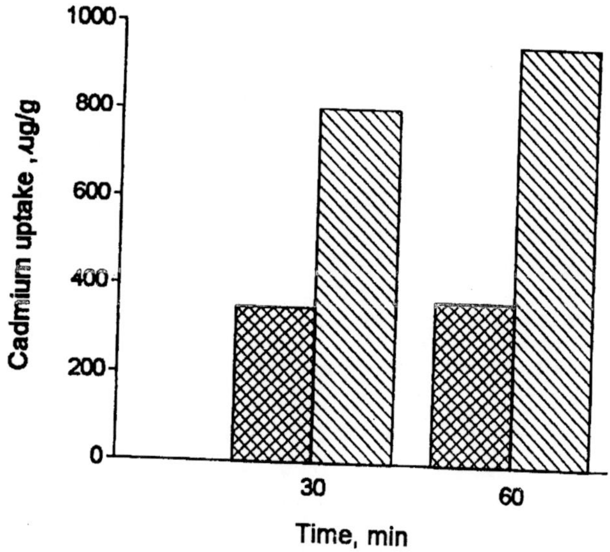

Fig. 12.4 : Effect of azide on cadmium uptake; with azide, control.

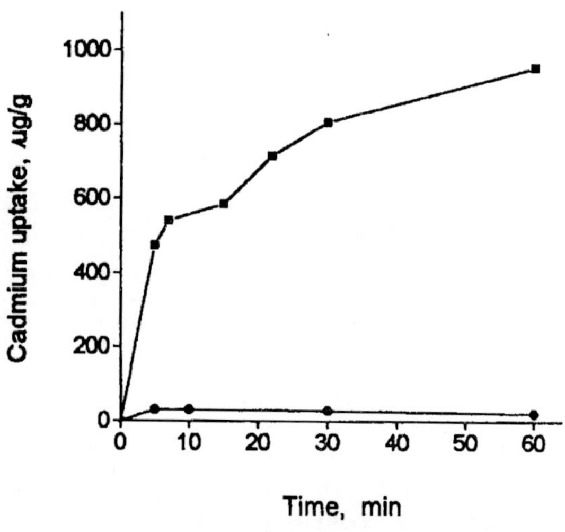

Fig. 12.5 : Cadmium uptake by viable and dead cells; —■— viable cells, —●— dead cells.

dinitro salicylic acid method (Miller, 1959).

Effect of azide on cadmium bioaccumulation

Sodium azide (10mM), a metabolic inhibitor was added to the yeast cell suspensions 10 min prior to the addition of metal. The methodology remains similar to the bioaccumulation assay.

Cadmium uptake by dead cells

The cells were killed by autoclaving for 20 min at 15 lb and used for cadmium uptake studies.

Effect of temperature on cadmium bioaccumulation

The cell suspensions were maintained at different temperatures using thermocontrolled water bath and the metal solution was added after acclimatizing the cell suspensions to the set temperatures. Samples were collected after 1h of incubation and were assayed as described above.

Effect of pH on cadmium bioaccumulation

pH of the cell suspensions were adjusted between 3 - 10 by addition of HCl or tetra methyl ammonium hydroxide. After incubation for 1h with 40mg/ml of metal, the samples were centrifuged, washed, digested in acid and analysed for metal content.

Effect of different metals on cadmium bioaccumulation

Different metal salt solutions (40 mg /ml) were added to the yeast cell suspensions along with cadmium chloride (40 mg/ml) solution and were incubated for 1h on an orbital shaker (125 rpm). Metal uptake was studied as described earlier.

Effect of ionic strength on cadmium bioaccumulation

Bioaccumulation assay was carried out in the presence of varying concentrations (0-500) of sodium chloride to determine the effect of ionic strength on the bioaccumulation of cadmium.

Results and Discussion

Yeasts have shown potentials as efficient biosorbents in metal uptake studies. Hence, a metal tolerant yeast isolate was investigated for its metal uptake capability. The yeast isolate could tolerate high concentrations of metals. The MIC for different metals of the yeast isolate is as shown in Table 12.1. It could tolerate 80 ppm of cadmium, 75 ppm of mercury and 275 ppm of cobalt. It also could accumulate cadmium in large amounts from aqueous solutions.

Cadmium uptake by the metal tolerant yeast is shown in Fig. 12.1. The uptake of cadmium was found to be significantly enhanced in the presence of glucose. Different authors have seen different responses towards the bioaccumulation of metals, which in general has been shown to be enhanced in the presence of a metabolisable substrate such as glucose. Thus Brady and Duncan (1994) using *S. cervisiae* reported that even though the bioaccumulation showed a biphasic response, it was not found to be effected by the presence of glucose. On the other hand Stoll and Duncan (1996) observed that bioaccumulation of heavy metals (Cu, Cd, and Cr) from industrial effluent was enhanced considerably when the yeast biomass was pretreated with glucose for 30 min prior to the addition of the metal. In the present study a different response was seen. The simultaneous addition of glucose and cadmium did not affect the bioaccumulation efficiency of the cells (Fig. 12.2) to a great extent.

Metal resistance has been attributed to number of factors. Major among them are the microbes developing certain potentials to prevent the entry of metal inside the cell and the second is the ability of the cells to take up metal inside and its safe storage in vacuoles. Suggested mechanism for transport into the microbial cell include complex permeation, carrier mediated ion channels/ pumps and endocytosis (Ford and Mitchell, 1992). Bioaccumulation of metals in the presence of a metabolisable substrate is primarily determined by active transport in response to electro-chemical proton gradient generated by a membrane-bound H^+ ATPase, across the cytoplasmic and vacuolar membranes (Jones and Gadd, 1990). The inability of *S. cervisiae* to take up metal when added together with glucose has been attributed to the inability of yeast cells to internalise the glucose in the presence of high metal ion concentration. (Stoll and Duncan, 1996) The glucose uptake by metal tolerant strain in the presence or absence of metal is shown in Fig. 12.3. It is evident that glucose uptake was not significantly inhibited upto 60 min of incubation, but after 180 min of incubation a cadmium concentration of 40ug/ml reduced only 40 per cent of its glucose utilising capability. Unlike *S. cerevisiae*, the ability of the metal tolerant strain to internalize glucose suggests that such strains might have developed resistance towards inactivation of its biochemical machinery for the uptake of glucose. This may be an added advantage for its practical use in the bioremediation of Cd^{2+} through bioaccumulation.

Various metabolic inhibitors that interfere with the membrane potential have been shown to reduce metal accumulation (Karna et al., 1996). In the present study the metabolic inhibitor like azide was found to reduce the metal uptake ability (Fig. 12.4.)

There is ambiguity concerning whether live or dead biomass is a better metal sorbent. Volesky et al., (1993) observed better cadmium uptake by cells of aerobic baker's yeast compared to dry non living cells. On the other hand *S. cerevisiae* when killed by drying and grinding increased their ability to take up a range of metals (Kuyucak and Volesky, 1988). In the present study it was observed that dead biomass was a poor metal accumulator. The cells killed by autoclaving showed very little accumulation of cadmium (Fig. 12.5.).

The effect of temperature on cadmium bioaccumulation is shown in Fig. 12.6. Maximum uptake occurred at 37°C, which is also the optimal temperature for the growth of this organism. Over the range of temperature studied (5 - 55°C), the metal uptake was significantly reduced at low temperatures. Since the bio-accumulation process is metabolism dependent, the observed decrease in metal uptake could be due to the decreased metabolic activities at low temperatures. Decreased uptake at higher temperatures has normally been attributed to loss of membrane integrity of the cell. A similar response has been reported by Brady and Duncan (1994) for *S. cerevisiae* when studying copper uptake.

pH of the medium also plays an important role in the amount of metal accumulated by the cells. Metal accumulation at different pH range (3 - 10) is shown in Fig. 12.7. The maximum accumulation

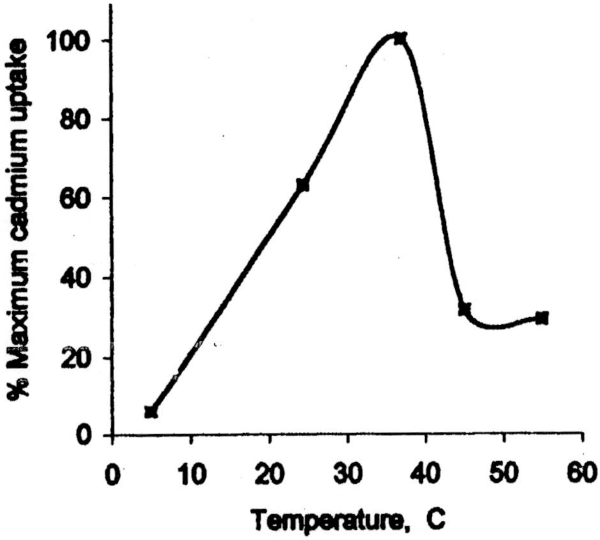

Fig. 12.6 : Effect of temperature on cadmium uptake.

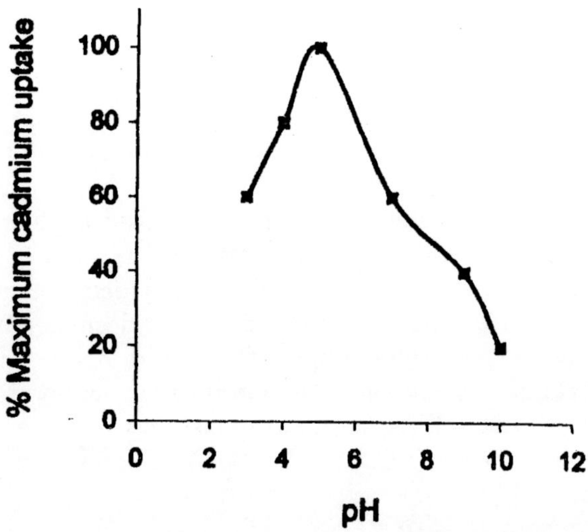

Fig. 12.7 : Effect of pH on cadmium uptake.

was found at pH 5. The dependence of metal accumulation on the pH of the medium has been attributed to the competition of metal cations with hydrogen ions or formation of metal hydroxides at higher pH thus reducing the availability of the metal cations. However, in a metabolically dependent accumulation process as in the present case, the effect of pH on the biochemical machinery itself can also play an important role.

Cadmium uptake was studied in the presence of other metals (40 ppm) by incubating the cells with 40 ppm of cadmium (Fig.12.8). Copper, Nickel and Calcium were found to reduce the cadmium uptake whereas monovalent cations like Potassium and Sodium had no effect. This indicates, a competition between divalent cations for the same ligands might have occurred for the cell surface or transport channel accumulating the metal ions into the cells interior. Different authors regarding bioaccumulation from a binary mixture of metal cations have seen different responses. Thus K^+ and Mg^{2+} were found to inhibit Zn^{2+} uptake by *S. cerevisiae* whereas Na^+ and Ca^{2+} did not (Blackwell et al., 1995).

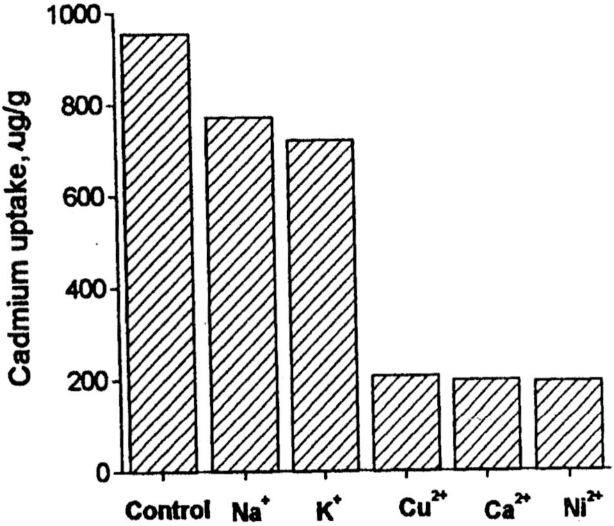

Fig. 12.8 : Cadmium uptake in the presence of other metal cations.

Table 12.1. Minimum Inhibitory Concentrations of different metals on the yeast isolate

Metals	*Minimum Inhibitory Concentration (MIC) (ppm)*
Cadmium	80
Mercury	75
Cobalt	275

Sr^{2+} and Ca^{2+} uptake by yeast was inhibited by monovalent cations K^+ and Na^+ (Roomans et al., 1979). Co^{2+} and Cd^{2+} uptake was found to be severely inhibited by cations of the same size. Thus Ca^{2+} was found to depress Cd^{2+} uptake indicating that Cd^{2+} was accumulated via the Ca^{2+} transport system (Norris and Kelly 1977, Mowll and Gadd 1984).

Higher ionic concentration of the medium was found to affect the bioaccumulation of cadmium. Higher ionic concentration may interfere with the initial adsorption of the metal on cell surface which is the first step in the accumulation of heavy metals by cells.

The above studies show that viable cells of the metal tolerant yeast can remove metal from aqueous solution through energy dependent bioaccumulation process. Dead cells or the viable cells in the absence of an energy source like glucose exhibited very low metal uptake capacity. Studies are under progress to understand the mechanism of bioaccumulation of cadmium as well as other heavy metals which may provide important information for the technological application of the yeast for the bioremediation of heavy metals.

ACKNOWLEDGMENTS

Authors thank Dr. P.S. Rao, Head Nuclear Agriculture and Biotechnology Division (NA&BTD) , BARC for his keen interest in this work ; Dr. K.N. Joishy, Somaiya College, Mumbai for his involvement in the isolation of the strain; Dr. V. Ramachandran, NA&BTD , BARC and Dr. S. Sudershanan, Dr. Tantri and Mr. Shahji, Analytical Chemistry Division , Bhabha Atomic Research Centre, for their help in the analysis of cadmium by Atomic Absorption Spectrophotometer.

REFERENCES

Blackwell, K.J., Singleton, I and Tobin, J.M. (1995) Metal cation uptake by yeast: A review. Appl. Microbiol. Biotechnol. 43, 579-584.

Brady, D. and Duncan, J.R. (1994) Bioaccumulation of metal cations by *Saccharomyces cerevisiae.* Appl. Microbiol. Biotechnol. 41, 149-154.

Ford, T. and Mitchell, R. (1992) Microbial transport of toxic metals . In: Environ Microbiol, Wiley-Liss, New York pp. 83-103.

Gadd, G.M. (1988) Accumulation of metals by microorganisms and algae. Biotechnology vol. 6B, ed. Rehm HJ, VCH Verlagsgesellschaft Weinheim 401-433.

Jones, R.P. and Gadd, G.M. (1990) Ionic nutrition of yeast-physiological mechanisms involved and implications for biotechnology. Enzyme Microb Technol 12: 1-17.

Karna, R.R., Sajani, L.S. and Mohan, P.M. (1996) Bioaccumulation and biosorption of Co^{2+} by *Neurospora crassa.* Biotechnol. Lett. 18, 1205-1208.

Kuyucak, N. and Volesky, B. (1988) Biosorbents for recovery of metals from industrial solutions. Biotechnol. Lett. 10, 137-142.

Mowll, J.L. and Gadd, G.M. (1984) Cadmium uptake by *Aureoasidum pullulans.* J. Gen. Microbiol. 130, 279-284

Murphy. Jr., C.B. and Spiegel, S.J. (1982) Bioaccumulation and toxicity of heavy metals and related trace elements. J. Water Pollut Control Fed 54: 849-854.

Norris, P.R. and Kelly, D.P. (1977) Accumulation of cadmium and cobalt by *Saccharomyces cerevisiae.* J. Gen. Microbiol. 99, 317-324.

Nriagu, J.O. and Pacyna, J.M. (1988) Quantitative assessment of worldwide contamination of air, water and soils by trace metals. Nature, 333: 134-139.

Panchapakesan, S. (1997) Yeast: A potential biosorbent for metal uptake. M.Sc dissertation, University of Mumbai, Mumbai, India.

Puranik, P.R., Chabukswar, N.S. and Paknikar, K.M. (1995) Cadmium biosorption by Streptomyces pimprina waste biomass. Appl. Microbiol. Biotechnol. 43, 1118-1121.

Roomans, G., Theuvenet, A., Van Den Berg, T., Borst - Pauwells (1979) Kinetics of Ca^{2+} and Sr^{2+} uptake by yeast : Effect of pH, cations and phosphates, Biochim.Biophys. Acta, 551, 187-196.

Stoll, A. and Duncan, J.R. (1996) Enhanced heavy metal removal from waste water by viable glucose pretreated *Saccharomyces cerevisiae* cells. Biotech Lett. 18, 1209-1212.

Volesky, B. and Holan, Z.R. (1993) Cadmium biosorption by S. cervisiae, Biotech. Bioengg. 41 : 826 -829.

Volesky, B. and Holan, Z.R. (1995) Biosorption of heavy metals. Biotechnol Prog. 11, 235-250.

13

Phytoremediation : A Feasibility Study with Trees in Florida, USA

ASHA SYED*, K. KOMAR**, D.L. ROCKWOOD*, AND L. Q. MA**

Phytoremediation, the process of using plants to remove or stabilize a contaminant in soil or water, provides a low cost, effective, and non-invasive method to deal with a contaminated area. This paper addresses some previous phytoremediation research. Examples are given of plant species that are effective hyperaccumulators of specific metals, such as copper, zinc, and lead. We also address our current research to determine the feasibility of utilizing trees to solve some of Florida's environmental concerns. Contaminated cattle dip and copper/chromium/arsenic (CCA) wood treatment sites are being evaluated to identify plant and tree species that may effectively accumulate heavy metals. A greenhouse study will be conducted to determine the effectiveness of high biomass producing trees in the removal of heavy metals from contaminated soil. Additional laboratory experiments will

* School of Forest Resources and Conservation, University of Florida, Gainesville, FL, USA.

** Soil and Water Science Department, University of Florida, Gainesville, FL, USA.

Table 13.1 : Golita Loam Soil Analysis

Constituents Tested	*Unite*
(Total)	mg/kg dw
As	5
Cd	1.6
Cr	41
Cu	31
Zn	130
Fe	14,000
Mn	730
Pb	26
B	4
Mo	1.1
Ni	35
P	1,000
Se	<10
Ca	90,000
Mg	3,400
Na	130
K	5,300
Hg	<0.3
Water Soluble (saturated paste)	***% saturation***
Ca	0.8 meq/100 gms
Mg	1.67 meq/100 gms
Na	0.36 meq/100 gms
K	138 meq/100 gms
Cl	0.18 meq/100 gms
NO_3	0.57 meq/100 gms
PO_4	0.35 meq/100 gms
SO_4	0.59 meq/100 gms
Electrical Conductivity (EC 10 mmhos/cm)	0.64 meq/1
Exchangeable-cations	
Ca^{++}	5.4 meq/100 gms
Mg^{++}	0.92 meq/100 gms
Na^{+}	0.07 meq/100 gms
K^{+}	0.26 meq/100 gms
Cation Exchange Capacity	16.5 meq/100 gms
Exchangeable Sodium Percentage	0.4 percent
pH on saturated paste	6.7

be conducted to determine methods to increase the availability of metals for removal by trees. Promising trees and methods will then be field tested.

INTRODUCTION

Throughout history, anthropomorphic activities, such as mining and agricultural activities, have been the sources of contamination of much of the world's soil and water. With the increase in public opinion about the state of the environment, the process of remediating these areas has increased during the last few decades. Methods have been established to handle many of the pollutants, but due to the unique nature of metal in soil and water, few techniques have been adopted to clean-up these areas. One possible method to remove the metal contaminant from a soil or water system is with the use of the treatment process known as phytoremediation.

Phytoremediation is the use of green plant-based systems to remediate contaminated soils, sediments, and water. Compared to most traditional remediation and engineering techniques, it is a fledging technology that addresses a wide variety of surficial contaminants. The word phytoremediation is formed from the Greek prefix 'phyto' meaning plant and the Latin suffix 'remedium' meaning to cure or restore. The term remediation often has more of a legal connotation than a technical one. Remediation can imply: (a) clean-up where either the contaminant is removed from the matrix (leaching, bioremediation, etc.) or the entire contaminated matrix is removed from the site (excavation and landfilling), or (b) stabilization where the physical or chemical form of the matrix or contaminant is transformed to a more inert condition. Stabilization and biological activity are the basis of plant-based approaches for the cleanup of inorganic and organic environmental contaminants. Phytoremediation uses a plant's natural ability to take in water and nutrients through roots, evaporate water through leaves, and act as a natural reactor to metabolize organic compounds like oil and pesticides or to absorb and concentrate toxic trace elements such as lead and selenium.

Phytoremediation can take place in many ways. Plants can be used to stabilize a contaminated soil by decreasing wind and

Table 13.2 : Nutrient Contents (Total) of Composts[a]

Constituents Tested	*Compost Type[b]*					
	A1	*A2*	*A3*	*A4*	*B1*	*B3*
	% Dry Weight (dw)					
Total N	1.7	1.5	1.7	1.6	1.3	1.2
Organic N	1.6	1.4	1.6	1.4	1.3	1.2
NH_3-N	0.002	0.0015	0.001	0.0011	0.0018	0.0016
NO_3-N	0.15	0.098	0.16	0.11	0.064	0.041
P_2O_5	0.65	0.63	0.61	0.61	0.63	0.61
K_20	1.1	1.8	1.1	1.6	1.6	2.5
Ca	2.4	2.7	4.0	2.7	2.0	2.4
Mg	0.6	0.78	0.9	0.84	0.6	0.99
SO_4-S	0.11	0.083	0.1	0.079	0.079	0.056
Cl	0.37	0.32	0.38	0.35	0.31	0.25
Na	0.29	0.27	0.29	0.25	0.20	0.17

[a] Soil analysis was performed by Soil Control Lab in Watsonville, CA.

[b] Key :

A1 - 100% MSW[c]
A2 - 53% MSW, 47% UPD
A3 - 96% MSW, 4% Bio
A4 - 52% MSW, 46% UPD, 2% Bio
B1 - 96% UPD, 4% Bio
B3 - 100% UPD

[c] MSW = municipal solid waste; UPD = urban plant debris; Bio = biosolids/UPD.

soil erosion while decreasing water infiltration and the leaching of the contaminant. This process is referred to as phytostabilization. Another method, phytoextraction, removes the contaminant from the rhizosphere and accumulates it into the plant's biomass. The contaminant is removed from the system when the plant material is harvested and treated. Other methods include rhizofiltration, the removal of contaminants from water and rhizodegradation,

Table 13.3 : Trace Element Contents (Total) of Composts[a]

Constituents Tested	*Compost Type[b]*					
	A1	*A2*	*A3*	*A4*	*B1*	*B3*
	mg/kg (dw)					
As	10	<10	24	22	20	22
Cd	6.6	3	8	5.6	2.2	1.5
Cr	52	26	64	44	22	24
Cu	200	98	260	172	70	58
Zn	800	380	940	680	220	240
Fe	17000	18000	17000	12000	13000	11000
Mn	460	240	500	420	220	280
Pb	240	100	280	200	30	42
B	4	50	32	32	30	50
Mo	4.4	4	5.4	8	13	9
Ni	56	26	74	56	26	28
Se	<10	<10	<10	<10	<10	<10
Hg	0.55	0.45	0.33	0.4	0.61	0.45

[a] Soil analysis was performed by Soil Control Lab in Watsonville, CA.

[b] Key :

A1 - 100% MSW[c]
A2 - 53% MSW, 47% UPD
A3 - 96% MSW, 4% Bio
A4 - 52% MSW, 46% UPD, 2% Bio
B1 - 96%% UPD, 4% Bio
B3 - 100% UPD

[c] MSW = municipal solid waste; UPD = urban plant debris; Bio = biosolids/UPD.

which is the symbiotic relationship between plants and microorganisms that results in the breakdown of soil contaminants. Much of this paper will deal with phytoextraction.

A small group of plants known as hyperaccumulators can take up, translocate, and tolerate levels of certain heavy metals that would be toxic to other organisms (Baker and Brooks, 1989). Some hyper-accumulating plants can tolerate shoot concentrations of heavy metals in excess of 0.1% Ni, Co, Cu, Cr, Pb, or 1% Zn on a dry weight basis (Baker and Walker, 1990) and have often evolved on metalliferous outcroppings. They are remarkable not only for their high levels of accumulation and tolerance, but also for the plant's tendency to concentrate these elements from even normal soil. Although these plants are taxonomically widespread, the trait is relatively rare, indicating a rather late appearance in the evolution of modern species. Some of the more effective metal accumulating plants includes Indian mustard, sunflower, corn, peanuts, broccoli, etc. Raskin (1996) reported that Indian mustard could absorb 2 metric tons (2.2 short tons) of lead/hectare. Because of slow growth and low biomass of these plants, they are insufficient to form the backbone of a technology useful for remediation or

Table 13.4 : Statistical Test and Variance Components estimates for Broccoli Yield Data

Source	*NDF*[1]	*DDF*[2]	*Broccoli*[3] *F-statistic*
Compost Type (Type)	5	12	197.34**
Compost Rate (C_Rate)	2	24	2531.03**
Type x C_Rate IA	10	24	84.99**
Nitrogen Rate (N_rate)	2	72	595.42**
Type x N_rate IA	10	72	2.55*
C_rate x N_rate IA	4	72	44.83**
Type x C_rate x N_rae IA	20	72	2.07*
	Main Plot		0.345
Variance Components	Split Plot		0.367
	Split-Split Plot		10.75

1 Numerator degrees of freedom.

2 Denominator degrees of freedom.

3 F-statistic for the overall test of the effect. Significance indicated by *=p<0.05, **=0.05<p<0.01.

Table 13.5 : Broccoli Yields by Compost Type for Combinations of Compost Rate and N Rate

		Compost Product					
		100% MSW[1] (A1)	*53%MSW 47%UPD* (A2)	*96%MSW 4% Bio* (A3)	*52%MSW 46%UPD 2% Bio* (A4)	*96%UPD 4%Bio* (B1)	*100% UPD* (B3)
Average Yield Dry Weight (g/plant)		*88.1*	*76.9*	*96.4*	*80.3*	*75.0*	*66.7*
C_Rate Mg ha^{-1}	*N_Rate kg ha^{-1}*						
	0	27 a	37 a	40 a	37 a	41 a	37 a
0	75	65 b	62 b	58 b	64 b	58 b	59 b
	150	77 c	70 c	65 c	75 c	67 c	69 c
	0	94 d	72 c	92 d	65 b	62 bc	54 b
37	75	99 d	78 d	99 e	80 c	73 c	65 bc
	150	103 d	84 de	102 e	87 d	82 d	76 cd
	0	100 d	90 ef	129 f	95 e	88 d	72 cd
74	75	102 d	95 f	138 g	105 f	98 e	80 de
	150	116 e	104 g	145 h	115 g	106 f	88 e
Plot variance		0.0	0.00	1.79	0.56	0.66	0.00
Split plot variance		1.56	2.28	0.00	2.22	0.00	0.00
Split-split plot var		17.22	9.72	6.21	7.44	8.78	10.33

[1] Means in the same column with similar letters are not significantly different via LSD at a Type I error of 0.01.

biomining. Thus far, phytoremediation research has focused primarily on trace elements because of their few remediation options and ease of tracking throughout process (Cunningham et al., 1997).

Although phytoremediation is a low-tech and low-cost technique, there are limitations and constraints associated with this process. Perhaps the most important one has to deal with the contaminated site; it must be capable of sustaining plant growth. Many contaminated sites will have metal concentrations that will inhibit plant growth. In these situations, the soil must be amended to dilute the contaminant so that plants can grow. Another limitation is the fact that plants can only accumulate contaminants that are accessible in the rhizosphere or are transportable via transportation driven water movement in the soil (Bolton and Gorby 1995). The type of accumulation mechanism in the plant, the soil type, and

Table 13.6 : Various Forms of Nitrogen Added to Soil by Different Compost Types and Rates

Compost Rate ($Mg\ ha^{-1}$) N	*Compost Types*	*Total N*	*Organic N*	*Estimated[1] mineralized*
Kg ha^{-1}				
	A1	629	592	59-237
	A2	555	518	52-208
37	A3	629	592	59-237
	A4	592	518	52-208
	B1	481	481	48-194
	B3	444	444	44-177
	A1	1258	1184	118-472
	A2	1110	814	104-416
74	A3	1258	1184	118-472
	A4	1184	1036	104-416
	B1	962	962	96-385
	B3	888	888	89-355

[1] Based on estimated 10-40% mineralization during the first year of application.

the origin of the contaminant can also limit the effectiveness of phytoremediation. The greatest limitation can be associated with the amount of time that is required for the plants to accumulate the contaminant and to reduce the soil contaminant to acceptable levels (Kelly and Guerin 1995). Lasat et al. (1998) demonstrated that the use of red pigweed (*Amaranthus retroflexus* L.) to reduce the levels of radiocesium to acceptable levels would require two croppings a year and approximately fifteen years.

There has been a good deal of research conducted on hyper-accumulating plants for many types of contaminants. Nyer and Gatliff (1996) reported that cottonwood trees are capable of removing 36-71 kg of nitrogen/hectare in a nitrogen contaminated aquifer in New Jersey. Negri et al. (1996) found high levels of Zn [4,200 mg/g (ppm)] in leaves of hybrid poplar growing as a cleanup system at a site contaminated with Zn and 55,000 ppm of Zn in roots of the same plant in a greenhouse experiment which is about 5% dry root weight (1-5% accumulation is considered hyperaccumulative). More than 20 per cent Ni accumulation was found in the latex of the New Caledonian tree *Serbertia acuminata* (Summary Report, 1994). Kabata-Pendias and Pendias (1991) reported greater than 10 per cent Zn accumulation by pennycress (*Thalspi calaminare*), 10 per cent Ni accumulation by alyssum (*Alyssum bretolonii*), upto 3 per cent Cr concentration in *Pimela* suteri and broom tree (*Leptospermum scoparium*), up to 3 per cent uranium by *Uncinia leptostachya* and Coprosma arborea, and up to 1 per cent mercury in paper birch (*Betula papyrifera*).

On the other hand, *Masunaga* et al. (1997) identified *Aporusa maingayi, Mallotus glaberrimus, Memecylon laurinum, Pellacalyx axillaris, Spathostemon javensis*, and *Madhuca sericea* as sulfur (170-13,600 mg/kg) accumulators; *Aporusa frutescens, Aporusa maingayi, Memecylon laurinum, Maschalocorymbus corymbosus*, and *Eurya acuminata* to be Al accumulators (10-46,000 mg/kg); Ficus ribies as a Ca (2300-100,000 mg/kg) accumulator; *Barringtonia pendula* and *Urophyllum macrophyllum* to be Cl (110-9,200 mg/kg) accumulators; *Mallotus affnis* and *Madhuca sericea* as Si (2-1700 mg/kg) accumulators; *Mallotus affinis, Lithocarpus elegans, Gonocaryum macrophyllum, Cryptocarya scotechinii* and *Eugenia* decipiens to be Na (130-700mg/kg) accumulators; *Gonocaryum macrophyllum*, *Eugenia opaca*, *Evodia latifolia*, and *Callicarp arborea* as B (4-70 mg/kg) accumulators; *Eugenia opaca* as a Cu (0.8-150 mg/kg) accumulator; Ficus lepicarpa to be an Fe (0.6-4100 mg/kg) accumulator; *Gonocaryum macrophyllum* and *Eugenia opaca* as P (19-2,540 mg/kg) accumulators; and *Aquilaria melaccensis* as a Zn (0.4-290 mg/kg) accumulator. Labrecque et al. (1995) reported that *Salix discolor* and particularly *Salix viminalis* can be used as filters for the purification of wastewater sludges as well as for biomass production purposes. They found 60 per cent Ni, 79.9 per cent Cu, 76.4 per cent Pb, 86 per cent Hg and 52 per cent Zn accumulation in roots and stems of both species. Turner and Dickinson (1993) reported metal resistance induction in woody species. Dickinson et al., (1996) found that trees in Prescot, Merseyside had become acclimated to elevated metal levels.

Punshon and Dickinson (1998) reported that chronic exposure appears to be more effective at eliciting a resistance response in willows, and the hardy species *Salix caprea* was most amenable to resistance acclimation in Florida

We are currently conducting research to determine the feasibility of phytoremediation to address some of Florida's environmental concerns. Many of the state's soils have become polluted over time by agricultural and industrial processes. Two processes that have resulted in areas of high contamination are the CCA wood treatment process which uses salts of copper, chromium, and arsenic and the dipping of cattle in a mixture of arsenic, toxaphene, and DDT which was adopted to eradicate a species of tick that was causing cattle fever. With both of these processes, there was a risk associated with the contamination of the surrounding soil from leaks, spills, and deliberate dumping. Another environmental concern that we are addressing is the remediation of sewage effluent disposal sites that can contain high levels of N and P.

In 1997, our project selected sewage effluent disposal, cattle dip, and wood treatment sites for study with the following objectives:

1. To determine the levels of concentrations of Cu, Cr and As in the contaminated sites,
2. To identify and analyze the existing plant species that effectively remove, accumulate, or tolerate these contaminants,

3. To determine the effectiveness of selected trees to hyper-accumulate heavy metals from a contaminated site, and
4. To improve the ability of these trees to accumulate contaminants by adopting chelating agents, fertilizers, amendments, pH, and microbial inoculations.

POTENTIAL OF SELECTED TREES

Few studies have investigated metal resistance in woody species for the purpose of land reclamation (Antoňovics et al., 1971; McCormack and Steiner, 1978; Baker, 1987; Watmough, et al., 1995; Punshon and Dickinson, 1997). Trees need to take up as much water as possible so that they can accumulate as much as possible (Hinchman, 1996).

Phreatophytic trees (plants known for fast growth and high water usage rates) root deeply and have potential impact on hydraulic control of ground water. For example, a single willow tree evapotranspirates more than 5000 gallons of water in one summer day (Miami Conservation District, 1991). Similarly, the pumping rate for cottonwood ranged between 50 and 350 gallons /tree/day (Nyer and Gatliff, 1996). According to Labreque et al. (1995) for trees a higher metal content is acceptable as long as normal physiological activities are not affected. Once the heavy metals are absorbed, they are sequestered in the tree roots or an organic compound maybe metabolized. Not all plants sequester metals in their roots; some have the ability to convert toxic metals into gas. For example, broccoli, cabbage, rice, and others can accumulate selenium in their roots and convert it to a nontoxic gas dimethyl selenide which is 500-700 times less toxic than soluble selenium (selenate) which is easily dispersable (Terry, 1996). Similarly, hybrid poplar treescan absorb trichloroethylene (TCE) and convert it into metabolites like trichoroethanol, trichloroacetic acid, dichloroacetic acid, and CO_2 (Newman et al., 1997). Burken and Schnoor (1997) reported that this tree can uptake, hydrolyze, and dealkylate atrazine to less toxic metabolites like hydroxyatrazine, diethylhydroxy atrazine, diethylatrazine, diisopropyl atrazine, didealkylated atrazine, and hydroxylated de-alkylated products. This tree also transforms TNT into 4-amino-2,6-dinitrotoluene, 2-amino-4,6-dinitrotoluene and to a number of unidentified compounds which were more polar than TNT (Thomson et al., 1998).

The distribution of heavy metals among the different plant parts (leaves, stems, roots) is of great importance in the evaluation of the capacity of trees to decontaminate soils fertilized with sludge (Labreque et al., 1995). According to them, the metals in stems and roots of trees could be of importance because it allows an immobilization of these elements for long periods of time.

For remediation of these contaminated sites, we are evaluating the potential of Eucalyptus amplifolia, *E. grandis*, *E. camaldulensis*, *Populus deltoides*, cypress (*Taxodium distichum*), laurelcherry, castorbean (*Ricinus communis*), Leucaena leucocephala, slash pine, longleaf pine, and loblolly pine. Almost all these species are very fast growing; many are grown widely, including in India. The Eucalyptus species are highly productive (height growth up to 6 m/year),resprout after harvest, are heavy consumers of nutrients and water, and have excellent potential for efficiently extracting toxic chemicals or heavy metals from soil or water in sites contaminated with a variety of hazardous wastes. *E. amplifolia* is suitable for the frost-frequent subtropics, while *E. camaldulensis* and *E. grandis* are suited to the warmer subtropics. No single species is the most productive in all regions of Florida nor most suitable for all phytoremediation applications.

Pisano and Rockwood (1997) reported that *E. grandis* could annually take up to 165 cm of water and 170 and 330 kg of N and P/hectare, respectively. *Populus deltoides* is easy to establish and grows quickly. Its high transpiration rate and widespreading root system make it ideal to intercept, absorb, degrade, and /or detoxify contaminants while reducing soil erosion. These trees are amenable to coppicing and short-rotation harvest that helps in maintaining sustained root vigor (Dix et al., 1997). Furthermore, widespread root systems of this species are more effective than confined root systems in uptake of nutrients (Licht, 1992). For example, O'Neill and Gordon (1994) in their simulated groundwater depletion studies reported that *Populus canadensis* roots substantially reduced the concentrations of nitrates continuously over a 2-month period. Cypress is also fast growing and effective in remediating sewage effluent applied to ponds. For example, baldcypress planted on muck soils in southern Florida were up to 20 cm DBH and 10 m tall in nine years. Various pine species are very productive. Rockwood and Dippon (1989) and Rockwood et al. (1997) found that slash pine growth could be more than doubled. *Ricinus communis*, *Leucaena leucocephala*, and laurelcherry also may have remediation potential due to fast growth.

PLANTING STOCK

As possible, the most productive genotypes of the candidate species are included in the study. Cuttings of the best cottonwood clones were secured from clone banks in Mississippi and Florida (Pisano and Rockwood 1997). Superior seedlots of *E. amplifolia*, *E. grandis*, and cypress were selected from ongoing tree improvement programs. Plantlets of three *E. camaldulensis* clones were obtained. Castorbean seed from specific collections in central and south Florida was arranged, as was seed from three varieties of Leucaena tested in Florida.

FIELD STUDY DESCRIPTIONS

The sewage effluent application site (Water Conserv II) in our project is located on sand hills west of Orlando. It totals 2.8 hectares consisting of 42 rows of 240 trees divided into four irrigation zones. Since March 1998, a split-split-block design with three replications has been employed to compare four silvicultural options (conventional practices, i.e., no compost and no mulch; compost only; mulch only; and compost+mulch) for enhancing the uptake of sewage effluent applied at the rate of 100 l/tree/day to seven species: *E. camaldulensis, E. amplifolia, E. grandis, Taxodium distichum, Leucaena leucocephala, Ricinus communis,* and *Populus deltoides*. To date, the fastest growing species are *Ricinus communis* and *Populus deltoides*, and the compost treatments have greatly enhanced growth.

The CCA wood treatment site in Archer, Florida, covers approximately 0.4 hectares and includes a small concrete enclosure and the former wood treating facility. It is listed as a Super fund Sites by the EPA. The whole site is subdivided into 18 subunits, four of which have high concentrations of metals. Although the area is contaminated, different plant species like grasses, herbs, weeds and trees are growing well in different seasons. These species were collected and screened for metal accumulation. At present, eastern red cedar, boxelder, red mulberry, sugerberry, Albizzia, and mockernut hickory in triplicate were collected with roots and preserved in polyethylene bags until they were transferred to the lab. In the lab, the plants were washed with tap water to remove soil, dust and contaminants on the surface of the plant, preserved in brown paper covers, and kept for drying at 60°C for three days. Then the whole plant without separation into stems, leaves and roots was ground in Laboratory Mill model-4. About 500 mg of powdered, ground plant samples were digested in 10 mL nitric acid using microwave digestion method AG6 and diluted to 100mL with distilled water and analyzed for metal accumulation by using AAS. If any plant shows metal accumulation, then plant parts will be analyzed and concentrations will be noted and that plant species will also be included in the list for remediating the soils. Apart from this, all plant species selected will be studied in greenhouse experiments for the capability of metals accumulation. Any species showing hyperaccumulation will be further studied for within species genetic variation.

Triplicate soil samples were randomly collected in three different areas of the same plot in different depths like 20 and 100 cm. All soil samples were mixed in the site itself and preserved in polyethylene bags. In the lab they were dried at 70°C temperature for three days. About 500 mg of soil were weighed, digested in 10 mL nitric acid in accordance to EPA method 3051A and diluted to 100 mL with distilled water and analyzed for metal concentration in atomic absorption spectra.

FUTURE RESEARCH

A greenhouse study will be conducted to determine the effectiveness of high biomass producing trees in the removal of heavy metals from contaminated soil. Four trees representing one genotype of each candidate species will be grown in 3.7 l containers of the CCA contaminated soil. After reaching a height of approximately 1 m, the trees will be destructively sampled to assess uptake of Cu, Cr, and As. Initial and post-harvest soil analyses will also be conducted to estimate contaminant removal.

Concurrently, laboratory experiments will be conducted to determine methods to increase the availability of metals for removal by trees. Chelating agents, fertilizers, amendments, pH, and microbial inoculations will be examined as techniques for improving the ability of these trees to accumulate contaminants.

Promising trees and methods will then be combined in a subsequent greenhouse study in early 1999. Additional genotypes (tested progenies or clones) of the best species will be included to expand the genotypes that could be used in future studies. The CCA contaminated soil used in the first greenhouse study will be reused and analyzed after the trees are harvested at a size of about 1 m and assessed for contaminant content.

Field testing of promising progenies or clones and methods will then be initiated at the Archersite and a cattle dip site near Gainesville in mid-1999. Using site preparation methods that minimally disturb the soil yet improve bioavailability and enhance tree growth, the most remediating will be established

The Water Conserv II study will be monitored through 1999 to estimate the species that best take up N and P and use high amounts of effluent. These species will then be planted in plots as large as one hectare employing the cultural treatment that maximizes growth so that operational scale costs and productivity can be documented for short rotation cropping of trees that combine effluent remediation with commercial production of fuelwood, mulchwood, or even more conventional timber products. Soil and water monitoring of the original and new plots is also planned to determine the fate of nutrients and water added by the effluent and compost.

Gasification procedures for concentrating contaminants contained in trees will be initiated in late 1999. Gasification properties and yields for biomass components of promising genotypes will be determined, and solid residues will be analyzed to evaluate if indirectly heated gasifiers offer a benign method of distilling off and utilizing organics while retaining metals or other contaminants in residual char.

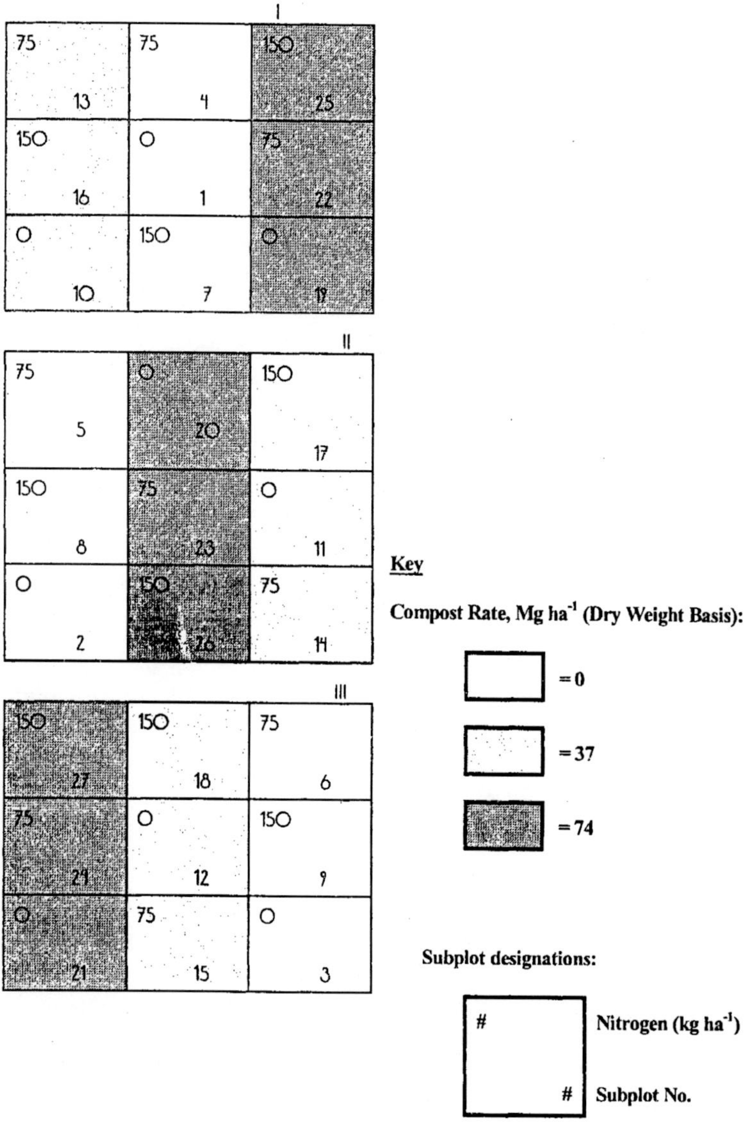

Fig. 13.1 : Compost Field Experiment Sample Plot Design

CONCLUSIONS

Trees have several attributes that are desirable for phytoremediation of soil and water contaminated with heavy metals, nutrients, and organic compounds. Research in Florida and elsewhere promises to identify the best species, develop cultural treatments for enhancing their effectiveness, and establish management systems for their long-term use.

REFERENCES

Antonovics J, Bradshaw AD and Turner RG. 1971. Heavy metal tolerance in plants. Advances in Ecological Research 7:1-75.

Baker AJM and 1987. Metal tolerance. New Phytol 106(Supp):93-111.

Baker AJM and Brooks RR. 1989. Biorecovery 1:81-126.

Baker AJM and Walker PL. 1990. In: Heavy metal tolerance in plants: Evolutionary Aspects. Shaw, AJ. ed. CRC Press, Boca Raton, FL. pp 155-177.

Bolton H. Jr., Gorgy YA. 1995. "An overview of the bioremediation of inorganic contaminants." In R.E. Hinchee, J.L. Means, and D.R. Burris (Eds.), Bioremediation of Inorganics, pp. 1-16. Battelle Press, Columbus, OH.

Burken JG and Schnoor JL. 1997. Uptake and metabolism of atrazine by poplar trees. Environ. Sci. Technol 31:1399-1406.

Cunningham SD, Shan JR, Crowley DE and Anderson TA. 1997. Phytoremediation of contaminated water and soil. In :Phytoremediation of soil water contaminants. Kruger EL, Anderson TA, Coats JR. eds. ACS Symp. Series 664 Amer. Chem. Soc., Washington, DC. pp 2-19.

Dickinson NM, Watmough SA and Turner AP. 1996. Ecological impact of 100 years of metal processing at Prescot, North West England. Environmental Review 4:8-24.

Dix ME, Klopfenstein NB, Zhang JW, Workman SW and Skim M. 1997. Potential use of Populus for phytoremediation of environmental pollution in riparian zones. In: Klopfenstein NB, Chun YW, Kim MS, Ahuja MR. eds. Micropropagation, genetic engineering and molecular biology of Populus. Gen. Tech. Rep. RM-GTR.297. Fort Collins, CO:U .S Dept. of Agric., Forest Service, Rocky Mountain Research Station. pp. 206-211.

Hinchman R. 1996. Cited by Gail Dutton. 1996. Stemming the toxic tide. Compressed Air 101(4) : 40.

Kabata-Pendias A and Pendias H. 1991. Trace elements in soils and plants. 2nd ed., Boca Raton, Fla: CRC Press.

Kelley RJ and Guerin TF. 1995. "Feasibility of using hyperaccumulating plants to bioremediate metal contaminated soil." In Hinchee R.E., Means J.L, Burris DR. (Eds.), Bioremediation of Inorganics, pp. 1-16. Battelle Press, Columbus, OH.

Labrecque M, Teodorescu TI and Daigle S. 1995. Effect of wastewater sludge on growth and heavy metal bioaccumulation of two Salix species. Plant Soil 171:303-316.

Lasat MM, Fuhrmann M, Ebbs SD, Cornish JE and Kochian LV. 1998. Phytoremediation of a radiocesium-contaminated soil: evaluation of cesium-137 bioaccumulation in the shoots of three plant species. J. Environ. Qual. 27: 165-169.

Licht LA. 1992. Salicaceae family trees in sustainable agroecosystems. For.Chron. 68:214-217.

Masunaga T, Kubota D, Hotta M. and Wakatsuki T. 1997. Nutritional characteristics of mineral elements in tree species of tropical rain forest, West Sumatra, Indonesia. Soil Sci. Plant Nutr. 43(2):405- 418.

McCormack LH and Steiner KC. 1978.Variation in aluminium tolerance among six genera of trees. Forest Science 24:565-568.

Miami Conservancy District. 1991. Aquifer Update no.11.

Negri MC, Hinchman R and Gatliff EG. 1996. Phytoremediation: using green plants to clean up contaminated soil, groundwater and wastewater. In: Proceedings of International Topical Meeting on Nuclear and Hazardous Waste Management, Spectrum 96. Seattle,WA, August 1996. American Nuclear Society.

Newman LA, Strand SE and Choe N, Duffy J, Ekuan G, Ruszaj M, Shurtleff BB, Wilmoth J, Heilman P, Gordon MP. 1997. Uptake and biotransformation of Trichloroetylene by hybrid poplars. Environ. Sci. Technol. 31:1062-1067.

Nyer EK and Gatliff EG. 1996. Phytoremediation. Ground Water Monitoring and Remediation 16(1):58-62.

O'Neill GJ and Gordon AM. 1994. The nitrogen filtering capability of Carolina poplar in an artificial riparian zone. J. Environ. Qual. 23:1218-1223.

Pisano SM and Rockwood DL. 1997. Stormwater phytoremediation potential of Eucalyptus. Proceedings 5th Stormwater Research Conference, Nov 5-7, 1997, Tampa, FL. Punshon T, Dickinson NM. 1997. Mobilisation of heavy metals using short rotation coppice. Aspects of Applied Biology 49:285-292.

Punshon T. and Dickinson NM. 1998. Acclimation of Salix to meta l stress. New Phytol 137(2):303-314.

Raskin, I. 1996. Cited by Gail Dutton.1996. Stemming the toxic tide. Compressed Air 101(4):39.

Rockwood DL and Dippon DR. 1989. Biological and economic potential of E. grandis and slash pine as biomass energy crops. Biomass 20(3&4):155-166.

Rockwood DL, Saucier JP and Clark A. 1997. Culture and genetic variation influence juvenile wood properties of Pinus elliotii and P. taeda in Florida. In: Proc. 24th. South. For. Tree Improvement Conf., June 9-12, 1997, Orlando, FL. p299-308.

Summary Report of a Workshop on Phytoremediation Research Needs, Santa Rosa, Calif. Dept. Energy, 1994. US Dept. of Energy Publication EM-0224.

Terry N. 1996. Cited by Gail Dutton. 1996. Stemming the toxic tide. Compressed Air 101 (4), 38.

Thomson PL, Araman L and Schnoor JL. 1998. Uptake and transformation of TNT by Hybrid poplar trees. Environ. Sci. Technol. 32(7):975-981.

Turner AP and Dickinson NM. 1993. Survival of Acer pseudoplatanus L. (sycamore) seedlings on metalliferous soils. New Phytol 123:509-521.

Watmough SA, Gallivan CC and Dickinson NM. 1995. Induction of nickel resistance in Acer pseudoplatanus L. (sycamore) callus cell lines. Environmental and Experimental Botany 35:465-473.

14

Sewage Sludge Hygienisation by Gamma Radiation and Recycling of Irradiated Sludge for Agricultural Applications : A Model Study

GAGAN A. PANDYA*

Sewage treatment an age-old practice adopted by human society, is the largest and one of the most important controlled applications of microorganisms. In the earlier days, it was mainly considered from the sanitary and hygiene point of view. The traditional aims of sewage treatment have been to reduce the concentration of organic matter which otherwise exerts an oxygen demand expressed as BOD (Biological Oxygen Demand) and to decrease the number of potential pathogens to a level where they are no longer a risk to human, animal or plant health. However, the problems involved in wastewater and sludge treatment have been magnified with acute industrialization, urbanization and a rise in population. There is no doubt, however, that today's sewage treatment processes are more scientifically based and are even aimed at resource conservation, ecological balance and economical gains, besides improving sanitation and hygiene. The growing importance of sewage treatment can be realized by the

* Wyeth Lederle Vaccines, Pearl River, New York, U.S.A.

fact that it has become one of the leading industries of the century. A US government technology report had estimated a \$ 400,000 million expenditure on pollution control between 1976-1986 (Wheatley, 1984). The European Community also has a policy that the spending on pollution control should continue to rise in accordance with increased expectations of environment quality (Wheatly, 1984). Its impact is so great that legally and socially imposed discipline in the form of rules and codes has become a basic necessity in the scheme of proper environmental management.

Sewage treatment processes have evolved scientifically also over the last two decades or so. Several methods, like conventional treatment plants (aerobic and anaerobic, mesophyllic and thermophilic digestion), composting, liming, lagooning and drying beds, pasteurization, thermal disinfection or chlorination are available at present (Strauch et al., 1985). All these methods aim at making the final effluent and sludge biologically and chemically safe. However, most of these methods have some disadvantage(s) and hence, a compromise is necessary between the availability of existing methods, objectives and economics.

A lot of research work is going on to either upgrade the existing methods or to find out better alternative sewage treatment processes. Ionizing radiation as a tool for sewage treatment is comparatively a recent concept. Researches carried out by several workers have indicated it to be a promising technology (Wizigman and Wursching, 1975; Epp, 1975; Wizigmann, 1976; Krishnamurthy, 1980, 1981; Kapila et al., 1981; Suess *et al;* 1983; Alexandre, 1983; Watanabe and Takehisa, 1983; Hashimoto et al.,1986).

The first pilot-scale radiation plant was established at the University Centre for Pollution Research in Florida in 1969. The plant employed Cobalt-60 as a gamma ray source and treated only 360 gallons of sewage effluent per day (approx. 1.5 m^3 Day^{-1}). However, the first large-scale sewage irradiation plant at Munich handles upto 150 m^3 of anaerobically digested sludge per day round the year since 1973. Since, the commissioning of the radiation treatment plant at Munich, radiation technology in sewage treatment has gained a lot of importance worldwide.

Successful disinfection of sludge with electron irradiation has been demonstrated at pilot scale operation at the Deer Island sewage treatment facility, Boston, U.S.A. since 1976. A large-scale irradiation facility employing Caesium-137 was in operation since 1979 at Sandia Laboratories, Albuquerque, New Mexico for the treatment of dried sewage solids till recently. One of the recently developed, Miamė Dade electron treatment plant at Virginia key, treats an average of 120 million gallons of wastewater per day since 1985 (SHRI Information Sheet - 2). A number of other irradiators, either at laboratory scale, plant level or for commercial applications of sewage and sludge treatment have been set up in different countries all over the world.

India also faces the increasing complexities in handling and treatment of sewage-sludge because of steep rise in population, rapid urbanization, industrialization and progress. Epidemics of human disorders are not uncommon in the country because of lack of adequate facilities of handling, treatment and disposal of sewage and sludge. Therefore, development of newer technologies for sewage treatment has become a priority. The Department of Atomic Energy took initiative in this regard by sponsoring research projects to develop radiation technology for sewage and sludge disinfection. The development of radiation technology for sewage and sludge treatment has special significance for India because of the following reasons:

- The availability of Cobalt-60 sources in megacurie, quantities with proven source technology capabilities.
- The possibility of employing Caesium-137 as an alternative source which will be available in future, as a by-product of nuclear power reactor operations.
- Beneficial utilization of nuclear wastes instead of costly and risky current disposal practices.
- The promise it holds, for recovering the resource value of nuclear waste to combat a public health problem of basic importance to the society.

  (From: Krishnamurthy, 1980)

Irradiation with ionizing radiation is very efficient method of rendering sewage and sludge free of pathogens. It results in the removal of viruses, bacteria, algae, protozoa and parasite eggs

(Krishnamurthy, 1980, 1981: Watanabe et al., 1981a, 1981b; Epp, 1975: Wizigmann, 1976; Wizigmann and Wursching, 1975; Suess et al., 1983; Alexandre, 1983; Watanabe and Takehisa, 1983). There are distinct advantages of using ionizing radiation for sewage sludge hygienisation. They are as follows:

1. The energy requirement is less than that for heat pasteurization.
2. Sludge can be disinfected continuously and completely.
3. There is an improvement of sedimentation of sludge.
4. Irradiated sludge has better dewaterability. This saves 50 per cent of the costs for the storage in the drying beds and 20 per cent of the transportation costs, making the process economical.
5. Irradiated sludge can be stored at room temperature without reinfection.
6. Irradiated sludge has a better fertilizer value and is safe to handle.
7. No residual chemical remains in sludge or effluent in contrast to chemical treatment process. Obnoxious gases are not generated.
8. The operation is simple.
   (From: Suess et al. 1983: Watanabe & Takehisa, 1983).

Because of the above reasons, a timely decision was taken by the Department of Atomic Energy to develop radiation technology for sewage-sludge treatment in India. Baroda was selected as a site to experiment and develop this technology. The relevance of this research project is justified considering the growing demand of sewage-sludge handling and treatment the indigenous technology to obtain Cobalt-60 or Caesium-137 in the country and partially solving the problem of nuclear waste disposal by using it for treatment of other waste. Radiation technology for sewage-sludge treatment can be of immense value from the point of view public health and hygiene, improved environment, resource conservation and even economical benefits.

The objectives and major achievements of the project undertaken during the period of 1980 to 1991 is outlined and summarised here. The results have been published independently (Pandya, et. al., 1987, 1988, 1989 & 1991).

Objectives

1. To analyze the sewage samples for seasonal variation in total bacterial and coliform counts.
2. To standardize the gamma ray dose requirements for inactivation of microorganisms commonly found in sewage.
3. To find a suitable radiosensitizer which (when added to sewage) will economize the treatment process and make it more efficient.
4. To check the fertilizer value of gamma irradiated sludge with regard to agriculturally important plants with a view to safely recycle the treated sludge for beneficial uses.
5. To scale up the process at the community or city level based on the results obtained in laboratory.
6. To replace the inefficient conventional sewage treatment processes with indigenously developed radiation technology throughout the country.

Achievements

1. All the sewage samples examined from Wadi sewage treatment plant, Baroda showed high bacterial and coliform counts, suggesting that the process of conventional treatment of sewage was inefficient.
2. The seasonal variation in the total bacterial and coliform counts in sewage samples have been monitored for a period of more than five years The results strongly indicate the inefficiency of conventional treatment and therefore urgency for a better alternative.
3. Gamma ray dose requirements have been standardized. A dose of 1000 krad could bring about a reduction of 5.3 to 6.7 log cycles in total bacterial counts of all the sewage samples.
4. A dose of 500 krad was sufficient to reduce the total coliform count in sewage samples by 5 log cycles.
5. The gamma radiation doses required for bacterial inactivation in sewage were found to be high compared to the doses required for inactivation of the same organisms in buffer, suggesting a protective effect of the sewage constituents towards organisms.
6. Sewage constituents were found to exert different radio-modifying effects on different organisms.
7. A number of chemicals have been screened to find an ideal radiosensitizer in sewage with following properties:

- The compound should be cheap and readily available.
- It should have a fairly broad radiosensitizing spectrum.
- The compound should not directly or indirectly produce any harmful compound or effect by reacting with sewage constituents.
- The radiosensitizing activity should be independent of sewage composition.
- The use of radiosensitizer should make the gamma radiation treatment of sewage more efficient and economical.

8. Potential of recycling gamma-irradiated sewage sludge for use as a fertilizer in agriculture has been assessed by studying the effect of irradiated sludge on rice (*Oryza sativa* Var. GR-3), chickpea (*Cicer arietinum*) and pigeonpea (*Cajanus cajan*). Results obtained from studies carried out in pot culture suggested gamma-irradiated sludge to be a better fertilizer compared to unirradiated sludge. Moreover, sludge obtained from conventional treatment process (unirradiated) was found to inhibit the plant growth indicating further treatment before its use as a fertilizer. No such inhibitory effect was observed in case of gamma irradiated sludge.
9. A sewage treatment plant employing gamma radiation (Cobalt-60) has already been set up at Gajrawadi, Baroda, which treats the Baroda city sewage since 1990. This is the first radiation treatment plant for sewage-sludge hygienisation in the country and probably in Asia. The technical specifications of the plant are as follows:
   - Maximum sludge treatment capacity: 110 M^3 per day
   - Cobalt-60 activity: 500 kilocuries.
   - Protection standard to house: 1000 kilo curies.
   - Treatment dose: 0.5 - 5.0 KGy (variable).
   - Pathogen reduction: To acceptable levels.
   - Process control: Fail-safe automated.

The results obtained clearly encourage the development of this technology in other parts of the region and compare the data obtained with the data obtained with sewage of Wadi Treatment plant. This would strengthen the study and point towards the refinement of the treatment technology with respect to the local sewage composition. Some of the future directions this project can point towards are as follows:

- To check the effect of gamma-irradiated sludge on other agriculturally important local plants.
- To analyze the radiation-induced changes in sludge, which modifies the plant growth.
- To carry out physico-chemical analysis of gamma-irradiated sludge in detail to understand the irradiation effects on sludge.
- To assess the possibility of recycling gamma-irradiated effluent for agricultural or industrial use.

ACKNOWLEDGEMENT

These studies were carried out during 1980-1991 under the guidance of Prof. V. V. Modi at the Department of Microbiology and Biotechnology Center of M.S. University Baroda, India. Thanks are due to Prof. V. V. Modi for his Financial assistance was provided by Department of Atomic Energy, India. Contribution of a number of other colleagues and co-workers on the project is also acknowledged.

REFERENCES

Alexandre, D. (1983). In: 'Disinfection of Sewage Sludge : Technical, Economio and Microbiological Aspects'. (A. M. Bruce, A. H. Havelaar and P. L'Hermite eds.), p. 212. D. Reidel Publishing Company, Holland.

Epp, C. (1975). In: 'Proceedings of a Symposium - Radiation for Clean Environment, p. 485. International Atomic Energy, Vienna.

Hashimoto, S., Nishimura, K., Kawakami, W. and Watanabe, H. (1986). J. Ferment. Technol. 64 299.

Kapila, S., Puranam, R.S., Rao, N.N., Modi, V.V., Ekbote, R. and Shah, M.N. (1981). In: 'Proceedings of the National Symposium on Applications of Nuclear and Allied Techniques in Public Health and Pollution Control' p. 44. Bhabha Atomic Research Centre, Bombay (Dept. of Atomic Energy), India.

Krishnamurthy, K. (1980). In: 'Management of Environment' (B. Patel, ed.), p. 149, Wiley Eastern Limited, New Delhi.

Krishnamurthy, K. (1981). In: 'Proceedings of the National Symposium on Applications of Nuclear and Allied Techniques in Public Health and Pollution Control'- p. 15. Bhabha Atomic Research Centre, Bombay (Dept. of Atomic Energy), India.

Pandya, G.A., Kapila, S., Kelkar, V. B., Negi, S. and Modi, V. V. Inactivation of bacteria in sewage-sludge by gamma radiation. Environ. Pollut. 43, 281-290, 1987.

Pandya, G.A., Prakash, L., Devasia, P, and Modi, V. V. Effect of gamma irradiated sludge on the growth and yield of rice (*Oryza sativa L. var.* GR3). Environ. Pollut. 51, 63-73, 1988.

Pandya, G. A., Sachidanand, S. and Modi, V. V. Potential of recycling gamma irradiated sewage-sludge for use as fertilizer: A study on chickpea (*Cicer arietinum*). Environ. Pollut. 56,101-111,1989.

Pandya, G. A'., Banerjee, S. and Modi, V. V. Perspectives of recycling gamma irradiated sewage-sludge in agricultural applications: A study on methi (*Trigonella foenum-graceum L.*) Environ. Pollut. 72, 225-237,1991.

SHRI Information Sheet-2 : 'Sludge Hygienisation Research Irradiator'. Radiation Technology Division, Isotope Group, Bhabha Atomic Research Centre, Bombay, India.

Strauch, D., Havelaan, A.H. and L'Hermite, P. (eds.) (1985). In: 'Inactivation of Microorganisms in Sewage Sludge by Stabilisation Processes'- Elsevier Applied Science Publishers, London.

Suess, A., Lessel, T. and Haisch, A. (1983)- In: 'Disinfection of sewage sludge : Technical, Economic and Microbiological Aspects' (A.M. Bruce, A.H. Havelaar and P.L. Hermite eds). p. 179. D. Reidel Publishing Company, Holland.

Watanabe, H. and Takehisa, M. (1983). Radiat. Phys. Chem. 2441.

Watanabe, H., lto, H., lizuka, H. and Takehisa, M. (1981a). Hakkokgaku, 59449.

Watanabe, H., lto, H., lizuka, H. and Takehisa, M. (1981b). Hakkokgaku, 59 455.

Wheatley, A. (1984). In: 'Biotechnology and Genetic Engineering Reviews' (G.E. Russell, ed.), Vol. I, p. 261, Intercept, New Castle, Upon Tyne.

Wizigmann, I. (1976). in: 'Radiation for Pollution -Abatement', Proceedings of the First International Conference of ESNA Working Group on, 'Waste-Irradiation, p. 87. Organised by European Society of Nuclear Methods in Agriculture.

Wizigmann, I. and Wursching, F. (1975). In: 'Proceedings of a Symposium Radiation for Clean Environment, p. 477. International Atomic Energy Agency, Vienna.

15

Microbes : Source of Energy for 21st Century

K.S. MANJA* AND H.S. MURALI

Energy crisis of 1970 has triggered a great interest in anaerobic digestion and methane production. Secondary goal of this process is to abate environmental pollution. Initially this was not found economically viable but due to increased cost of energy and awareness of environmental pollution, the interest in this field is still sustaining. Further the sale of digester solids as fertilizer for organic farming, production of electricity from methane added credence to this technology. Production of alcohol from microbes is an established source of tapping energy from microbes. Recently in our laboratory we have developed a technology to produce microbial oil from *Fusarium* spp. employing food factory waste which could act as a source of energy and spent material could be used as animal feed.

Microbial technology presents a viable alternative technology for energy intensive conventional pyrometallurgy technology for mining, mineral and waste water technology. Biohydrometallurgy technology is cleaner, consumes less energy than those used in pyro and hydrometallurgical industries. Commercial application of this technology appears to be viable for the recovery of copper,

* Division of Microbiology, Defence Food Research Laboratory, Mysore-570 011, India.

uranium and detoxifying the fly ash. Laboratory scale leaching plants using *Thiobacillus thiooxidans* and *Thiobacillus ferrooxidans* have been successfully developed, so also a large scale reactor for leaching of gold reducing the energy consumption. There are many more areas like agriculture, food, health care, automobile industry where the microbial energy could be employed thereby reducing the demand for scarce conventional energy. Microbes provide great promise as a source of energy in the 21st century when the oilgut gets depleted.

INTRODUCTION

Energy consumption is an index of a country's economic development. There is a close link between the availability of energy and the future growth of a nation. As it impinges on poverty, jobs and incomes, access to social services, gender disparities, population, agricultural production and food scarcity, health, land degradation, climate change and environmental quality, and economic and security issues, it is directly related to the most pressing social issues affecting sustainable development. The large proportion of humanity does not enjoy the benefits of modern energy sources and devices. It is estimated that about 2 billion people still use traditional fuels and about 1.5 - 2 billion people are without access to electricity (Reddy, 1997).

India's energy situation has been grim for several years. The demand for electricity in the country has been rising at the rate of 9.7 per cent and that in the residential and agricultural sectors have been increasing at an alarming rate of 14.03 per cent and 14.58 per cent respectively (Anonymous, 1996). It has been estimated that there would be a ten fold increase in energy requirement within next 20 years, if the national economic growth rate is maintained at around 6 per cent.

(1) Population increase
(2) Rise in the standard of living, and
(3) The relationship between energy growth and economic growth.

In addition to non-renewability and exhaustion, conventional energy sources have become a major threat to the clean environment. For example, the continued use of firewood leads to deforestation and soil erosion. The concentration of carbon dioxide has significantly increased due to ever increasing use of fossil fuels. In the last decade alone, it has increased by about 100 ppm resulting in the warm-up of upper layer of the oceans which has caused a rise in the sea level. In the case of fissionable nuclear fuels, the raw materials cause a plethora of hazards both to man and his balanced environment. Therefore, there is an urgent need for tapping energy from alternate sources to meet the increasing energy demand in the coming years. These sources should be energy efficient, cost effective and environment friendly.

It is in this context, microbes can be contemplated as a major source for the production and conservation of energy for different purposes. In fact biogas and ethanol which are produced by biological route from the biomass have been in use for many years. They have already made inroads in the domestic and industrial sectors. At present a serious attempt is being made to widen the scope of these processes by making the technologies more cost effective and readily available.

ENERGY FROM BIOMASS BY MICROBIAL PROCESS

India has a vast potential of renewable energy sources. The Ministry of Non-conventional Energy has designated a number of national programmes to harness energy from non-conventional sources (Anonymous, 1996). Among the various alternatives, the use of biomass energy is an option which is receiving increasing attention in many parts of the world. As an energy source, biomass has several important advantages. Renewability is a key factor. In biomass, solar energy is trapped as organic molecules synthesized by photosynthetic mechanisms. Out of the total solar energy on earth (3×10^{24}J), the plant life utilizes about 0.1 per cent annually leading to an annual net production of 2×10^{11} tonnes of organic matter which has an energy content of 3×10^{12}J. The total annual energy use, however, is of the order of 3×10^{20}J. In India, there is an abundant supply of sunshine in the country. The total solar radiation received in the country is about 60×10^{13} MWH with 250-300 days of useful sunshine per year in most parts (Vimal and Tyagi, 1984).

The main factors which will determine whether a biomass scheme can be implemented in a given country are: (a) the biomass resource, (b) the available technology and infrastructure for conversion, distribution and marketing, and (c) the political will combined with social acceptance and economic viability of the project with the array of conversion processes that now exist, a whole range of more convenient gaseous and liquid fuels can be produced from biomass. They would become the potential substitutes for fossil fuels. Biomass programmes are in action in several countries including U.K., Ireland, France, Germany, Denmark, Sweden, U.S.A., Canada, Mexico, Brazil, Sudan, Kenya, Australia, Botswana, Tanzania, Papua New Guinea, New Zealand, India, China, Philippines, Thailand, Israel and South Korea (Barnard and Hall, 1983).

Biogas from Microbes: Although the use of biogas technology, popularly known as "Gobar Gas" in India started long back, official interest was spurred only after fuel and fertilizer crisis in 1970. India being predominantly an agricultural country with 80 per cent of its population living in 5.7 lakh villages, major stress has been laid on production and supply of biogas and manure from biogas generators. Under this programme, as many as 23.59 lakh family type reactors have been installed by the end of 1995-96. About 1400 institutional plants have been installed upto 1994-95 (Anonymous, 1996).

Biogas is a clean and efficient fuel drawn from cowdung and hosts of other organic materials. It is produced by the anaerobic digestion technology involving hydrolytic bacteria, hydrogen producing acetogenic bacteria, homoacetogenic bacteria and methanogenic bacteria. The efficiency of the process is dependant on the content of the substrate with microbial cells, the type of the organism, the number of cells and type of reactors. Currently, a high rate reactors such as upflow anaerobic sludge blanket (USAB), expanded granular sludge bed (EGSB), fixed film (FF), fluidized bed (FB), disc bed and plug flow are being used extensively in developed countries for the treatment of waste waters and solid wastes for food processing and agro industries (Krishna Nand, 1995). A variety of substrates are available for anaerobic digestion. The studies carried out on the use of willow-dust, a textile mill processing residue and bagasse and press mud generated in sugarcane crushing industries by solid state fermentation showed that about 600 m^3 of biogas was produced from one tonne of willow-dust and 1:1 mixture of bagasse and press-mud (Balasubramanya, et al. 1995). Organic matter with potential for methane generation are listed in Table 15.1.

The gas produced from anaerobic digesters typically contains between 60 per cent and 70 per cent methane, the remainder being largely carbon dioxide and hydrogen (Khandelwala and Mahidi, 1986). Typical gas yields from selected feed stocks is given in Table 15.2. The energy content of biogas varies from 20,000 - 26, 000 KJ/m^3 depending on the amount of methane it contains, compared to 35,000 KJ/m^3 for pure methane.

Table 15.1 : Organic matter with potential for methane generation

Crop wastes
Sugarcane thrash, weeds, corn and related crop stubble, straw, spoiled fodder.
Wastes of animal origin
Cattle shed wastes (dung, urine, litter), poultry litter, sheep and goat droppings, slaughter house wastes (blood, meat), fishery wastes, leather, wool wastes.
Wastes of human origin
Feces, urine, refuse (night soil)
By products and wastes from agriculture based industries
Oil cakes, bagasse, rice brass, tobacco wastes and seeds, wastes from fruit and vegetable processing, press-mud from sugar factories, tea waste, cotton dust from textile factories.
Forest litter
Twigs, bark, branches, leaves
Wastes from aquatic plants
Marine algae, sea weeds, water hyacinths.

Source : Garg et. al 1971.

Table 15.2 : Typical gas yields from feedstocks

Feed Stock	*Gas yield (m^3/ton of organic matter fed)*	*Gas composition (% methane)*
Cow dung	90-300	65
Chicken manure	300-600	60
Pig manure	350-480	65-70
Farm Wastes	300-420	60-70
Elephant grass	420-540	60
Chicken manure/paper pulp	420-480	60
Chicken manure/grass clippings	350	68
Sewage sludge	600	68

Source : Barnard and Hall (1984).

As a fuel this gas has a variety of possible applications. It provides a convenient clean-burning flame for cooking and other heating purposes. It can also be used effectively in lamps, in duel fuel diesel engines or in modified spark ignition engines. In Zurich, the biogas produced in the compogas facility is processed to comply with natural gas quality standards and fed into municipal gas supply net work (Claude, 1998). In Bulgaria, experiments conducted under COGERSA to assess the quality and quantity of biogas generated in the landfill showed that there a real possibility exists to recover biogas from land fills (Martin et al, 1997). Besides energy it provides, rich organic manure for raising the yield of crop and also maintains the fertility of the soil.

Liquid fuel from Microbes: Providing alternative liquid fuels is probably the most challenging task. Ethanol production from biomass is one of the few viable options that is currently available. Several countries have been manufacturing fuel grade ethanol as a substitute for petrochemical fuels. The prominent ones among them are : Brazil, the U.S., Australia, Costa Rica, Indonesia, Philippines, Sweden and Zimbabwe.

Traditionally, alcohol for industrial use is manufactured from microbial fermentation of molasses. Grains, excess grapes or other fruits, potatoes, starchy roots like cassava, mahun flowers and palm juice may be used depending on their value in relation to human and animal foodstuffs. In Sweden, Finland and Canada, the sulphite liquor with 2-3 per cent fermentable sugar constitutes a suitable substrate for alcohol production. In Brazil, sugarcane is directly converted into alcohol. The attractive feature of low cost, abundance, and a renewable nature of a variety of cellulosic residues such as corn-stove, wheat straw, and rice straw results in their frequent consideration as substrates for fuel grade ethanol production. A novel single-step microbial conversion of cellulosic biomass to ethanol using *Cl. thermocellum* and *Cl. thermosacche-rolyticum* has been designed for the direct production of ethanol from agricultural cellulosics (Avgerinos et al., 1981). In Japan, an integrated bench-plant has been developed for this purpose (Miyakawa et al., 1986). A list of feed stocks for ethanol production and alcohol yields are given in Table 15.3. Regarding the fermentation process, much interest has been focussed on continuous operation of the fermenter and immobilization of microorganisms (Robinson et al., 1981; Davison and Scott, 1986; Sola et al, 1986). Continuous fermentation gives high productivity per unit volume of fermentor along with simple computer controlled operations. Immobilization provides a high concentration of cells in the fermentor and simple separation of products and cells.

Ethanol is employed for a wide range of end-uses. It can be used as a liquid-fuel substitute for internal combustion engines as a blend with gasoline or as the entire fuel. Brazilian tests with dual fuel systems indicate that as much as of 50 per cent substitution of diesel oil may be achieved in combustion engines (Harris et al, 1979). In addition, it is more economically used in chemical industries and as substitute for naptha as feed-stock.

Hydrogen from Microbes : Several blue-green algae containing hydrogenase are capable of producing hydrogen from water and sunlight. Since the substrate water and the energy source are abundant, hydrogen can be considered to be an economically viable energy source. The whole process is renewable because when hydrogen is consumed the substrate water is regenerated. It is also environment friendly as it operates at normal ambient temperatures without yielding any toxic intermediates and by-products. Having very simple nutritional requirements, algae can be cultivated on dilute waste materials.

India has a vast coastal area which abounds in many forms of algae. It may be worthwhile to explore the possibilities of using unicellular blue-green algae like *Synechococcus* spp., *Chlorella fusca, Cl. vulgaris f tertia, Spirulina platensis, Anabaena azollae* and *A. cylindrica* for the production of fuel hydrogen. To make the biophotolysis a viable alternative to conventional forms of energy, a great deal of research is needed on the following aspects :

(1) Modification of algae by genetic engineering to increase productivity.
(2) Reactor design.
(3) Optimum conditions for efficient hydrogen production, and
(4) Micro-capsulation and stabilization of sub-cellular fractions capable of hydrogen evaluation.

Table 15.3 : Feedstock for ethanol production

Crop	*Ethanol yield (L/ton)*
Sugar Crops	
Sugar Cane	70
Sugar beet	95
Sweet Sorghum	86
Molasses	280
Cellulosic Materials	
Wood	160
Stanch Crops	
Cassava	180
Sweet Potatoes	125
Jerusalem artichoke	82
Potato	100
Corn (Maize)	370
Wheat	355
babassu Palm Nut	80

Source : World Bank (1980); Kardos and Mulcock (1977).

Microbial Oil : It has been well established that a number of yeasts and molds are potential sources of oils and fats, which is a store house of energy. The oil content in the dry cell mass varies from 50 to 85 per cent. Recently in our laboratory microbial oil has been produced from *Fusarium* spp. employing food factory wastes (Murali, 1997). The possibilities of utilizing microbial oil in diesel engines directly or as a mixture with diesel fuel deserves more attention.

MICROBES IN ENERGY CONSERVATION

Microbial technology presents a viable alternative technology for various energy intensive processes. Biohydrometallurgy and biodegradation of toxic materials are the two important areas where microbes can be employed as a viable substitute for expensive physical and chemical agents. The role of microorganisms in the production of value-added products in agriculture and food processing cannot be overlooked.

Biohydrometallurgy : The biotechnology of microbial leaching has advanced technologically over the past years to commercial application. Microbially mediated chemical reactions result in significant levels of metal accumulation or trapping by microorganisms. Microbes could be an alternate process to energy intensive pyrometallurgical processes. Commercial application of these biological processes are of potential value to the metal industries. Microorganisms are already exploited in the mining industry where bioleaching is used to process certain ores, containing Cu and U for example which are too lean to be economically mined by conventional methods. Microbial processes for recovery of metals from process or waste streams for economical and environmental benefits are also under active study (Brinckman and Olson, 1986). The interaction of microorganisms with metals is of particular interest with regard to the base metals including vanadium, chromium, manganese, iron, cobalt, nickel, copper, zinc, molybdenum, silver, cadmium, and lead; the precious metals gold and silver; and the metaloides arsenic, selenium, and antimony (Ehrlich, 1997). An integrated technological circuit for gold production incorporating biological processes is under active consideration (Natarajan, 1995). It has been predicted that within the next decade, the frontier technology is set to revolutionize gold production, making it economically viable to recover the precious metal even from wastes and unconventional lean ore deposits. The types of microorganisms important in biohydro-metallurgy are listed in Table 15.4.

Table 15.4 : Types of microorganisms important in biohydrometallurgy

Microorganisms Promoting metal solubilization
Thiobacillus ferroxidans
Leptospirillum ferroxidans
Sulfolobus spp.
Thiobacillus thiooxidans
Microorganisms promoting precipitation of metals
Arthrobacter
Bacillus
Pseudomonas fluorescens
Pseudomonas dechromaticans
Gallionella spp.
Microorganisms binding metal ions
Rhizopus arrhizus
Bacillus subtilis
B. licheniformis
Pseudomonas aeruginosa

Biohydrometallurgy is a novel promising technology for recovering valuable metals from industrial waste materials (e.g. bottom and fly ash, galvanic sludge, and filter dust) and for detoxifying these materials for environmentally safe deposition. Biohydrometallurgical processing of solid waste allows recycling of metals, similar to natural biogeochemical metal cycles, and diminishes the demand for resources, such as ores, energy and landfill space (Brombaucher et al., 1998). This technology is cleaner and consumes less energy than technologies used in the pyro and hydrometallurgical industries.

Detoxification by Microbes : As a result of recent advances in bioremediation, several microbial processes have been designed for neutralizing specific chemical pollutants in agriculture and sewage wastes and landfills. The list of chemicals that can be removed from our environment by microbes include trichloroethylene, polychlorinated biphenyls, polyvinyl chlorides, nitroaromatic compounds and pesticides like DDT and isoproturan. In addition, these technologies can also be employed for neutralizing chemical warfare agents. For example, species of *Altermonas* bacteria are able to detoxify a whole range of chemical warfare agents and nerve gases like Sarin. These methods are simple, economical and safe when compared to conventional physical and chemical methods. A bacteria discovered at the Horticulture Research International (HRI), U.K. is adept at breaking down isoproturan, one of the widely used herbicides in U.K.

Microbial Enhanced Oil Recovery (MEOR) : Crude oil produced by current methods leads to recovery of only about 8-30 percent of the total oil present in the reservoir. The oil recovery process can be made more efficient by the use of microorganisms which also increase the productivity and longevity of the present oil fields.

There are several reports of successful use of microorganisms for the microbial enhanced oil recovery (MEOR). *Bacillus licheniformis* JF-2 and *Clostridium acetobutyricum* have been used successfully in experiments mimicking the reservoir conditions. The workers were able to recover an additional 17-19 per cent of oil by this process. Japanese scientists have shown that a strain of the thermophilic anaerobe *Cl. thermosaccharoliticum* can recover the residual oil from wells earlier flushed clean with water. Several field trials of MEOR have been conducted in USA, Russia, Poland, Hungary, Romania, Netherlands and Canada. In the field experiments at Bondyzchskoe oil fields in Russia, MEOR increased oil-production by 20-100 per cent. Improvement of strains by genetic engineering will help in successful recovery of oil out of not only the current producing wells, but even from out of those wells which have been abandoned as non-productive. The MEOR has some advantages over the chemical methods: They are; (1) low cost; (2) highly specific; (3) the activity of microbes easily controllable; and (4) applicable to both heavy and light crude oil.

Microbes as biofertilizers and biopesticides : Biofertilizers are preparations containing microorganisms, in sufficient numbers helping plant growth and nutrition. Biofertilizers are widely accepted as low cost supplements to chemical fertilizers and have no deleterious effects either on soil health or environment. Further, the production of chemical fertilizers is energy intensive. Biofertilizers, which can supply nitrogen and phosphorous, two major plant nutrients, are available currently. Free-living nitrogen fixing bacteria like *Azotobacter* not only fix nitrogen but produce growth promoting substances. Rhizobia living in symbiosis with legumes fix atmospheric nitrogen. *Azorhizobium caulinodans* forms stem nodules in the green manure plant *Sesbaria rostrata.* The other nitrogen fixing organisms are *Herbaspirillum spp, Acetobacter diazotrophiens*, Cyanobacteria like *Nostoc, Anabaena* and *Tolypothrix.* Similarly, phosphate solubilizing bacteria and fungi, solubilize unavailable forms of phosphorous in soil through the production of organic acids and make it available for plant growth. Some of the phosphate solubilizing organisms are: *Bacillus polymyxa, Pseudomonas striata, Aspergillus awamori* and *Penicillium digitatum.* Certain mycorrhizal fungi like *Pisolittus, Laccaria, Amanita, Scleroderma, Russula* and *Tricholoma* which live in symbiotic association with roots of crop plants and forest tree species improve uptake of phosphorous (Bagyaraj and Indira, 1995)

The Integrated Pest Management (IPM) concept being promoted world over recommends biopesticides instead of conventional chemicals to control pests and insects in crop fields. The present focus is on microbes which can grow on pests and insects without posing any danger to man and his environment. Two Bio-insecticides, *Bacillus thuringensis* var *israelensis* and *Bacillus thurengensis* var. *Krustaki*, which can control mosquito larvae and agricultural pests respectively have been developed by the Hindustan Insecticides Limited keeping in view the IPM concept.

CONCLUSION

Microbes a store house of biochemical processes can produce variety of byproducts without employing heat. Further, they can produce fuels like ethanol, methane and H_2 which can be used as energy source directly. Microbial degradation of toxic chemicals, biopesticides and biofertilizers would greatly reduce the energy requirement in future years. Therefore, microbes provide great promise as a source of energy in the 21st century when the oilgut gets depleted.

REFERENCES

Anonymous (1996). Energy In "India 1996 - A Reference Annual" Ed. Research, References and Training Publication Division, Ministry of Information and Broadcasting, Govt. of India. pp. 374-395.

Avegerinos, G.C., Fang, H.Y., Biocic, I and Wang, D.I.C. (1981). A novel, single-step Microbial conversion of cellulosic biomass to ethanol. In "Advances in Biotechnology" Vol.II. Eds: M. Moo-Young and C.W. Robinson. Pergamon Press, Toranto pp. 105-111.

Bagyaraj, D.J. and Indira (1995). Microbes - a goldmine of biofertilizers. In "Microbes for better living". Eds: R. Sankaran and K.S. Manja Association of Microbiologists of India, India. pp. 51-55.

Balasubramanya, R.H., Gangan, H.U., and Khandeparkar (1995). Biogas from cellulosic wastes by solid state fermentation. In "Microbes for better Living". Eds: R. Sankaran and K.S. Manja. Association of Microbiologists of India, India, pp. 235-240.

Barnard, G.W. and Hall, D.O. (1983). Energy from renewable resources. In "Biotechnology - a comprehensive treatise in 8 Vol." Vol.3 Ed. H.J. Rehm and G. Reed. Vol. Ed: H. Dellweg. Verlag Chemil GMBH, Weinheim, pp. 597-625.

Barnard, G.W., and Hall, D.O. (1983). Energy from renewable resources In "Biotechnology" Vol.3 Ed: H. Dellweg Verlag Chemil, Weinleim, Deerfield Beach, Florida, Basel. pp. 593-625.

Brinckman, F.E. and Olson, G.J. (1986). Chemical Principles underlying bioleaching of metals from ores and solid wastes, and bioaccumulation of metals from solutions. Biotechnology and Bioengineering symposium, 16, 35-43.

Brombacher, C., Bachofen, R., and Brandl, H. (1998). Development of a laboratory-scale leaching plant for metal extraction from fly ash by *Thiobacillus* strains. Applied and Environmental Microbiology, 64, 1237-1241.

Claude, W.J. (1998). Use of biogas in the public natural gas network, illustrated by the examples of compogas deliveries at Samstagern, Switzerland. Gas, Wesser, Abwasser, 78: 79-83.

Davison, B.H. and Scott, C.D. (1986). Ethanol production from an industrial feedstock by immobilized *Zymomonas Mobilis* in a fluidized-bed reactor. Biotechnology and Bioengineering symposium, 17, 629-632.

Ehrlich, H.L. (1997). Microbes and metals. Applied Microbiology and Biotechnology, 48, 687-692.

Garg, A.C., Idnani, M.A. and Abraham, T.P. (1971). Organic manures, Bulletin No. 32 Indian Council of Agricultural Research, New Delhi.

Harris, G.S., Leamy, M.L., Fraser, T., Dent, J.B., Brown, W.A.N., Earl, W.B., Fookes, T.W. and Gilbert, J. (1979). The potential of energy farming for transport fuels in New Zealand. New Zealand Energy Research and Development Committee Report No. 46, University of Auckland, New Zealand.

Kardos, N. and Mukock, A.P. (1977). Ethanol from agricultural crops : A literature survey. New Zealand Energy Research and Development Committee Report No.28. University of Auckland, New Zealand.

Khandelwala, K.C. and Mahdi, S.S. (1986) Benefits of Biogas. In "Biogas Technology - a Practical hand book". Tata McGraw Hill Publishing Co. Ltd., New Delhi, pp 2-3.

Krishnanand (1995). High rate biomethanation process for energy generation from agricultural and food industry wastes. In "Microbes for better living". Eds: R. Sankaran and K.S. Manja Association of Microbiologists of India, India. pp. 209-217.

Martin, S, Compos, F, Sastre, H and Fernandez, S (1997). Evaluation of biogas production from the landfill of Bratoro (Burgas, Bulgaria). Hung. J. Indo. Chem. 25: 241-243.

Miyakawa, H., Moriyama, S., Ishibashi, H., Shikasaka, Y., Mizutani, N., Michiki, H. and Saida, T. (1986). Fuel ethanol production from cellulosic biomass. Biotechnology and Bioengineering Symposium. 17, 345-353.

Murali, H.S. (1997). Factors influencing fat production in *Fusarium* spp. Ph.D. thesis, University of Mysore, Mysore.

Natarajan, K.A. (1995). An integrated biotechnological approach for gold processing. In "Microbes for better living". Eds: R. Sankaran and K.S. Manja Association of Microbiologists of India, India. pp. 241-247.

Reddy, A.K.N., Williams, R and Johansson, T (1997). Energy : Sustainable strategies. The Hindu Survey of the Environment, 27-33.

Robinson, C.W., Moo-Young, M. and Lamptey, J. (1981). Ethanol production by immobilized yeast cells in a packed tower. In "Advances in

Biotechnology Vol.II. Eds: M. Moo-Young and C.W. Robinson. Pergamon Press, Toronto, pp. 105-111.

Sola, C., Godia, C.F., Poch, M. and Seira, A. (1986). Continuous ethanol production by immobilized yeast cells and ethanol recovery by liquid-liquid extraction. Biotechnology and Bioengineering Symposium, 17, 519-532.

Vimal, O.P. and Tyagi, P.D. (1984). Biomass as a source of energy - Chapter 1.2 In "Energy from Biomass - an Indian Experience". Agricole Publishing Academy, New Delhi, pp. 9-14.

World Bank (1980). Alcohol Production from biomass in the developing countries. Washington, D.C.

16

Biotechnological Approaches for Ethanol Production from Lignocellulosic Biomass

T. Vijaya* and D.V. Dev

INTRODUCTION

The energy crisis of the seventies has aggravated the problem of continuous supply of energy at reasonable costs. The limitations of fossil fuel to meet the increased demand of the world have accelerated the search for new source of energy.

Biotechnology has received increasing attention in scientific circles during the past two decades because of its great potential in many fields. These potentials are being realized first in biomedical areas, such as in the production of human insulin and human growth hormone in bacterial cells. The attractiveness of biotechnology lies in its potential to provide processes and products that are impractical with non-biological chemistry, to increase specificity in reactions, to provide less environmentally deleterious processes and by virtue of the foregoing to decrease costs.

Forest and agricultural wastes present a great resource for the generation of energy and the production of fibres, chemicals, feed

* Dept. of Biotechnology, S.V.University, Tirupati - 517502

and other products. The plant biomass contains approximately 70-80% carbohydrates, i.e. cellulose, hemicellulose and 20-30% lignin and various small amounts of ash and protein. Cellulose is a linear hydrophilic polymer made up of glucose subunits linked by β 1-4 glycosidic bonds. The hemicelluloses are heteropolymers of pentoses (xylose, mannose, arabinose), hexose and glucoronic acid. Lignin is three dimensional globular irregular, insoluble, high molecular weight (<10,000) polymer made up of Phenylpropane subunit, methoxylated and linked in various ways. Lignocellulosic biomass is not digestible by humans and without pretreatment does not even provide adequate nutrition for ruminant animals. The major physico-chemical feature limiting the direct fermentation of biomass is its polymeric nature.

Energy is obtained from biomass by direct combustion or by microbial action. Amongst the microbial conversion of biomass to fuels, production of ethanol has a great potential and is one of the best ecofriendly technologies for biomass conversion in to energy (Vijaya and Dev, 1999). The abundance low cost and high carbohydrate content make biomass an attractive feed stock for biological conversion to ethanol. Biologically produced fuels and feedstocks have the apparent advantage of being renewable and raw material for their production one assumed to be cheap, free or even off negative value. Since 1973, the concept of producing ethanol as a fuel from agricultural biomass has been extensively investigated (Singh and Garg, 1996; Vijaya and Dev, 1998).

Biomass based alcohol offers immediate prospects of substituting petroleum products at least partially by providing a premium liquid fuel.

India is one of the largest sugar cane growing countries producing about 240 million tonnes of cane per anum. India is also the largest crystal sugar producing country in the world with per capita consumption of 12 kg. There are about 11 million tonnes of sugar and about 40 million tonnes of bagasse. Bagasse contains 49% moisture, 48% fibre, 2 to 3% soluble solids. The fibre is insoluble in water and consists of a cellulose with 2000-3000 units of glucose, 45%, pentosan 18%, lignin 23% and ash 2% advance in process increasing quantities of bagasse are being used in manufacturing news print, all grades of paper including bag, wrapping, printing, writing, toilet tissue, towelling, corrugating etc. It is also burnt as solid fuel in sugar factory boilers, have been used in limited ways as animal feed after various pretreatments. Sugar cane bagasse is the highest amongst the growing crops besides having an extremely high solar energy utilizing capacity which can be exploited for ethanol production. In recent years, considerable interest has focused on converting bagasse to ethanol largely because of the need to extend or enhance transportation fuels.

PHYSICAL AND CHEMICAL PROPERTIES OF ETHANOL

Ethanol is an excellent fuel for spark ignition engines as no soot or carbon built up occur on engine parts. It burns relatively cool, so the engine is less likely to over heat, exhaust valves are less likely to burn and lubricant oil may last longer. Octane is measure of the ability of a fuel to resist detonation during combustion. Ethanols significantly higher octane rating makes it a better fuel than gasoline for spark ignition engines.

Formula	:	CH_3CH_2OH
Molecular weight	:	46.1
Composition (wt%)		
Carbon	:	52.1
Hydrogen	:	13.1
Oxygen	:	34.1
Specific gravity	:	0.79
Boiling temperature °C	:	78
Flash point	:	13
Auto ignition temperature °C	:	423
Flammability limits (volume %)		
Lower	:	4.3
Higher	:	19.0
Octane number	:	106-111
Cetane number	:	0.5
Calorific value	:	11,500

ETHANOL PRODUCTION

Ethanol is produced by chemical as well as biological routes. Through the chemical routes, synthetic alcohol is produced by catalytic hydration of ethylene (C_2H_2) which is derived from petroleum or natural gas with water using petroleum phosphoric acid at 70 atmosphere pressure at 300°C. More than 70% production of ethanol in the world is by synthetic mode. Biologically ethanol can be prduced from three main types of biomass raw materials: (a) sugar bearing materials, (b) starches such as corn, potatoes etc., (c) lignocelluloses such as agricultural residues and wood etc. For biological production of ethanol sugar, starch and lignocellulosic materials are hydrolysed using amylase, cellulase, hemicellulase, pectinase, lignases producing microorganisms upon enzymatic hydrolysis of these materials sugars are liberated in which inturn are fermented to ethanol.

Starch → hexoses → glucose → ethanol
cellulose → hexoses → glucose → ethanol
hemicelluloses → hexoses → glucose, galatose, fructose, mannose → ethanol
pentoses → xylose, arabinose → ethanol

Lignocellulosic biomass hydrolysing microorganisms

Microorganism	**substrate**
Volvariella sp	Lignocellulose
Lentinus edodes	Lignocellulose
Pleuortus sp	Lignocellulose
Trichoderma reesi	Cellulose or pretreated lignocellulose
Thermoactinomycetes sp	Cellulose
Phanerochaete chrysoporium	Lignocellulose
Clostridum thermocellum	Cellulose or pretreated lignocellulose
Pseudomonas fluorescens var *cellulose*	Cellulose, hemicellulose
Thermophilic sporocytophage	Cellulose, pentose, xylose

However, the rising cost of petroleum has renewed the need for production of ethanol by fermentation for use as a fuel. In

India ethanol production by fermentation is gaining importance. By 1980, India became almost self sufficient in ethanol production. The growing demand for ethanol in industries such as organic chemicals and pesticides, pharmaceutical, paints and varnishes and several other laboratory requirement envisage a need for approximately 1280 million tonnes of ethanol by 2000. This is more than double the requirement in 1980.

Brazil has been producing large quantities of ethanol since 1975 from sugarcane and cassava. Currently several other countries including Costa Rica, Africa, Venezuela and Philippines and also exploring sugar cane and cassava for production of ethanol. India has also been producing millions of hectoliters of ethanol from the fermentation of molasses.

Possible routes for biological ethanol production

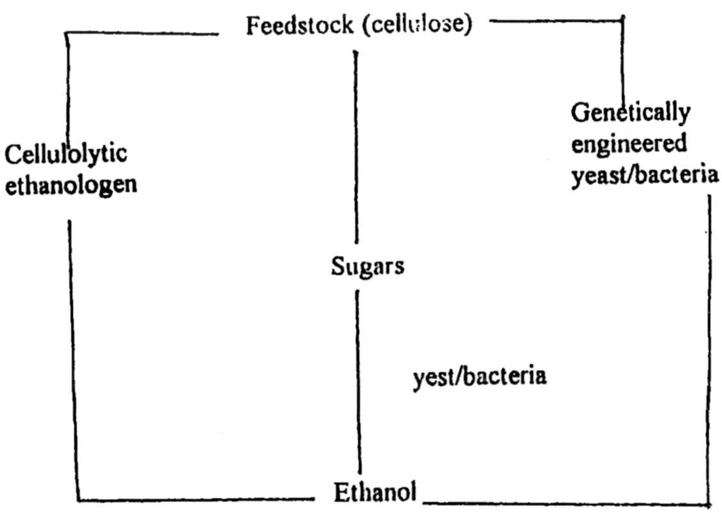

Microorganisms useful for the production of ethanol

Bacteria

Clostridium thermocellum
C. thermosaccharolyticum
C. papyrosolvans
C. stercorarium

C. acetobutylicum
Klebsiella pneumoniae
Leuconostoc mesenteroides
Sarcina ventriculi
Zymomonas mobilis
Thermomonospora curvata

Fungi

Aspergillus oryzae
Endomyces lactis
Khuyveromyces fragilis
Neurospora crassa
Saccharospora crassa
Saccharomyces beticus
S. cerevisiae
S. ellipsoideus
S. oviformis
S. saki
Trichosporium cutaneum
Candida utilis

More than 95% of the fermentation ethanol produced world wide employs the yeast *Saccharomyces cerevisiae* (Rosillo-Calle & Hall, 1987) by conventional two step process. Wherein the first step the lignocellulosic biomass is hydrolysed by enzyme or acid to liberate free sugars. The sugars are then fermented to yield ethanol. Recently ethanol is being produced by single step process either by using monoculture i.e. cellulytic ethanologen (Mori, 1990; Sairam et al., 1991; Sudharani et al., 1996; Vijaya & Dev, 1996) or by coculture i.e. cellulolytic organism + ethanolgen (Grahmann et al., 1992) or by simultaneous saccharification and fermentation (Karel Grohmann, 1993) where the substrate is hydrolysed enzymatically in one vessel and the resulting sugar solution is either separated from residual solids or transfered without purification to the second vessel where the fermentation takes place. Simultaneous saccharification and fermentation approach eliminates one reactor vessel and the need for a solid-liquid separation device. More importantly, it addresses one of the serious shortcomings of current cellulolytic enzymes i.e. extreme sensitivity to end product inhibition. Solid State Fermentation

(SSF) of ethanol has gained much importance in recent years because of its advantages over conventional submerged fermentation.

Solid state fermentation (SSF)

Both solid state and submerged fermentation has been exploited in earlier years. SSF technique was neglected upto 1980 in whole of the world (Lonsane et al., 1985). Fortunately, a few dedicated groups, however continued to work on SSF in spite of prevailing situation and their efforts finally proved sufficient enough to stimulate interest on SSF even among those who were strong advocators of submerged fermentation process. In spite of resurgence and extensive efforts, the SSF even today is not well studied and fully developed as in submerged fermentation process.

Solid state fermentation is defined as the growth of the microorganisms on solid substrates in the absence or near absence of free water.

The energy input in SSF process is much lower due to the absence of medium agitation (Lonsane et al., 1994). This assumes critical importance in the context of present energy crisis throughout the world. There is no water output in SSF (Lonsane et al., 1992). Control of foam formation in many submerged fermentation process is extremely difficult and is a cost intensive proposition (Ghildyal et al., 1988). No such problem exist in SSF due to absence of foam formation. The microbial products which were produced so far by the conventional submerged fermentation process can be produced as efficiently as by SSF system by employing fungi, bacteria, yeast and actinomycetes culture. Simplicity of the system, simpler fermentation media and low cost of SSF may prove useful to the fermentation industry in whole of the world for ethanol production.

Genetic engineering and transgenic microbes for ethanol production

Currently researchers are testing a number of technological strategies including biotechnological schemes to deal with large scale wastes such as lignocellulosics and toxic substance. Wide range of bacteria and fungi are naturally capable of degrading lignocellulose through the concerted action of several enzymes collectively referred to as cellulose. The enzymes required for degradation of lignocellulosic biomass are located in the chromosomal DNA or on the plasmids. In some organisms, the genes that contribute the biodegradation occur on both chromosomal and plasmid DNA. In a number of microorganisms, cellulose activity is found as a part of multiprotein complex that is called a cellulosome. This complex lies on the external surface of the cell and contains several copies of each of the following enzymes. Endoglucanase, exoglucanase, cellobiohydrolase and β-glucosidase or cellobiase.

The breakdown of cellulose by cellulolytic organisms is slow and often incomplete. Therefore, genetic engineering strategies have been employed to create organisms that have more effective cellulase activity. For this purpose, genes coding for the individual enzyme functions of cellulase activity can be isolated from prokaryotic organisms and a clone bank of DNA can be constructed. The genes coding for endoglucanase were isolated from *Streptomyces, Clostridium, Thermoanaerobacter, Erwinia, Pseudomonas, Cellvibrio, Ruminococcus, Cellulomonas, Fibrobacter* and *Bacillus* species. In future there will be a variety of uses for cloned cellulose genes to express cellulolytic activity in organisms that will convert waste cellulose into a commercial product ethanol.

For convenience, most cellulase genes are initially cloned and expressed in *E. coli*, but these genes can be introduced in *Saccharomyces cerevisiae*, *Zymomonas mobilis* and other ethanol producing organisms can be used as hosts for the expression of cellulase genes which efficiently convest lignocellulose directly to ethanol.

REFERENCES

Ghildyal NP. Lonsane BK and Karanth N.G 1988. Adv Appl Microbiol. 33: 173-222.

Grohmann K., Wyman C.E. and Himmel ME 1992. Potential for fuels from biomass and wastes. *In* : Rowell RM, Schultz TP and Narayan R (eds), Emerging Technologies for Materials and Chemicals from Biomass. American Chemical Society, Washington DC, USA, 354-392.

Grohmann K., 1993. Simultaneous saccharification and fermentation of cellulosic substrates to ethanol. *In* : Saddler JN (ed) Bioconversion of forest and agricultural plant residues. Colset Private Limited. CAB International, UK, 183-209.

Lonsane BK., Ghildyal NP, Budiatman S. and Ramakrishna SV. 1985. Enzyme Microbiol Technol.7:156-161.

Lonsane B.K. Vijayalakshmi G., Krishnaiah MM 1994. *In*: Spencer JFT, Spencer DM (eds), Yeasts – Their World and Ours. Springer Verlag, Heidelberg.

Lonsane B.K., Krishnaiah MM. 1992. *In*: Doelle HW, Mitchell DA, Rolz CE (eds) Solid substrate cultivation. Elsevier Science Publisher Ltd., London, 191-209.

Mori Y. 1990. Characterization of a symbiotic coculture of *Clostridium thermohydrosulfurium* YM3 and *Clostridium thermocellum* YM4. Appl. Environ. Microbiol 56:37-42.

Rosillo-Calle, F. and Hall D.O. 1987. Brazilian alochol. Food Versus feul. Biomass, 12:97-128.

Sai Ram M. and Seenaya G. 1991. Production of ethanol from straw and bamboo pulp by primary isolates of *Clostridium thermocellum*. World J. Microbiol Biotechnol. 7:372-378.

Singh SK. and Gary AP. 1996. Bioconversion of Pea Crop residues using *Gliocladium Virens* Miller et al. J. Indian Bot Soc 75 : 245-250.

Sudha Rani, K. Swamy, MV. and Seenayya G. 1996. High ethanol production by new isolates of *Clostridium thermocellum*. Biotechnol Lett 18 : 957-962.

Vijaya, T. and Dev DV. 1998. Studies on factors affecting ethanol Production from cellulosic wastes by a mesophilic *Clostridium* species. J. Indian Bot. Soc 77 : 75-77.

Vijaya, T. and Dev DV. 1999. Effect of pretreatment of lignocellulosic materials on ethanol and other by - products production by a mesophilic *Clostridium* sp. J. Microbiol World 1(2) 47-50.

17

Biodegradation of Bagasse for Production of Alcohol Mixture as Automobile Fuel Using Soil Microbes

S. ANURADHA, R. ABBHIRAMI, NANDINI ARUNKUMAR, UTPAL ROY, S.K. RAY*

INTRODUCTION

An ever growing population, energy and fuel demand has been very hazardous to the environment. Limitation in demand by energy thrift and protection of environment from further depletion is extremely important. In this project we harness an easily available and cheap natural source – bagasse from sugar mills for production of ethanol-methanol mixture[4] through the process of fermentation by culturing microbes from soil in these bagasse disposal areas.

MATERIALS AND METHODS

Six soil samples and two bagasse samples from various depths in bagasse disposal areas were taken from a sugar mill in

* Biological Sciences Group, Birla Institute of Technology and Science Pilani, Rajasthan.

Haryana. Soil samples were cultured in a formulated medium containing: glucose (1.5%), Beef extract (0.5%), Dipotassium hydrogen phosphate (1%), magnesium sulphate (0.1%), ferrous sulphate (0.001%), Sodium chloride (0.5%), manganese sulphate (0.1%). The mixture was left in the shaker for 48 hours[1]. Bagasse was pretreated by autoclaving at 15 psi for 30 minutes to soften it [2,5]. The pretreated bagasse was added to the soil culture and the mixture incubated at 40°C and biomass observations were taken every two days. Biomass was recorded by premobilising the cells in 0.05% Triton - X and centrifuging and recording the cell mass. Observations were repeated after two days and increase in biomass was noted.

The centrifuged cell mass was serially diluted and cultured in Nutrient agar and incubated at 40°C for 24 hours. The colonies were transferred to Nutrient broth and in order to screen the bacterial strains gram's staining was done. Finally plasmid DNA isolation was done on 18 hour cultures prepared using M-17 broth[3].

DISCUSSION AND CONCLUSION

The biomass determination experiments indicated a maximum percentage increase in biomass in samples of bagasse itself. Trial 1 indicated 0.0042 gms of cell mass in 1 ml of microbial culture and after two days the cell mass increased to 0.0122 gms in 1 ml. The spread plating observations indicated large and few small white colonies decreasing in number with increasing dilution. At high dilutions the large colonies were more predominant. Grams staining observations indicated mainly gram positive cocci, rods and bacilli. None of the colonies were pure. The twenty isolates were subjected to plasmid DNA isolation and none of them were found to harbour any extra chromosomal DNA. This shows that bio-degradation capacity may not be associated with plasmids. However, further works in this line are extremely essential to draw any concrete conclusion. In the near future, the parameters including extensive screening and characterisation of the bacterial and fungal strains will be done.

The biochemical and environmental optimisation of the production of the biodegrading principles will also be extensively investigated in order to realise the full potential of the industrially important microbes. Further studies at the molecular level will also be carried out.

REFERENCES

1. Stainer R.Y., Ingraham, J.L., Wheelis M.L., Painter P.R. : General microbiology (1995), Macmillan Press Ltd.
2. Kamra DN., Kewalramani N., Lall D., Pathak N.N. : Effect of substrate pretreatment on degradation of sugar cane bagasse by white rot fungi. Indian J. Microbiol (29), 4, 1989, 391-393.
3. O Sullivan J.D., Klaen Hammer RT: Rapid miniprep isolation of high quality plasmid DNA from Lactococcous and Lactobaccilus sp. App Environ Microbiol. 59:2730-2733.
4. Goodgear EM, Alternate Fuels, The Macmillan Press Ltd. (1980).
5. Aspmall. G.O. (1980), Chemistry of cell wall polysaccharides in The Biochemistry of Plants, a comprehensive treatise. Vol.3, Preiss J (ed.) Academic Press, pp. 437-500.

18

Utilization of Distillery Spentwash for the Production of Biogas as a Renewable Energy Source Using Hybrid Upflow Anaerobic Sludge Blanket (HUASB) Reactor

C.B. SHIVAYOGIMATH, T.K. RAMANUJAM*

INTRODUCTION

Distillery spentwash is a high-strength acidic waste that presents significant disposal and treatment problems. In India there are more than 200 distilleries (Godbole et al., 1995) producing about 10,000 million litres of spentwash or vinasse annually. Hence spentwash is recognized as one of the worst pollutants produced in India both in magnitude and strength. Since spentwash is highly biodegradable and includes large amount of protein (3-4 g/L) and because large amount of spentwash is produced, it is a potential energy resource. Hence indigenous methods should be developed and implemented to exploit this potential energy resource and to combat the pollution at the same time.

Over the last decade, increases in energy prices have made anaerobic wastewater treatment (AWT) processes very attractive.

* Department of Chemical Engineering, Indian Institute of Technology, Madras, Chennai - 600 036, India.

Particularly after the development of high-rate anaerobic reactors, AWT is becoming an accepted simple technology not only for the treatment of a variety of wastewaters but also for environmental protection and recovery of renewable resources. In developing tropical countries like India direct utilization of AWT technology becomes even more attractive.

Anaerobic systems studied in India for the treatment of the spentwash include conventional digester (Parthasarthy et al., 1967), open lagoons (Subba Rao, 1980), anaerobic filter (Gadre and Godbole, 1986) two staged fixed film reactor (Kaul and Badrinath, 1984) and UASB reactor (Jayantha and Ramanujam, 1994).

Hybrid upflow anaerobic sludge blanket (HUASB) reactor which combines the advantages of UASB and anaerobic filter (AF) design is among the newer designs. Feasibility of HUASB design for the treatment of spentwash is not reported. In view of this, present work was undertaken to study the optimum conditions for the treatment of spentwash and production of usable biogas.

MATERIALS AND METHODS

Reactor design

The laboratory-scale reactor was made of plexiglass tube of internal diameter 10 cm and a height of 76 cm as shown in Figure 18.1. The empty reactor volume was 5.97 L and effective liquid volume was 4.96 L. The liquid height was 69 cm; and a 5 cm column of liquid separated the screen and effluent overflow line. The top third of the reactor was filled with 350 polypropylene pall rings (diameter, 15 mm; specific surface area, 341 m^2/m^3; and density, 910 kg/m^3). The reactor was operated under ambient conditions.

Reactor inoculum

The reactor's inoculum (seed) was obtained from an anaerobic digester of a sewage treatment plant. Initial concentration of the seed was 13-15 g VSS/l and VSS/SS ratio was in the range of 0.30 to 0.41.

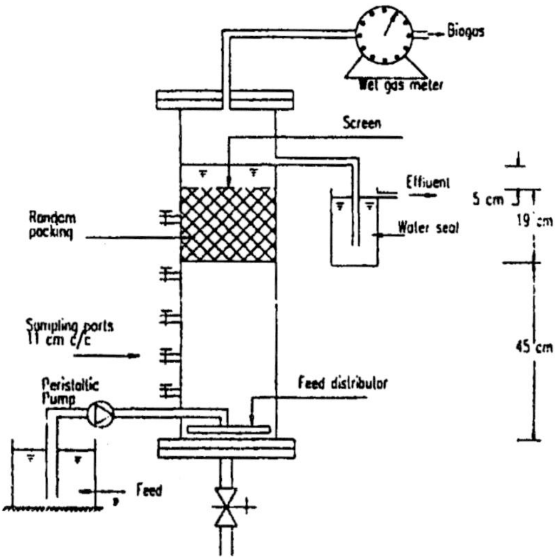

Fig. 18.1 : Experimental setup.

Reactor feed

Spentwash was collected from the molasses based distillery. The major characteristics of the spentwash are shown in Table 18.1. The spentwash was diluted with demineralised water to a desired concentration and pH was adjusted to 7.0 by adding $NaHCO_3$. The required amount of Urea and NaH_2PO_4 were added to the substrate to maintain a COD:N:P ratio of 100:5:1.

Sampling and analysis

Gas production rate, and pH of the effluent were measured daily. Chemical oxygen demand (COD), volatile fatty acids (VFA) and alkalinity were measured on alternate days. The reactor profile was studied once per month by taking samples from the various sampling ports of the reactor. All the analyses were carried out according to Standard Methods (APHA, 1995).

Gas production was measured with wet gas meter and its composition was determined by using gas chromatograph equipped with thermal conductivity detector (TCD) and porapack Q column. Injector and detector were at 130°C and column was at 50°C.

Table 18.1. Major Characteristics of Spentwash (all parameters except pH and colour are in mg/L)

Parameters	*Concentration*
Colour	Dark brown
pH	4 - 4.5
Total solids	60000 - 85000
Volatile solids	40000 - 65000
Suspended solids	800 - 1500
Acidity, as $CaCO_3$	8000 - 13500
COD	80000 - 125000
BOD_5	35000 - 45000
Total nitrogen	1400 - 1800
Total phosphate	300 - 800
Sodium as Na	120 - 250
Potassium	7000 - 7500
Calcium	2200 - 2600
Sulphate	2000 - 3200
Magnesium	1400 - 1600

For a scanning electron microscopy (SEM) of bacterial granules, the sample was fixed in 6% gluteraldehyde fixative (pH 7.1), washed with 0.1 M phosphate buffer, dehydrated in graded ethanol and air dried. It was then sputter-coated with gold and examined with JEOL-JSM 5300 (Japan) model.

REACTOR OPERATION

Two distinct phases occurred during the operation of HUASB reactor.

Phase 1: Start-up

The reactor was seeded with 2.5 L of anaerobically digested sewage sludge (1.5% VSS) and started at an initial sludge loading rate (SLR) of 0.13 kg COD /kg VSS.d. Corresponding organic loading rate (OLR) and hydraulic retention time (HRT) were 1.0 kg

COD/m^3.d and 48 h. Then OLR was increased progressively by stepwise decrease in HRT until start-up was successfully completed with formation of granules at an OLR of 8 kg COD/m^3.d.

Phase 2: Increased loading phase

(a) A following successful start-up with granulation OLR was increased in steps, to 12, 16, 20 and 24 kg COD/m^3.d.
(b) The OLR was increased in steps, with an increment of 2gCOD/L.d, to 36 kg COD/m^3.d.

RESULTS AND DISCUSSION

Phase 1: Start-up (days 1-118)

After 14 days of the operation COD removal efficiency was only 46%. This low efficiency was because of the time required for acclimatization and poor mixing in the system. Then OLR was steadily increased in steps by decreasing HRT (Figure 18.2). This type of procedure washed out poor sludge and improved mixing conditions. Hence COD removal efficiency increased to 67% at a loading of 3 kg COD/m^3.d. Around day 50 the black granules, 1-1.5 mm in size appeared. Beyond this period granulation proceeded rapidly, which improved the reactor performance significantly and COD removal of 80% at an OLR of 4 kg COD/m^3.d was achieved. Further 56 days were required to reach OLR of 8 kg COD/m^3.d at an HRT of 6 h. At this loading COD removal was 80%. During this phase of operation, the gas production rate increased with OLR from 0.5 m^3d to 15.15 m^3 /d. At the end of this phase the reactor was filled with distinct granules with clear blanket zone. Sludge also got trapped in the filter matrix.

Phase 2 : Increased loading phase

(a) Increment of 4 kg COD/m^3.d (day 119-240)

Following successful granulation, the loading was increased in steps of 4 kg COD/m^3.d to 24 kg COD/m^3.d. The feed concentration was increased without changing the HRT (6h). The increase in loading was followed by decreased COD removal efficiency (Figure 18.3). This kind of behaviour was due to the fact that increase in feed concentration inhibited the biomass till they acclimatized to new environment. After few days (12-15 days) COD removal efficiency improved (80-82%). At the end of this phase the average COD of the feed, biogas production and methane content of the biogas were 6000 mg/l, 8.95 m^3/m^3.d and 80.5% respectively.

(b) Increment of 2 kg COD/m³.d (day 241-352)

During this phase of operation loading was increased in steps of 2 kg COD/m^3.d to an OLR of 36 kg COD/m^3.d. However on day 275 when OLR was increased by 4 kg COD/m^3.d, COD removal efficiency dropped to 66% and did not recover even after 2 weeks. At the end of this phase the average COD of the feed, biogas production and methane content of the biogas were, 90000 mg/l, 71.8 m^3/d and 79% respectively.

Throughout the study pH in the reactor was between 7.1-8.0 except for the first few days (Figure 18.4) where acidogenesis prevailed over the methanogenesis. This is attributed to the neutralization of the spentwash before feeding as well as buffering ability of the system.

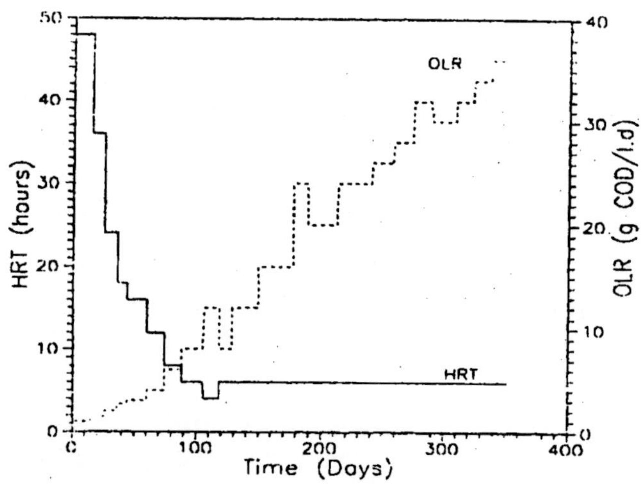

Fig. 18.2 : Stepwise increase of organic loading with time.

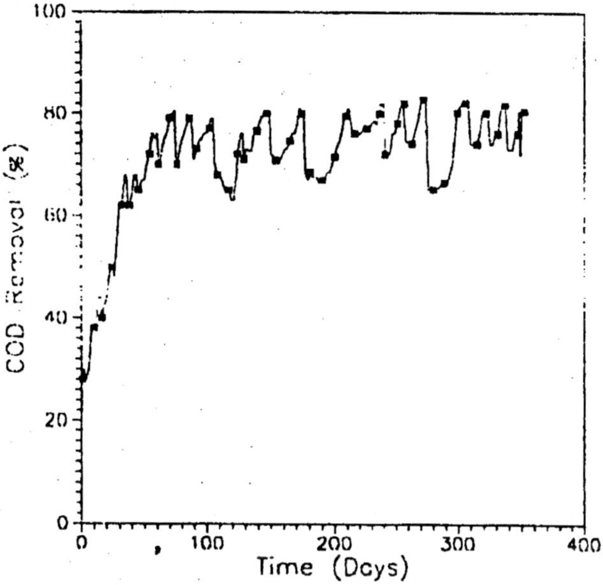

Fig. 18.3 : Variations of COD removal efficiency with time.

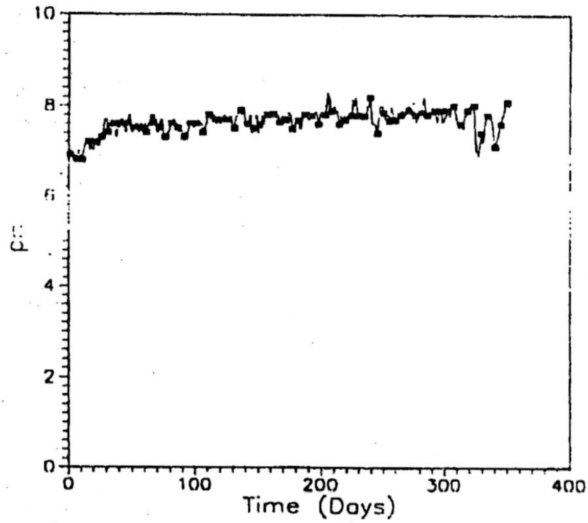

Fig. 18.4 : Variations of effluent pH with time.

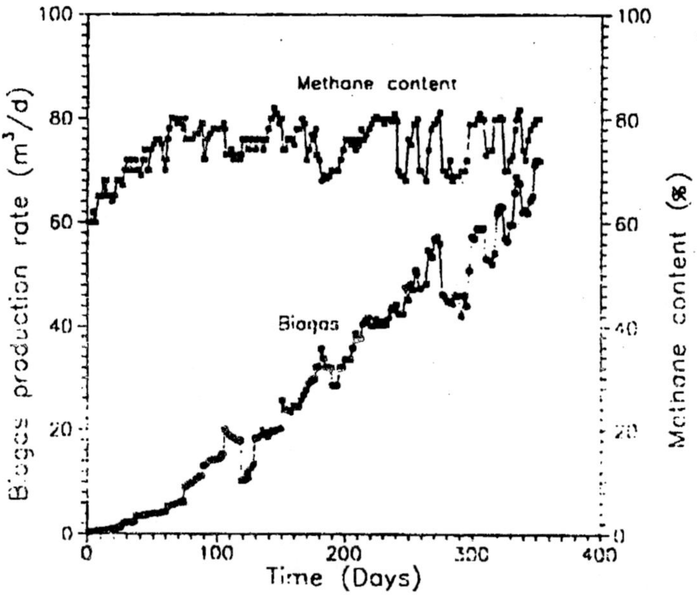

Fig. 18.5 : Variations of biogas production rate and methane content with time.

Variations of gas production rate and methane content of the biogas are shown in Figure 18.5. It is clear from this figure that as production rate was closely related to the applied OLR, Methane content in the biogas was found to be between 70 - 80% throughout the study with minor fluctuations. The specific biogas yield was about 0.40 m^3CH_4/kg COD and specific methanogenic activity of the granular sludge was found to be 0.85 kg CH_4 COD/kg VSS.d indicating highly active granules.

Scanning electron microscopy (SEM) of granules

The granular biomass in the reactor was black in colour and most granules were approximately 2-3 mm in size. The surface was found rough and uneven (Figure 18.6). The predominant bacteria in the granules, observed by SEM, were filamentous *Methanothrix*-like species. *Methanosarcina*-like species were also found; in few granules. A high percentage of *Methanothrix*-like filaments in the form of skeleton (Fig. 18.7) confirm their importance

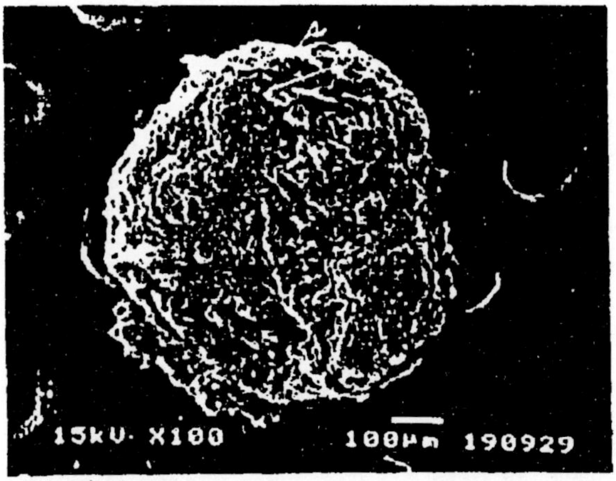

Fig. 18.6 : SEM Picture of Typical Sludge Granules.

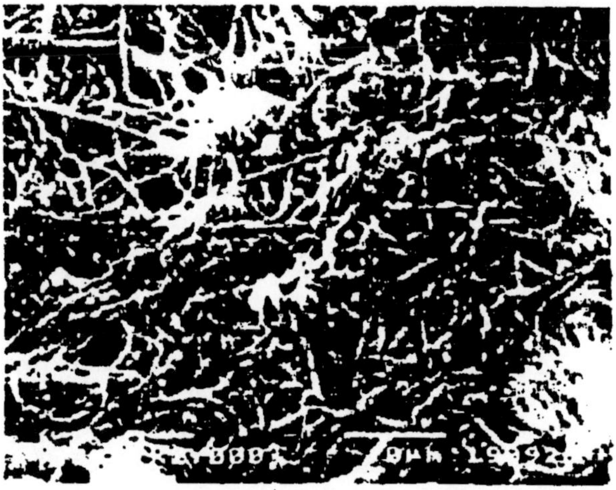

Fig. 18.7 : SEM Picture of Granule showing *Methanothrix*-like Microorganisms in the form of Skeleton.

in the formation and maintenance of the granular structure as suggested previously (Forster, 1991).

CONCLUSIONS

1. The HUASB reactor was able to treat high- strength wastewater like spentwash successfully at a high OLR of 36 kg/COD.m^3.d at a short HRT of 6h under ambient conditions.

2. Pall rings packing in the top third of reactor was an efficient gas-liquid solid (GLS) separator. This design is simple compared to that of conventional UASB.
3. The reactor was able to retain high biomass concentration without serious sludge washout even at higher loadings.
4. Large quantity of usable biogas (Methane 80%) with a high specific biogas yield (0.4m^3 CH_4/kg COD) can be obtained even at a low HRT, besides treating wastewater efficiently.
5. Present studies indicate HUASB reactor could be a very feasible alternative for distillery wastewater treatment.

REFERENCES

American Public Health Association (APHA). (1995). Standard Methods for Examination of Water and Wastewater. 19th Ed., APHA, Washington, D.C.

Forster, C.F. (1991). Anaerobic Upflow Sludge Blanket Reactors : Aspects of their Microbiology and their Chemistry". J. Biotechnol., 17, 221.

Cadre, R.V. and Godbole, S.H. (1996). Treatment of distillery waste by upflow anaerobic filter. Indian J. Environ. Hith. 28, 54-59.

Godbole, S., Zende, N.A and Dhamankar, V.S. (1995). Distillery spentwash (vinasse) treatment with water hyacinth plant, 33rd Intl. Conf. Appropriate waste management technologies for developing countries. NEERI, Nagpur., 25-26.

Jayantha, K.S. and Ramanujam, T.K. (1994). Applicability of upflow anaerobic sludge blanket (UASB) reactor for treatment of distillery spentwash. Indian J. Environ. Protec., 15,106-110.

Kaul S.N. and Badrinath, S.D. (1984). Treatment of distillery spentwash by two stage fixed film reactor - IAWPC (India) Tech. Annual XI 61-67.

Parthasarathy, T., Thangadurai, D. and Basheer Ahmed, N. (1967). Anaerobic Stabillization of distillery waste from Cane Sugar Industry., Environ. Hlth. 9, 110-117.

Subba Rao, B. (1980). Distillery effluent treatment. IAWPC (India) Tech. Annual. 7, 83-101.

19

Environmental Hazards of Sugar Industry and By-Product Utilization in Distilleries

RAJALAKSHMI N.*

Environmental impact of sugar industry is the environmental cost of this economic process. It gives the amount of an agency external to the eco-system when intruded upon tends to degrade its capacity of self-adjustment. A few growing environmental economists today challenge unlimited growth. To them continued output of any industry is the environmental cost of a given economic progress. Industrial Impact assessment is of major concern today. Sugar industry has an impact on agriculture, industry as well as environment. Apart from the economic contribution what is relevant is the environment impact and the measurement of economic cost of environment. The price of the product AR=MR=MC. This is the basic market theory when price of the firm is fixed where Average Revenue is equal to Marginal Revenue is equal to the Marginal Cost of production. But when EC (Environment Cost) also has to be taken into account, then Marginal Cost i.e., the cost of producing last unit of production input has to include the environmental cost or pollution abatement also. Pollution costs are essentially opportunity cost or real cost.

* Department of Economics, University of Madras, Chepauk, Chennai, India.

Pollution cost = Pollution avoidance + Welfare damage avoidance cost. After all, production process of sugar implies increase in per capita consumption of sugar and thereby to raise the utilities and welfare of the people.

The normative theory of Externalities lays foundation for the use of Market Based Instruments (MBIs) in Pollution Control, goes back to the pioneering work of Pigou (1920). The divergence between social and private benefits or costs are what economists call externalities. Several writers, Baumol and Oates, have shown that, under the assumption of perfect knowledge; marketable emission permits are an equivalent alternative to the Pigouvian taxes. Weitzman (1974) analysed the choice between price and quantity instruments in the presence of uncertainty about the benefits and cost of pollution control. Coase (1960) attributed the absence of an economic transaction in an externality situation to absence of well defined property rights for environmental resources and high transactions costs in effecting exchanges through markets.

How dangerous are toxic chemicals to humans? The answer to this question is given by John O' Connor. Industry, Government and citizens all agree that toxic chemicals pose a significant problem. Moreover, some statistics suggest that the spread of toxic chemicals into the environment is causing a public health crisis of major proportions. Each year in the United States, 100,000 deaths are attributed to occupational exposures. It has been estimated that each weak as many as 500 people in the United States die from diseases related to asleestos exposure alone. The birth defect rate had doubled in the last twenty five years, and scientists believe that part of this increase is due to exposure to toxic chemicals. The sperm count has declined by roughly 20 per cent since the emergence of synthetic petrochemicals in the 1940s[1].

This paper attempts to study by-product utilization of sugar industry in the state of Tamil Nadu. Molasses is used in Distilleries. Hence the distillery units are studied and the pollution abatement cost of Trichy Distilleries at micro level has been taken into account to analyse pollution abatement cost.

The Sugar Industry in Tamil Nadu which showed a good record during the entire plan period. In 1950-57, there were only two factories in Tamil Nadu. Tamil Nadu had a major share in the Indian Sugar Industry during the above period. In 1975-76 there were, 16 Sugar Mills in the State. Out of these 16 mills, nine were in the private sector and seven were in the co-operative sector. It has increased to 32 Sugar Mills in the State. For 1994-95 the Sugar Mills in Tamil Nadu crushed about 144.30 lakh tonnes of cane and produced 13.02 lakh tonnes of Sugar. The production of molasses has reached 80,455 tonnes in the year 1994-95.

India became self-sufficient in Sugar with number of unit going up from 33 to 137. The Sugar industry in Tamil Nadu is 3rd largest in the country. 1977-78 was the record year of Sugar production. Tamil Nadu's output was 7.59%, Maharashtra's share was 32.45 per cent and U.P.'s share was 28.83 per cent. There are 32 Sugar Mills in Tamil Nadu of which 15 are in co-operative, 3 in public and 14 are in private sector. The Sugar Industry has given rise to subsidiary industry like distilleries. Fifteen molasses based distilleries are functioning in the State thus leading to environmental hazards.

Hazardous substance, is defined as any substance or preparation which, by reason of its chemical or physio-chemical properties or handling, is liable to cause harm to human beings, other living creatures, plants, microorganisms, property or the environment.[2]

The effluent from sugar industries contains a large amount of dissolved organic matter. This organic matter is readily decomposed by biological action, consequently discharge damage to aquatic life in stream. One of the most serious threats faced today by mankind is the pollution of our environment. In fact, most of the developed countries have already realised that the very existence of life on earth may be endangered, if suitable steps are not taken for the control and abatement of air and water pollution. That is why the industrialized nations are already spending vast sums of money to control the environmental pollution and the developing nations are beginning to follow their example within the constraints imposed by the limited financial resources.

The quality of water in general and industrial wastes in particular may be described by their physical, chemical and microbiological characteristics. However, the number of such characteristics (known usually as water quality parameters) necessary to completely specify the quality of polluted water is quite large.

MOLASSES is the main by-product of sugar industry. Its production depends on the total quantity of cane crushed as well as quality, which varies from region to region. The Molasses Control Order of 1961 regulates the sale distribution and prices of molasses in Tamil Nadu. The bulk of molasses produced is being consumed by the distillery industry and is also used in tobacco, in foundries, and also used by the cattle food industry.

Due to lack of proper storage facilities, substantial quantities of molasses undergo deterioration, particularly in the years of large production. Bulk of molasses produced in the country is being consumed by the distillery industry for the production of industrial alcohol. Molasses is also used in small quantities for curing of tobacco cattle feed industry and foundry. Molasses is also exported in a limited scale through which we earn foreign exchange. Molasses can be used as motor fuel in admixture of 33 per cent with petrol.

Molasses is a very valuable by-product of sugar industry and as such more attention should be paid to adequate and proper utilization of it. The main difficulty is the proper utilization of this by-product from factories. Distilleries should build steel tanks of adequate capacity for proper storage of molasses which otherwise causes pollution problem.

"Toxic chemicals may be released at numerous stages in the production and manufacturing process. They may be fugitive, such as emission from compressors and pump seals, valves, spills, storage tanks, and pipelines, or they may be intentional releases into air, water or by discharging solid wastes of land".[3]

BAGASSE is another important by-product of sugar industry, the production of which depends on the fibre content of sugar cane. There, mostly consumed is as fuel for generating steam. Bagasse produced by five mills are supplied to the Tamil Nadu News Prints Limited.

An important form of utilization of bagasse, discovered by the National Sugar Institute is its conversion into bio-gas and bio-manure. Bagasse has become an important raw material for the manufacture of some commercially important chemicals such as alpha, cellulose and plastic materials. It is also useful to refining vegetable oils.

PRESSMUD is another subsidiary of the sugar industry and this is almost used as manure. Efforts are being made to develop the industry for manufacture of cane-wax. During the course of clarification of sugar cane juice most of its impurities are precipitated. These are filtered in filter presses or content. The cake which is discharged after washing, is called pressmud (or) filter mud. It is also called filter cake. The resultant product of pressmud contains certain mineral and organic matter including wax.

The major source of air pollutants in a sugar industry is the Bagasse-fired Boiler; the products of combustion of bagasse in the boiler furnaces are the air pollutants. The combustion of bagasse leads to pollutants like fly ash carbon monoxide (CO); Sulfurdioxide (SO_2) emissions are virtually negligible. Nitrogen Oxides (NOX) emissions are also considerably less as compared to that from coal/ oil fired boiler:

The firing of bagasse results in significant emissions of particulate matter to the atmosphere. By comparison, emissions of other pollutants such as sulphur dioxide and nitrogen oxides are very much lower than the amount produced by boilers fired with oil or coal. The factors influencing particulate emissions include boiler design, boiler operation and moisture content of bagasse. Several technologies are currently available for controlling particulate emissions for bagasse fired boilers.

The sugar mills and distilleries send their wastes under the ground and they never control the fly ash. These both effluents severely contaminate water and air. These environmental problems are not suitable for human beings as well as animals. So that people may get lung cancer, some may lose their vision. These problems are caused because of the carelessness of the particular sugar mills. According to Atkins & Lowe (1977:24-29), the impact of pollution abatement in firms and industries must begin with the effect on the firm's cost curve and on industry supply curve.

Table19.1 : Overall Molasses Production in Tamil Nadu

Year	*Tamil Nadu (tonnes)*
1980-81	2,20,073
1981-82	3,79,026
1982-83	3,15,351
1983-84	1,88,658
1984-85	2,66,440
1985-86	3,48,382
1986-87	3,63,676
1987-88	3,65,906
1988-89	4,33,606
1989-90	4,25,022
1990-91	5,88,159
1991-92	5,05,326
1992-93	4,40,959
1993-94	5,23,233
1994-95	9,48,922

SISMA, Handbook of Sugar information, pp. 25-26 and final manufacturing reports for the reason from (1986-87) to (1994-95).

Pollution can be defined from many angles. Economic definition of pollution is depending upon both some physical effect of waste on environment and a human reaction to the physical effect. In the real world, no good can be produced without any pollution, only the degree and its nature varies. So what level of pollution is to be tolerated is determined by the economic considerations. As Ogden (1966) contended that "economic considerations are necessary to determine which pollution sources are to be controlled and to what degree control measures should be initiated. Application of the marginal principle enables a judgement concerning the degree of control which ensures the cost of the final increment damages prevented".[4]

UTILIZATION OF MOLASSES

Molasses is the main by-product of sugar industry. Its production depends on the total quantity of cane crushed as well as quality which varies from region to region. The Molasses Control Order of 1961 regulates the sale, distribution and prices of molasses in Tamil Nadu. The prices of molasses are fixed by the union Ministry of Petroleum and Chemicals from time to time. The bulk of molasses produced being consumed by the distillery industry and is also incurring tobacco, in foundries, and also used by the cattle food industry. The price of molasses has 'been fixed at Rs. 120/- per M.T. The industry is pressing for decontrol of molasses.[5]

The amount of molasses produced in Tamil Nadu has been given in Table 19.1. Production of molasses depends upon the quantity of cane crushed and sugar produced. During the period from 1980-81 to 1994-95, there had been fall in the production of molasses in the year 1983-84 and 1984-85, due to the failure of monsoon and drought condition. Moreover there was diversion of cane for the production of Khandsari, gur and jaggery as the prices of these products went up. In 1981-82, cane crushed and sugar produced had been the maximum for the period and as a result molasses production also had gone up. There was increase in sugar cane acreage, cane production and the duration of the crushing season had been extended. In the remaining years, there had been variation in the production of molasses due to variation in the cane production and cane crushed.

The prices of molasses are controlled under the Molasses Control Amendment Order, 1976, issued in exercise of powers conferred by section 18 (G) of the Industries (Development and Regulation) Act, 1951. Molasses is fermented by yeast to produce ethyl alcohol and carbon dioxide. Ethyl alcohol apart from being a source of potable alcohol is also used as a raw material for various alcohol-based industries e.g. Synthetic rubber, plastics, solvents etc. Table 19.1 shows the rising trend of molasses production. The molasses production was 1,88,658 tonnes in 1983-84, it went up to 9,48,922 tonnes in 1994-95. Thus, molasses below ancillary industries in India seem to have bright scope in future given proper opportunities. Some of the states like Bihar, Uttar

Pradesh, Maharashtra, Tamil Nadu and Madhya Pradesh had considerable high prospect on the basis of increasing molasses production during the period 1979-80 to 1994-95. There are at present about 70 distilleries in the country, with installed production capacity of 3.5 lakh kilo litre of alcohol. For production of one kilo litre of alcohol about 4.5 tonnes of molasses are required.

Due to lack of proper storage facilities, substantial quantities of molasses undergo deterioration, particularly in the years of large production. The sugar industry contends the price of distribution of molasses are controlled by the central and state Government and the price is very low being Rs. 6.70 per tonne up to 5th February, 1972 which increased thereafter to Rs. 10 per tonne. Alcohol can be mixed with petrol to reduce, to some extent, our dependence on imports for petroleum products. Molasses is also used as raw materials for production of lactic acid, oxalic acid, citric acid, butanol and chemical (or) biochemical process. Therefore, due to low return some of the sugar factories are not willing to put up proper molasses storage tanks.[6]

Table 19.2 : Utilization of Molasses

Molasses Utilization

Direct	*Distillery Industry*	*Other Fermentation Industries*	*Miscellaneous*
1. Export	1. Rum	1. Vinegar and Acetic Acid	1. Aconitic Acid
2. Fertilizer	2. Ethyl Alcohol	2. Acetone Butanol	2. Mono Sodium gultanate
3. Cattle	3. Rectified Spirits	3. Citric Acid	3. Cattle feed
	4. Power Alcohol	4. Lactic Acid	
	5. Alcohol Derivatives	5. Yeast	
	6. Spent wash	6. Food yeast	
	7. Industrial Alcohol		

Bulk of molasses produced in the country is being consumed by the distillery industry for the production of industrial alcohol. Molasses is also used in small quantities for curing of tobacco cattle feed industry and foundry. Molasses is also exported in a limited scale to earn foreign exchange. Molasses can be used as motor fuel in admixture of 33 percent with petrol.[7]

Molasses the by-product of sugar industry is used in distilleries and as such more attention should be paid to adequate and proper utilization of it to save it from being wasted. The main difficulties in the proper utilization of this by-product are concerned with its proper storage and transport from factories to the ultimate utilizer industries. Both sugar factories and distilleries should build steel tanks of adequate capacity for proper storage of molasses.

Table 19.3 : Districtwise Distilleries in Tamil Nadu

Sl. No.	*District*	*Name of the Distilleries*	*No. of Distilleries*
1.	Cuddalore	i. Aruna Sugars ii. Chemplast Indl. Ltd. iii. E.I.D. Parry (I) Ltd.	3
2.	Tiruchirapalli	i. Kothari Sugars & Chem. ii. Trichy Distilleries & Chem Ltd.	2
3.	Erode	i. Sakthi Sugars Ltd. ii. Bannariamman Sugars Ltd.	2
4.	Dharmapuri	Chemplast Sanmar Ltd.	1
5.	Salem	Salem Co-op. Sugar Mills Ltd.	1
6.	Madurai	Rajashree Sugars & Chem. Ltd.	1
7.	Coimbatore	Amaravathi Sugars Ltd.	1
8.	Tirunelveli	Dharani Sugars & Chem. Ltd.	1
9.	Thanjavur	Thiru Arroran Sugars	1
10.	Chengalpattu	Mohan Breweries & Distilleries Ltd.	1
		Southern Agrifurance (I) Ltd.	1
		Total	15

Source : Handbook of Sugar Statistics - 'SISMA' - 1997 pp. 149-151.

Table 19.3 shows the growth and districtwise distilleries in Tamil Nadu it is significantly more distilleries domicile in Cuddalore District each two distilleries in Tiruchirapalli and Erode districts. The other districts have only one distillery in each. Table 19.3 further explains private sector as well as cooperative sector's dominance is higher, and have not even only one distillery in public sector.

Table 19.3 indicates districtwise growth of distilleries in Tamil Nadu.

There are 15 distilleries situated in 10 Districts of Tamil Nadu out of 15, more distilleries are in Cuddalore District. Next to Cuddalore the concentration is more in Trichirapalli and Erode. These three districts occupied half of the total distilleries in Tamil Nadu. Eight Distilleries are situated at various Southern districts of Tamil Nadu.

Table 19.4 : Growth of Distilleries and Capacity Utilization of Distilleries in Tamil Nadu

Year	*No. of Distilleries*	*Total Installed Capacity (K.L.)*	*Average Installed Capacity (K.L.)*	*Capacity Utilised (%)*
1985-86	3	25450	8483.33	70.66
1986-87	8	99470	12433.75	64.39
1987-88	8	100130	12516.25	67.82
1988-89	8	100130	12516.25	73.29
1989-90	8	101480	12685.00	73.76
1990-91	8	127244	15905.50	73.27
1991-92	8	114194	14274.25	78.11
1992-93	12	212900	17741.66	46.06
1993-94	12	201480	16790.00	55.88
1994-95	13	210480	16190.76	42.39

Source : Handbook of Sugar Statistics 'SISMA' - 1997, p.151.

Table 19.4 clearly defines the capacity utilization of Distilleries in Tamil Nadu from 1985-86 to 1994-95. Period from 1987-88 to

1991-92 there was not entered any new distilleries started in Tamil Nadu but the total installed capacity (K.L.) has increased. This has been shown in input as well as in out turn which has increased. Table 19.4 further explains average installed capacity utilization which has simultaneously increased. The capacity utilized per centage also has increased after 1991-92 it was on decline.

Table 19.4 shows Growth of Distilleries and capacity utilization of Distilleries in Tamil Nadu during 1985-1997. In 1985-86, there were only three distilleries in Tamil Nadu. It has increased to 15 in 1996-97. Total installed capacity of sugar mills during 1985-86 was only 25450 KL. It increased to 101480 KL in 1989-90, i.e. it increased by 4 times, whereas from 1990-91 to 1994-95, capacity have increased only 2 times i.e. from 127244 to 210480 respectively.

In 1985-86 average capacity was only 8483.33 KL within that decade it raised to 16190.76 KL. Except in 1990-91 to 1992-93 all years have experienced only moderate increase in its capacity. From 1985-1995, on an average the capacity utilization of sugar mills in Tamil Nadu was 65%.

Table 19.5 : Average No. of Days for Distilleries in Tamil Nadu Worked Each Year (During 1985-86 - 1994-95)

Year	*No. of Distilleries Worked*	*Average No. of Days Worked*
1985-86	3	281.00
1986-87	8	295.00
1987-88	8	296.75
1988-89	8	305.37
1989-90	8	295.12
1990-91	8	299.88
1991-92	8	311.50
1992-93	12	209.91
1993-94	12	233.66
1994-95	8	251.25

Source : "SISMA" - Handbook of Statistics, 1997. Figures computed average value.

Table 19.5 explains clearly average number of days worked for distilleries in Tamil Nadu during 1985-86 to 1994-95. It has been noted clearly during 1991-92 significantly Tamil Nadu distilleries are worked well. During 1992-93 it was lower as the level of work stood at 209.91 days. The table also indicates average number of days distilleries worked each year in Tamil Nadu during (1985-86 to 1994-95).

Over a period of ten (10) years 10 distilleries were established on an average of each in one year. The average working days of these distilleries range from 209 days to 311 days. It is observed that the minimum days (209) were worked in 1992-93, and maximum days (311) in 1991-92.

Table 19.6 : Alcohol Production and Exports of Distilleries in Tamil Nadu

Year	*No. of Distilleries Worked*	*Alcohol Production (KL)*
1985-86	3	16666
1986-87	8	61767
1987-88	8	63836
1988-89	8	70155
1989-90	8	69302
1990-91	8	75332
1991-92	8	89479
1992-93	12	92371
1993-94	12	89068
1994-95	13	108548

Source : Handbook of Sugar Statistics "SISMA" - 1997.

Table 19.6 explained Alcohol production and exports to other states in Tamil Nadu during 1985-86 to 1994-95. It was in 1989-90 Alcohol production has come down compared to the previous years and after the period it has increased. During 1994-95 the alcohol production was significantly very high. But alcohol exports to other states was without exports after 1991-92.

In 1987-88 alcohol exports was very high compared to following years.

Table 19.6 represents alcohol production, number of Distilleries in Tamil Nadu and export to other states from Tamil Nadu during the period of 1985-86 to 1994-95.

It reveals the fact that the number of distilleries have increased from 3 in 1985-86 to 13 in 1994-95 over a period of ten years. 10 Distilleries were established in an average of one in each year.

From 1985-86 to 1988-89 there was tremendous progress in Alcohol production i.e. increased from 16666 KL to 70155 KL. The very next year it is observed that a slight reduction in production. Next six (6) subsequent years, data showed that production has increased at moderate rate.

In 1985-86, since the alcohol production was much less, no exports is registered. From 1986-87 – 1989-90, exports to other states were significantly higher than the other two subsequent years 1989 to 1991, inspite of growth of Alcohol production its exports were not encouraging. From 1992 to 1995 no account of exports of Alcoholic production is registered in Tamil Nadu.

MICRO LEVEL STUDY OF DISTILLERY UNIT AND POLLUTION ABATEMENT COST OF DISTILLERIES

Trichy Distilleries industry growth from 1985-86 to 1994-95 has been estimated. The installed capacity has increased to 24000 KL. Amount of Molasses consumed has increased from 50,670 tonnes to 65,452 tonnes. Alcohol produced has been 74625 KL by 1994-95. The capacity utilisation was highest by 1991-92 at 101%. Mandays worked also was highest by 359. Table 19.7 shows that alcohol production has been at high demand with no stock. Hence of the by produced molasses utilisation is made in more safe manner without environmental hazards. The by-products, production itself has a wide market and profit generation. When pollution abatement technique has been installed, the efficient level of pollution which can be allowed can be explained by the Figs. (19.1 and 19.2). The intersection of marginal Damage cost and Marginal control cost leads to arrive at optimum level. But the total cost also includes total damage costs.

As already stated, the TDCL has been originally letting out the effluent in the river Cauvery, after fulfilling the norms of Pollution Control Board. However, the pollution caused by the effluent was intolerable and the TNPCB has ordered to treat the effluent fully before it is let out. Accordingly the sample industry has installed an effluent treatment plant with the total cost of Rs. 200 lakhs in 1989 with the collaboration of UK Technology and the Kaveri Engg Limited who have helped to design the plant.

The factory has so far spent more than Rs. 6.5 crores in the effluent treatment. Recently it has also allotted Rs. 4.5 crores in upgradation and maintenance of the plant. The recurrent expenditures in the form of wages and salaries and others may exceed the initial expenditure, in the long run. This may be reduced when the factory succeeds in connecting the factory and solar evaporation ponds with pipe lines, as this would considerably reduce the transport incurred in taking the effluent from factory to field.

Table 19.7 : Capacity Utilisation of Trichy Distilleries and Chemicals Ltd. (During 1985-86 to 1994-95)

Year	*Installed Capacity (KL)*	*Molasses Consumed (Tonnes)*	*Alcohol Produced (KL)*	*Capacity utilised (%)*
1985-86	16500	50670	10079	74.00
1986-87	16500	58469	13970	85.00
1987-88	16500	60233	13968	84.70
1988-89	16500	57074	12834	77.80
1989-90	18000	71065	15176	84.30
1990-91	18364	72190	16187	88.00
1991-92	18364	85294	18711	101.90
1992-93	24000	69678	14733	61.40
1993-94	24000	51082	11364	47.40
1994-95	24000	65452	14625	60.90

Source : Hand Book of Sugar Statistics, "SISMA", Chennai - 600 035. pp. 151-152.

Table 19.8 : No. of Days for Trichy Distilleries & Chemicals Ltd. worked Each Year (During 1985-86 to 1994-95)

Year	*1985-86*	*1986-87*	*1987-88*	*1988-89*	*1989-90*	*1990-91*	*1991-92*	*1992-93*	*1993-94*	*1994-95*
No. of Days Worked	264	332	340	333	350	359	351	321	282	..

Table 19.9 : Alcohol Production Particulars of Trichy Distilleries and Chemicals Limited in Tamil Nadu (During 1986-87 to 1994-95)

Year	*Opening Stock*	*Production*	*Total Availability*	*Disposal Potable*	*Disposal Industries*	*Export to Other states*	*Other Uses & shortage*	*Total disposal*	*Closing stock*
1986-87	230	13970	14200	2124	8370	2778	30	13302	898
1987-88	912	13968	14880	-	9143	2293	1793	13229	1651
1988-89	1670	12834	14504	-	11459	1201	1437	14097	407
1989-90	422	15176	15598	2376	12005	53	-	14434	1164
1990-91	786	16187	16973	3728	10303	81	1281	15393	1580
1991-92	1580	18711	20291	1133	16785	12	9	8569	168
1992-93	1877	14733	16610	2352	13536	-	46	15934	676
1993-94	750	11364	12114	1917	8082	-	7	10006	2108
1994-95	750	14625	15375	816	14687	-	-	15503	-128

Source : Handbook of Sugar Statistics 1997, "SISMA" Chennai-35, pp.153-157.

CONCLUSIONS AND POLICY SUGGESTIONS

Depending upon the method used for cane harvesting, large quantities of soil may be delivered to plant. In addition, solid waste consisting of packaging materials, pieces of cane, solids removed by screens and trash collection devices, filter muds, sludges, boilers and fly ash, is also generated in the plant.

C.B. Christainsen and H. Haveman (1981: 381-390) attributed the contribution of environmental regulations to the decrease in productivity growth. According to them increase in Economic Welfare is not reflected in marketed or measured output rather it should include improved health, longer lives, expanded outdoor opportunities and reduced demand for cleaning and other related activities.

R.E. Kohn (1985) has demonstrated that the Pigovian tax increased the prices of polluting good: reducing inelastic output and as a consequence, the relative price of the input in which polluting firms are relatively intensive. Resides, the Pigovian tax on pollution promotes an efficient allocation of inputs, including an efficient no. of firms in the industry.[8] This type of pollution tax may me suggested for sugar and distillery units.[9]

The number of processes that have been adopted to treat sugar industry effluent are:

Prevention at source
Disposal on land for irrigation
Trickling filter and activated sludge process
Anaerobic treatment
Disposal in lagoon or stabilization pond.

Eventhough in Tamil Nadu some Distilleries industries have excellent pollution control facilities, they never use it properly, because it will consume more money. So that most of the industries are simply sending their wastage under the ground.

The argument related to the market failure is the divergence between price and social value of environment as individual preferences and public benefits invariably clash. This requires that enable policy should bring the difference to the minimum (Perrings,

1996). The economy-environmental nexus is so inexonerably intimate that every decision in the socio-economy has some impact on environment. Thus in all socio-economic policies and programmes Environmental valuations are important. In the non-market - market techniques of extended Cost Benefits Analysis (CBA) , the value of instal or damage should find reflection. CBA may not dictate decisions, but no one denies that it aids the decision making process (Bards and Pearce, 1991).[10]

The suggestion would be not the closure of the entry. All we want is that the mills should take sufficient steps to stop polluting the environment. Because not only sugar mills contribute to National Income but also, generate employment opportunities. A threat to the environment is tag attached to any industrial development. Any good medicine, is bound to cause side effects, simultaneously treatment for the side effects may go a long way. Because the quality of life is closely linked with the quality of the environment.

Anaerobic Rotating Biological drum contractor (RBC) is reported to be used successfully for the treatment of sugar mill effluent necessarily.

A large number of factories had effluent treatment plants, but were either all treatment plants, (or) some of them were never used. The pollution control board should be equipped with enough teeth to regulate and to redefine the ethos of industrial operations.

The government should give licence to a factory only after it submits the proof of purchase orders of pollution control equipments and their installation. The development of industries has disturbed the "Ecological balance" in our economy and therefore an action planning for environmental protection to be taken into consideration in the Industrial and Economic Planning.

Robert and Spencer (1976) recommended a mixture of marketable permits supplemental by an effluent fee and a subsidy, when a regulator is uncertain about the marginal abatement cost function of polluters.

Pollution abatement in these industries consists of treating waste water in order to bring the BOD, the chemical oxygen

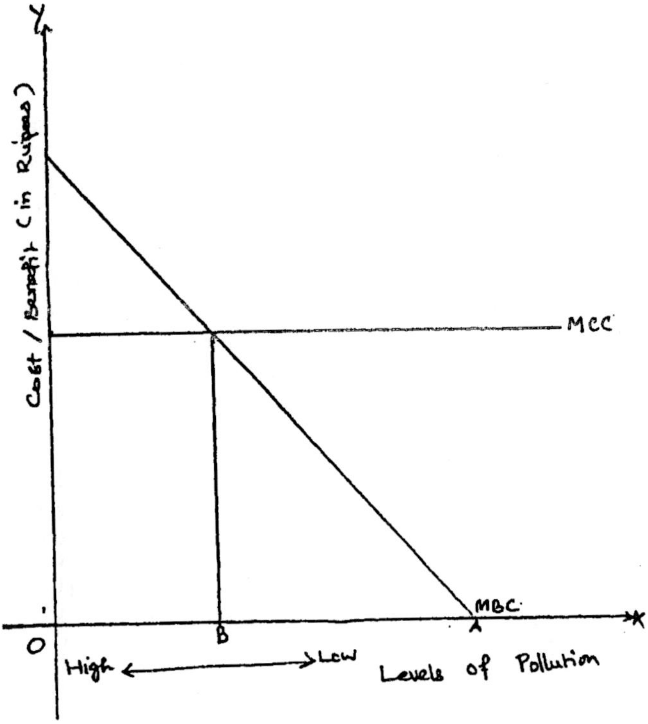

MCC – Marginal Control Cost

OB – The Efficient level of Pollution Control

A – Pollution is Zero

MBC – Marginal Benefit from Control of pollution

Fig. 19.1 : Efficient Level of Pollution.

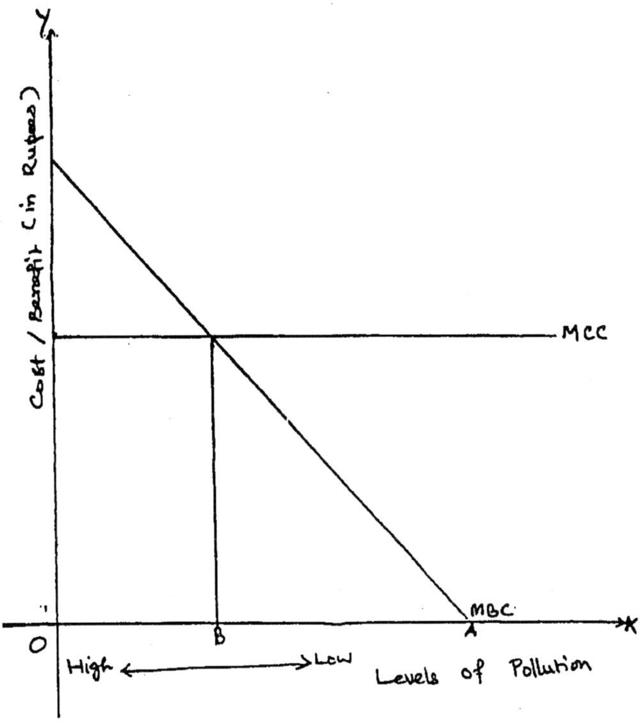

MCC – Marginal Control Cost

OB – The Efficient level of Pollution Control

A – Pollution is Zero

MBC – Marginal Benefit from Control of Pollution

Fig. 19.2 : Efficient Level of Pollution.

demand (COD) and alkaline content within specified tolerance limits. The annualised capital cost of effluent treatment plant for sugar is calculated as (0.26%) for sugar. (Sweker Mehta, Sudip Mundle and U. Sankar, 1994) . Abatement cost and incentives for waste water treatment. Fruas and Munley (1984), Subrahmanyam (1990) provides information about production processes and waste water treatment alternatives in Indian paper and pulp industry. A general abatement cost function for distillery industry can be worked out. Hafford and Harrington (1990) and Russel (1992) report some empirical evidence in favour of voluntary compliance of the part of many large industrial pollution in U.S.A. and Western Europe. U. Sankar (1994) classified taxes and charges and suggested different taxation with tax rebates for ecofriendly outputs/inputs.

REFERENCES

1. John O'Connor and Gary Cohen, "Fighting Toxic", Island Press, Washington D.C., 1990, p. 13.
2. Pandys C.G. "Hazards in Chemical Units" Oxford and IBH Publishing, New Delhi, 1988.
3. Andrews Richard N.L., and Turner G. Alvis, "Toxic Chemicals Health and the Environment", John Hopkins University Press, Baltimore, USA, 1987.
4. Christainsen, Gregory, B., and Haveman Robert, H., (1981), "The Contribution of Environmental Regulations to the Slowdown in Productivity Growth" Journal of Environmental Economics and Management, 8(4): 381-390.
5. SISMA, Handbook of Sugar information, 1997 MS. p. 3
6. Prof. Ram Vichar Singa, Sugar Industry in India, p. 200 and p. 208.
7. Sugar Industry in India Prof. Ram Vichar Singa, p. 218.
8. C.B. Christainsen, Gregory. B and H. Haveman Robert, H., (1981) "The contribution of Environmental Regulations to the Slowdown in productivity growth" Journal of Environmental Economics and Management; 8(4) : 381-390.
9. Kohn, R.E. (1989) "Effluent Scale of the pollution abating firm – A Comment and Reply" Land Economics, 65(4) 413-416.
10. Perrings, (1996); "The value of bio-diversity for sustainable economic development". Background papers and abstracts of the Indo-British workshop on Bio-diversity; Conservation and Evaluation, TBGRI Thiruvananthapuram, Kerala, India, Feb. 15-17, 1996, pp.30-45.

20

Municipal Ash/Fly Ash as a Substitute to Natural Resources in the Construction Industry

D.V. Reddy, T. Pierro, A. Yemeniciler*, S. Kastury**

OVERVIEW

The utilization of ash in structural concrete applications contributes to environmentally sound disposal and cost/energy effective use of a waste product. When used as partial replacement (15-50%) of the cement volume, fly ash reduces the amount of cementitious material needed, while contributing to significant improvements in strength and durability of the hardened concrete. Municipal ash may be incorporated in mix designs as a portion of fine and coarse aggregates, thus reducing the large amounts of these natural resources that are normally required by the construction industry. In both cases, ash replacement in small percentages of the entire concrete volume provides the opportunity to preserve natural resources and save millions of dollars in many concrete applications.

* Center for Marine Structures and Geotechnique, Dept. of Ocean Engineering, Florida Atlantic University, Boca Raton, FL, 33431 (USA).

** Florida Department of Environmental Protection (FDEP), 2600 Blair Stone Rd., Tallahassee, FL 32399 (USA).

INTRODUCTION

Coal Ash

The increasing demand for electricity is responsible for the massive consumption and combustion of coal, producing over 80 million tons of fly ash per year in the United States. Generally, there are two physical types of ash that result from the incineration of coal; bottom ash (collected from the bottom of the incinerator) and fly ash (collected from the flues of the incinerator). Accounting for nearly 2/3 of the total annual production of utility wastes, intensive research has promoted many safe, economical and environmentally effective uses for the waste material, which is otherwise stockpiled or disposed of in landfills, (Reddy, 1996). Although research has been performed on both types of coal ash (i.e. N. Ghafoori and J. Bucholc on bottom ash and D.V. Reddy on fly ash), applications generally employ the flue collected fly ash due to its fine particulate qualities and the properties discussed in this paper.

When considering utilization for concrete applications, coal fly ash is the best artificial pozzolan. The pozzolanic reaction,

$$CH + S + H \rightarrow C - S - H$$

allows the ash to react in water with lime to form low soluble compounds with cementitious qualities. This leads to longer lasting concrete products with low permeability, low heat of hydration, and improved strength and workability.

Considering that the most desirable mode of disposal is utilization, coal fly ash has a very important role in the economy of developing countries. In most of these situations, cement is a very expensive material and is always in short supply. However, coal fired electric power plants provide a steady source of readily available fly ash which can realistically reduce cement consumption by up to 50%. Therefore, by incorporating this inherent waste product as a material source in the construction industry, a number of environmental and economical benefits can be achieved.

MUNICIPAL SOLID WASTE (MSW) ASH

The combustion of municipal waste in the modem trash-to-energy facilities is one of the primary components of a fully integrated and environmentally sound approach to solid waste management. There are many extremely valuable energy and cost-effective benefits that arise from trash-to-energy conversion. Primarily, garbage is always in abundance and easily collected. Furthermore, its efficient and safe convertibility to energy makes it a renewable source of fuel and material.

Upon combustion of municipal solid waste (MSW), there are also two types of ash produced. A waste material referred to herein as "MSW bottom ash," is produced as a secondary bottom-ash waste product along with an airborne residue MSW fly ash. Both products can be safely managed in a number of ways. Like coal fly ash, MSW bottom ash may be utilized as a partial replacement material in concrete. However, due to the heterogeneous origin, the ash resulting from mass-burned municipal waste does not meet the chemical requirements for either Class C or Class F pozzolans, (Hamernik and Frantz, 1991). Therefore, it is more effective to add the ash as a part of total aggregate rather than cement.

Proportions of 10-25% ash replacement significantly reduce the necessary amounts of coarse and fine aggregates. In major construction jobs that require massive amounts of concrete, such as the highway infrastructure, small percentages can contribute to savings of many tons of coarse and fine aggregates.

PROPERTIES

1. Pysical Properties

The primary comparison of ash materials is based on the physical characteristics. With reference to concrete applications, these include physical appearance, particle size and shape, microstructure, moisture content, and water requirements of the mix. Although MSW bottom ash and coal fly ash derive from similar incineration processes, their physical properties vary quite extensively. The shape, fineness, particle-size distribution and density greatly affect the properties of the resulting concrete.

Generally, fly ash is divided into two classes. *Class C* ashes are derived from sub-bituminous and lignite coals, whereas *Class F* ashes come from bituminous and anthracite coal sources,

American Coal Ash Association (1995). Furthermore, fly ashes produced at different power plants or with different coal sources tend to vary in color from tan to light gray or black. In contradistinction, MSW bottom ash tends to be very black in color and possesses a texture similar to topsoil. In addition, municipal ash that has been collected from the bottom of the incinerator often contains small pieces of melted glass and particulates. Besides the indication of a variation and lack of homogeneity of sources, chemical contents, or burning conditions, different colored ashes can influence the final color of the hardened concrete.

Fly ash collected from the flues of coal-fired incinerators possesses extremely fine particulate qualities similar to unmixed cement, which is an advantage when it is used for partial replacement in a mix. The smooth surfaced spherical particles aid in the production of a cementitious material when combined with activators such as lime, cement or kiln dust, American Coal Ash Association (1995). The typical electron microscope image shown in Figure 20.1 (a) depicts the spherical microstructure of coal derived fly ash magnified 1000x. For comparison. Figure 20.1 (b) shows the 1000x magnified image of a MSW ash. Bottom ash samples have greater surface areas and show the existence of a variety of shapes and surface textures, such as "shredded sponge," "rolled paper," and "corral," (Hamernik and Frantz, 1991).

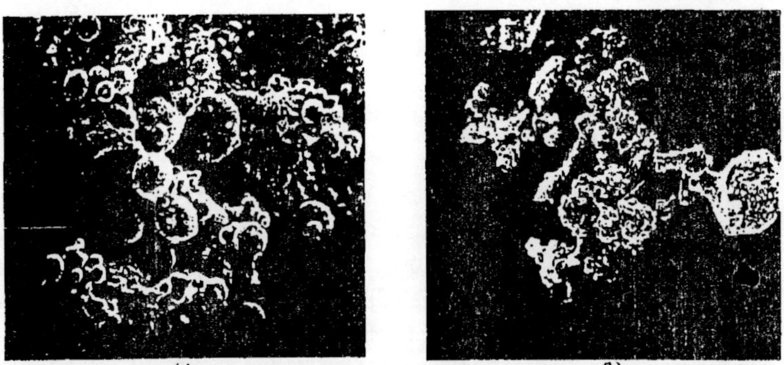

(a) (b)

Fig. 20.1 : Scanning Electron Microscope (SEM) images of the microstructures magnified 1000x of (a) coal fly ash and (b) MSW bottom ash (Reproduced from Hamernik and Frantz, 1991).

Normally, coal is highly pulverized before firing and subsequently burned at temperatures near 1500°C, resulting in a homogeneous material that fuses and condenses into the cenospheres found in the resulting fly ash. In contrast, municipal waste is highly heterogeneous and not pulverized prior to incineration at 1000°C. Therefore, the resulting bottom ash lacks a clearly defined matrix and the spherical attributes of coal fly ash. These variations in the microstructure of MSW bottom ash are part of the properties that detract from the cementitious qualities of municipal ash-modified concrete.

Furthermore, the relatively large surface areas associated with MSW bottom ash contribute to the tendency for this material to absorb moisture from the atmosphere. It has been shown by Hamernik and Frantz (1991) that with sufficient time, MSW ash has the ability to absorb 20% moisture from the surrounding air. Coal derived fly ash, on the other hand, does not absorb any moisture from the atmosphere. These are very important aspects because they indicate how each ash will affect the water/binder ratio needed for proper mix design. Coal fly ash does not require any increase in water; in fact, it can provide equal or better performance with a lower w/b ratio. However, due to the natural affinity to water of MSW derived ash, any applications of municipal ash-modified concrete must provide an increase in w/b ratio to cope with the problem of hydration.

II. Chemical Properties

Whenever using ash modifications in concrete applications, it is imperative to understand how these admixtures will react with and affect the surrounding environment. The factors of concern with coal fly ash generally include the reduction of heat generated by hydration, the reactions with the products of hydration, and the resistance to permeation. However, municipal waste sources necessitate the need for chemical investigations of the leaching abilities of various toxic compounds and heavy metals that may exist in the ash, as well as their effects on the cementing process.

The chemical aspects of coal fly ash greatly improve the properties of hardened and fresh concrete. When portland cement reacts to form concrete, there tends to be an early rise in temperature that presents a major problem when pouring massive structures.

However, the ash has the ability to reduce the high heat of hydration normally associated with the cementing process by slowing the rate of reaction. By achieving a slowed reaction rate, there is less heat generated in the early stages of cementing and the result is more structurally durable concrete.

Upon mixing, solutions of calcium and alki-hydroxide are released into the pore structure of the paste and combine with the pozzolanic particles of the coal derived fly ash to form a cementing medium, (Reddy, 1996). Combined with ordinary portland cement, the resulting paste is highly cementitious and workable. Furthermore, the reaction products have the ability to fill the pore spaces between the cement particles reducing the permeability of the hardened concrete. The advantage of reduced permeability is the protection of imbedded steel rebar by inhibiting the infiltration of aggressive corrosion promoters such as chlorides and carbon dioxide. In the marine environment, where cracking and spalling occur due to internal corrosion, this chemical property strongly supports the utilization of coal fly ash-modified concrete.

Unfortunately, the chemical properties of municipal bottom ash are not as advantageous as coal fly ash. The only MSW ash that showed signs of pozzolanic activity (similar to Class C fly ash) was that collected from refuse-derived fuel, which requires removal of all ferrous material, glass, and shreds of the remaining solids prior to burning, (Hamernik and Frantz, 1991). However, municipal ash from the as-received condition does not meet the requirements of a pozzolanic classification and therefore, does not react with portland cement in a cementitious fashion. Furthermore, according to Hamernik and Frantz (1991), MSW ash does not exhibit cementitious properties and does not set when mixed with only water. This is the reason for adding MSW bottom ash to the mix design as a fraction of total aggregate, while holding the cement content constant.

Without the pozzolanic reaction, there is no reduced permeability and no reduced heat of hydration. In fact, due to the increased particle surface areas previously discussed, municipal ash-modified concrete demands a significant increase in the w/b ratio of the mix design to cope with hydration.

The major problem with incorporating MSW bottom ash in mix design is the leaching capabilities of the inherent organic compounds and heavy metals that may be present in the resulting concrete. Of particular and obvious concern are the cadmium and lead levels that are significantly higher than the toxicity limits of the Toxicity Characteristic Rule published in the Federal Register. However, *Wheelabrator Environmental Systems Inc.* has developed a technology that effectively immobilizes the leaching capabilities of these and other toxic metals found in municipal ash through dissolution and precipitation reactions (Lyons, 1995). This process is known commercially as the WES-PHix® Ash Stabilization Process and guarantees that the chemical contaminant contents of the treated ash are below the federal limits through a test termed the Toxicity Characteristic Leaching Procedure (TCLP), thereby providing safe and environmentally sound ash for utilization in concrete applications. The results of this test for lead and cadmium are listed in Table 20.1.

Table 20.1 : TCLP Results for Untreated vs. WES-PHix® Treated Ash (mg/L)

Metal	*Federal Limit*	*Untreated Raw Ash*	*Treated WES-PHix®*
Lead	5.0	6.20	0.276
Cadmium	1.0	1.30	0.078

Sources: Lyons and Toxicity Characteristic Rule.

Clearly, the WES-PHix® process effectively reduces the levels of toxic materials to well below the federal limits. Therefore, the treated ash is a realistically safe and viable material for utilization in environmental applications.

Research Performed at Florida Atlantic University

The Department of Ocean Engineering at FAU has been actively involved in evaluating and comparing the influence of ash substitution of cement and aggregates on the mechanical properties (compression, tension, bending) of hardened and fresh concrete. Typical results from an experiment comparing coal fly ash and MSW bottom ash are presented in the following sections.

Mix Design

The test specimens were 39 in total and covered a wide variety of comparative mix designs that are represented in Tables 20.2(a)-(d). The control specimens (9 total) were comprised of an unmodified classical-type mix represented by *Mix A*. There were also nine specimens of *Mix B* containing varying proportions of coal fly ash. In order to provide a direct comparison between the behaviors of different ash modifications, *Mix C* was designed to exactly replicate *Mix B*, except that MSW bottom ash was substituted for the coal fly ash proportions. It is important to note that both of these mixes (*B* and *C*) were used to test the effect of reducing cement content with ash replacement while everything else is held constant. However, due to the lack of cementitious qualities that was expected of MSW bottom ash, twelve specimens of *Mix D* was used to test the concept of holding the cement content constant and replacing part of the aggregate (sand and pearock).

Compressive Strength Tests

Compressive load tests of 4×8 inch cylindrical specimens were performed for all mixes in accordance with ASTM C469, the standard for uni-axial compression testing. The compressive strength (C) was then be computed from the following formula and the results are included as Figures 20.2(a)-(c):

$$C = 4P/\pi D^2$$

where, P is the ultimate axial load and D is the diameter of the specimen.

Split-Cylinder Tensile Strength Testing

Based on the results of the compressive strength tests, the most efficient mixes from categories A, B and D were identified and then tested in tension (Mix C was omitted due to extremely poor compression results). The tensile strength of the chosen mixes was determined by way of indirect compression whereby a cylindrical specimen is loaded diametrically until fracture by lateral split is experienced. The test was performed according to ASTM C496, the standard for splitting tensile strength tests.

Table 20.2(a) : *Mix A*: Control (Unmodified)

Mix Number	*Cement (lb/cu yd)*	*Ash (lb/cu yd)*	*Water (lb/cu yd)*	*Sand (lb/cu yd)*	*Pearock (lb/cu yd)*	*W/B Ratio*
A-1	587	NA	195	1409	1722	0.33

Table 20.2(b) : *Mix B* : Fly Ash - Modifed
(Partial replacement of cement)

Mix Number	*Cement (lb/cu yd)*	*Ash (lb/cu yd)*	*Water (lb/cu yd)*	*Sand (lb/cu yd)*	*Pearock (lb/cu yd)*	*W/B Ratio*
B - 1	410	176	195	14-9	1722	0.33
B - 2	381	205	195	1409	1722	0.33
B - 3	352	235	195	1409	1722	0.33

Table 20.2(c) : *Mix C* : Municipal Ash - Modified
(Partial replacement of cement - Replicate of Mix B)

Mix Number	*Cement (lb/cu yd)*	*Ash (lb/cu yd)*	*Water (lb/cu yd)*	*Sand (lb/cu yd)*	*Pearock (lb/cu yd)*	*W/B Ratio*
C - 1	410	176	195	1409	1722	0.33
C - 2	381	205	195	1409	1722	0.33
C - 3	352	235	195	1409	1722	0.33

Table 20.2(d) : *Mix D* : Municipal Ash - Modified
(Partial replacement of aggregate)

Mix Number	*Cement (lb/cu yd)*	*Ash (lb/cu yd)*	*Water (lb/cu yd)*	*Sand (lb/cu yd)*	*Pearock (lb/cu yd)*	*W/B Ratio*
D - 1	520	200	208	1309	1622	0.40*
D - 2	500	400	250	1209	1522	0.50*
D - 3	450	800	270	1009	1322	0.60*
D - 4	450	1000	315	909	1222	0.70*

* W/B Ratio increases for municipal ash mixes to cope with the problem of hydration.

The tensile strength (T) of each specimen was then calculated from the following formula and the results are represented by Fig. 20.3:

$$T=2P/\pi LD$$

where, P is the ultimate diametrical load, and L and D are the length and diameter of the cylindrical specimen, respectively.

Slab Testing

Continuing research on this project involves flexural testing of conceptual highway pavements with ash modifications. As in the split-cylinder experiments, the specimens are comprised of three specific mixes providing a comparison between unmodified, coal ash-modified, and MSW bottom ash-modified concretes: *A-1*, *B-l*, and *D-2*. However, based on the theory of an *infinite slab on an elastic foundation*, these specimens were determined to be 4 ft square slabs with no reinforcement. Primarily, the highway infrastructure can be modeled as a uniform slab that extends to infinity in both directions and upon a point loading at the center, the two directional strains can be measured. A schematic of the experimental test setup and the actual test setup are shown as Fig. 20.4(a) and (b), respectively. Currently, slab testing for two mixes (B-1 and D-2) has been completed and the third mix (A-1) is in the testing process. The resulting strain data from the coal fly ash-modified and MSW bottom ash-modified concretes are graphed in Fig. 20.5(a) and (b).

Results

The test results indicated that MSW bottom ash does not possess the high-strength characteristics found in coal fly ash-modified concrete. The cement altered municipal ash-modified concrete (*Mix C*) was relatively weak compared to the coal fly ash-modified concrete *(Mix B)*. However, Fig. 20.2(c) shows that when MSW bottom ash is added as a 10-25% replacement of total aggregate *(Mix D)*, the strength values are comparable to those of unmodified concrete *(Mix A)*. This is a notably important observation because it promotes the utilization of MSW ash in applications where extremely high strengths are not needed.

LEGEND

A – No Ash B – Fly Ash C – Municipal Ash D – Municipal Ash

Fig. 20.2a : FLY ASH-MODIFIED (Partial replacement of cement)

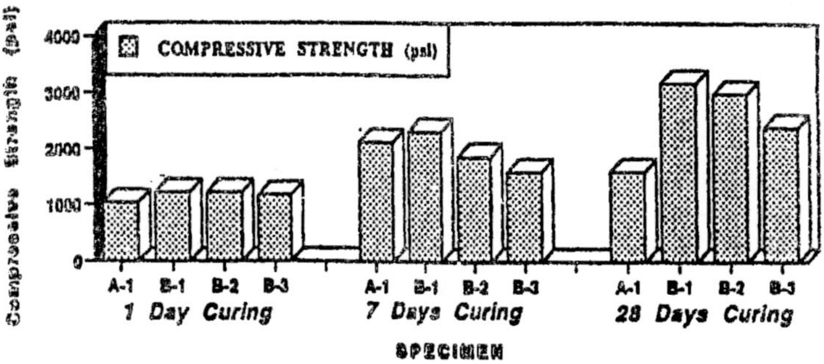

Fig. 20.2b : MUNICIPAL ASH-MODIFIED (Partial replacement of cement)

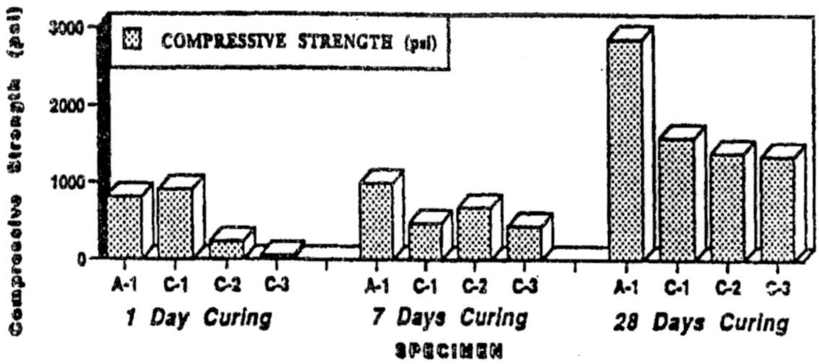

Fig. 20.2c : MUNICIPAL ASH-MODIFIED (Partial replacement of aggregate)

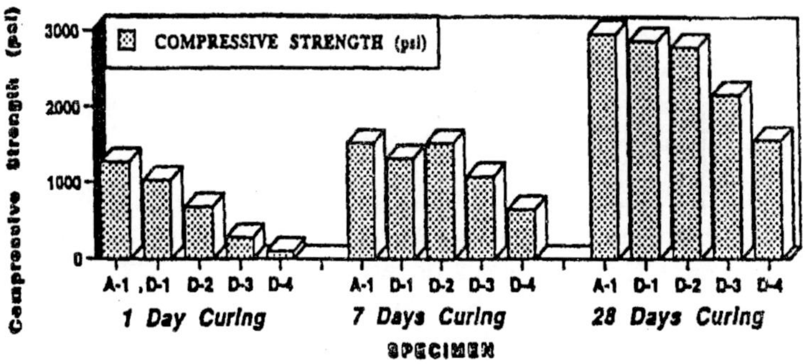

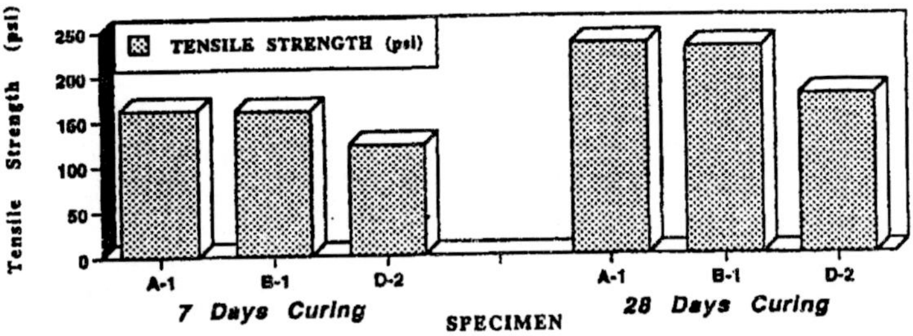

Fig. 20.3: Tensile strength comparison of three mixes: *A-1* (unmodified concrete), *B-1* (partial replacement of cement with coal fly ash), and *D-2* (partial replacement of total aggregate with MSW bottom ash).

However, the slab test results depicted in Fig. 20.5 show that the strains experienced by the MSW bottom ash-modified pavement are comparable to those that occurred in the coal fly ash-modified slab. This realization suggests that the both ash-modified mixes are reasonable designs for highway pavement applications. Furthermore, by using the ash as opposed to stockpiling it, environmentally sound disposal of the waste can be achieved while providing a cost-effective material to the construction industry.

Conclusions and Applications

In conclusion, it is clear that ash utilization as a substitute to natural resources in the construction industry is an energy efficient and environmentally safe alternative to the disposal of a waste product. Coal fly ash is highly acceptable for a wide range of applications from the high strength needs of highway infrastructure to the corrosive elements of the marine environment. MSW bottom ash is inherently restricted to lower strength applications, but still may be utilized in a number of ways.

The applicability of different fly ash mixes is linked to the level of concrete technology required for the concerned structural components. The applications can be divided into three categories: Low Technology (less than 20 MPa), Medium Technology (20-50 MPa) and High Technology (greater that 50 MPa). It is to be pointed out that MSW bottom ash is excluded from High Technology applications due to the relative lack in strength. The following is a list of some examples of these applications:

Low Technology Applications (< 20 MPa)

(a) Pavement base and sub-base courses

Fly ash is combined with lime or portland cement and aggregate and the pozzolanic reaction of fly ash results in an increase in strength with time.

(b) Subgrade stabilization

Fly ash, in combination with lime, is used to reduce plasticity and increase strength.

Medium Technology Applications (20-50 MPa)

(a) Ash-modified ready-mixed concrete

Many concrete producers use fly ash to overcome deficiencies in aggregate grading or have developed mixtures specifically for pumping. This takes advantage of the capacity of concrete containing fly ash to pump higher and further at faster rates, and with less segregation. MSW bottom ash may be used in this application to reduce the amounts of aggregates that are normally included. Fly ash is increasingly being used for the following reasons:

- Technical benefits.
- Reduced cement requirement with partial replacement of fly ash.
- Promotes high technology applications of fly ash-modified concrete.
- Increasing availability of fly ash meeting industry needs.

(b) Concrete pavements

Fly ash reduces map cracking and abnormal expansion and many states have conducted experiments on fly ash-modified concrete to evaluate the strength, crack resistance, finishing, and long term wear resistance. MSW bottom ash reduces the amount of coarse and fine aggregates used in pavement applications.

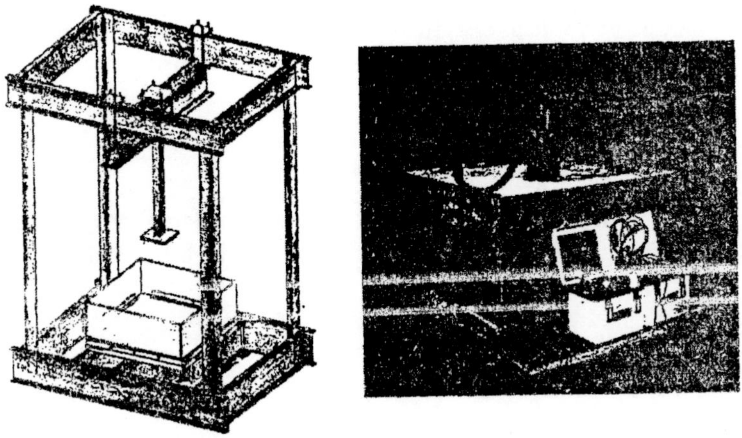

(a) (b)

Fig. 20.4: Experimental Slab Test Setup : (a) Schematic and (b) Actual.

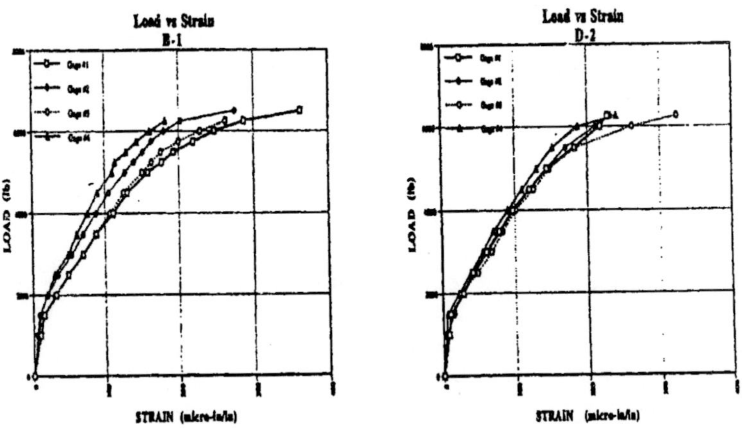

(a) (b)

Fig. 20.5 : Load vs. Strain plots for (a) coal fly ash-modified and (b) MSW bottom ash-modified 4 ft square concrete slabs.

(c) Mass concrete

The use of fly ash can reduce thermal stresses by providing a reduction in the large heat of hydration generated in the mixing process for massive concrete structures. Fly ash effectively reduces the temperature rise (relieving thermal stresses) of these massive structures and MSW bottom ash can reduce the sizable amounts of aggregates normally required.

(d) Concrete masonry units

Using fly ash in concrete mixtures improves the finish, the texture, and enables sharper corners. Most importantly, it provides added plasticity to the relative harsh mixtures associated with molding process of concrete masonry units.

(e) Concrete pipes

Pipes containing fly ash are more resistant to weak acid and sulfates due to decreased permeability of the modified concrete. Furthermore, there is an increase in cohesiveness and a reduction of hairline cracks.

(f) Precast/Prestressed products

Due to improved workability, products precast with fly ash provide sharp distinctive corners and edges and result with a better surface appearance. Applications of this nature are particularly desirable for products with intricate shapes or heavy reinforcement.

(g) Hollow-core slabs

Use of fly ash increases the workability of the harsh mixes associated with the "no-slump" conditions required by pretensioned hollow-core slabs.

(h) Bricks, blocks and paving stones

Both fired and unfired, bricks can be made with fly ash and bottom ash. The coal research bureau of West Virginia University has developed a method for molding and firing bricks from a mixture of 72% fly ash, 25% bottom ash, and 3% sodium silicate, (Naik, Singh, and Patel, 1992).

High Technology Applications (> 50 MPa)

(a) Bridge decks and substructures.
(b) Offshore and coastal structures.
(c) High rise buildings.
(d) Shell structures.

Other Applications

(a) Grouts and mortars

The benefits of using fly ash and MSW bottom ash in these applications are the same as those achieved upon utilization in concrete.

(b) Controlled low strength materials (CLSMs)

CLSMs are made up of cement ash, fine aggregate and water. They are ideal for backfilling, pipe bedding and protection, foundation sub-bases, spread footings, paving bases, floor files, culverts, steamlines and sewer lines.

(c) Soil cement

Soil cement is used as a base for road, street, and airport paving. Other uses include slope protection for dams and embankments, and liners for reservoirs. Both coal fly ash and MSW bottom ash can be utilized in these applications.

It is quite apparent that there exists an abundance of uses for fly ash and municipal ash. Furthermore, the benefits of applying wasted ash as a source of material are limitless. By utilizing these waste materials, the construction industry can not only reduce costs but also contribute to the effective disposal of a large volume waste product and save valuable natural resources.

ACKNOWLEDGEMENT

The authors would like to thank Dr. L. Berry, Distinguished Professor and Director, Center for Environmental Studies, Florida Atlantic University for funding the work described within this paper.

Further thanks and gratitude are expressed to Duos Engineering, Jacksonville, FL, for gifting the MSW bottom ash and technical advice that supported the project in a significant manner.

REFERENCES

ACI Committee 232, "Use of Fly Ash in Concrete," *ACI 232.3R-96,* Farmington Hills, MI, April 1996, pp.21-26.

Highway Administration, U.S. Department of Transportation, Washington, D.C., August 1995, pp. 1-8.

Ghafoori, N. and Bucholc, J., "Properties of High-Calcium Dry Bottom Ash Concrete," *ACI Materials Journal*, V. 94, No. 2, March-April 1997, pp. 90-101.

Hamernik, J.D. and Frantz, G.C., "Physical and Chemical Properties of Municipal Solid Waste Fly Ash," *ACI Materials Journal*, V. 88, No. 3, May-June 1991, pp. 294-301.

Lyons, M.R., "The WES-PHix® Ash Stabilization Process," Wheelabrator Environmental Systems Inc., Hampton, New Hampshire, 1995.

Naik, T.R., Singh, S.S., and Patel, V.M., "State of the Art: Fly Ash, Silica Fume and Slag Utilization in USA," Report No. REP-150, Univ. of Wisconsin, Milwaukee, May 1992.

Reddy, D.V., "Use of Fly Ash in the Transportation and Building Infrastructure," Florida Atlantic University, Boca Raton, September 1996, pp. 1-7.

Toxicity Characteristic Rule, Federal Register, 40CFR 261.24,1990.

21

Green Manufacturing—Technological Challenges for Innovation

A. DASGUPTA*

INTRODUCTION

Green Manufacturing can be defined as the intersection of manufacturing product and process practices with environmental aspects as shown in Fig 21.1. Greater the overlap between these areas, the greater the extent to which manufacturing practices can attend to the environmental issues, concerns and practices.

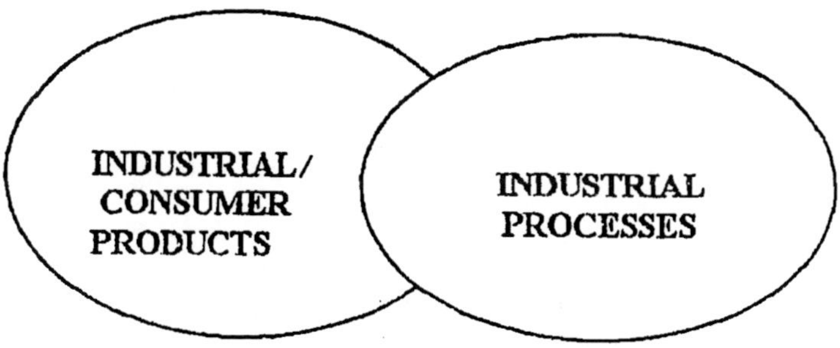

Fig. 21.1 : Intersection between Product and Process on Environmental aspects.

* Central Manufacturing Technology Institute, Bangalore

Green Manufacturing is also referred as clean technology or manufacturing, Green production, environmentally conscientious manufacturing, ecologically responsible manufacturing etc. Approach to green manufacturing means adopting such techniques and technologies which have positive or neutral impact on the environment.

GREEN MANUFACTURING TECHNOLOGY

As a whole Green Manufacturing encompasses almost all the areas of manufacturing. As a matter of fact both the existing design of the product as well as the manufacturing practices need to be given a fresh look to explore the possibility of pollution control and energy conservation and to achieve both tangible and intangible benefits.

The environmental issue remaining the main concern, innovative development or redesign of the product and processes, substitution of raw material to reduce or eliminate environmental waste, increasing life cycle of the products, recycling of waste product etc. come in the purview of the Green Manufacturing. The environmental issues related to manufacturing industries in past and present decades are shown in Fig 21.2.

The above concept is comparatively a new concern worldwide, which pose challenges to all design and manufacturing organizations. The new regulations and legislations are to be adhered to, the norms are becoming more stringent day by day, hence technological innovation is becoming a must to address this concept of Green Manufacturing and is likely to play an important role in the areas of manufacturing technology. Different aspects of Environmental issues in present days products and processes are given in Fig. 21.3.

A new international standard for environmental responsibility has already been laid down as ISO 14000.

GREEN MANUFACTURING IN METAL WORKING INDUSTRIES

In the Metal Working Industries, there are large no. of issues which need to be dealt with the concept of green manufacturing. In a very broad sense it can be classified into 3 groups:

Aspects	*Pre 1970s*	*1970s - 1980s*	*1990s*
General approach	Reactive	Compliant	Proactive
Environmental awareness	Very Limited	Limited to Specific Areas	Environmental Awareness in various sectors
Management Controls	Addressing Remedies	Inspection	Environmental Audit
Pollution and Waste	Not an Issue	Measure to Control	Process Innovation Life Cycle Approach
Legislative Controls	Few Regulations	Control on Emission & Waste	Integrated pollution control

Fig 21.2 : Industrial Response to Environmental Issues.

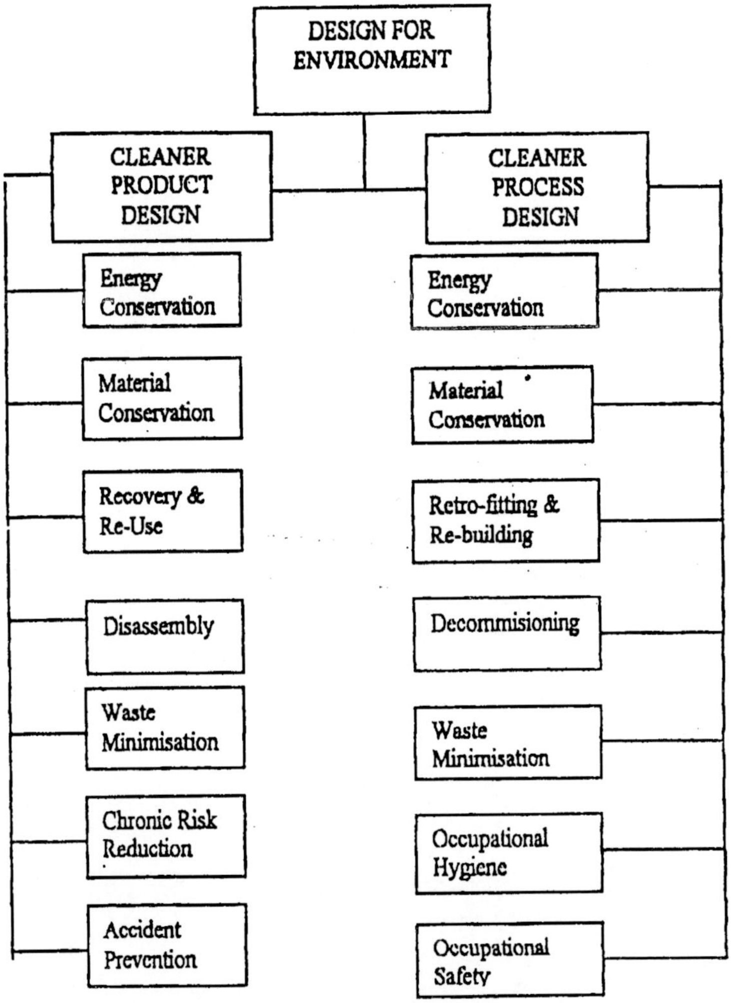

Fig. 21.3 : Design for Environment (Product and Process).

1. Improved or alternate Product and process design.
2. Additional measures in the existing products and processes.
3. Technological pursuit in Waste management.

Any new product and process which can reduce or eliminate the negative impact on the environment over the existing process may be considered belonging to first gap. As an example the Vibratory Stress Relieving equipment used for relieving stresses in structures, can be cited as an example for first category. Redesign of IC engines, for less pollution Catalytic converter manufacturing technique, etc. may also be considered in this group.

In the second category any improvisation on the existing processes like providing facilities for fume extraction, sorting and disposal of toxic chips, coolant cleaning, reducing exposure to noise pollution and other hazardous working condition etc. are included.

In the third category, conversion of pollutant waste into usable product like Briquetting of agro and forest waste for renewable source of energy may be included.

Similarly the following areas can be identified related to Green Manufacturing vis-a-vis Metal Working Industries viz.:

1. Propagation and development of renewable sources of energy–microhydel, solar, wind etc.
2. Conservation of energy by increasing the efficiency of mechanical transmission system.
3. Safety in manufacturing as a system.
4. Technology for reducing power consumption, water consumption etc.
5. Material handling, with Robot in hazardous area.
6. Technology of machining of toxic materials, Pyrophoric material.

It becomes essential that we not only become concerned regarding the impact of product and process to the environment but we should keep on searching for the new techniques to be adopted to effectively address the requirements. In many cases it becomes necessary to assimilate the established knowledge of different branches of engineering and technology.

WORK DONE AT CMTI

In the process of developing new technologies, Central Manufacturing Technology Institute, Bangalore, has carried out several works which can also be categorised under Green Manufacturing. Some of the examples are given below:

1. Special Purpose Machines for Hazardous Environment

CMTI has designed several Special Purpose Machines. One of those is for machining various items in hazardous environment, where human operators cannot be positioned. The Operator has to operate the machine remotely standing outside the work cell. CNC control is employed for operating the machine both in manual mode and auto mode.

Machine for trimming (facing) of special mass of inflammable material has also been developed with a remotely controlled semi automatic special purpose machine. There are no electrical elements or components on the machine hence no electrical sparks are generated. After the machine is set, the machine is totally operated remotely for safety, as the machining operation is hazardous for the operator. All control elements, electrical equipments and hydraulic power system are housed in adjoining control room for remote operation of the machine.

2. Vibratory Stress Relieving Equipment

After some detailed experimental study in CMTI, Vibratory Stress Relieving equipment was developed to reduce the residual stresses in casting and weldments. The conventional process of thermal annealing is high energy intensive, since it is required to be carried out approximately at 600°C for prolonged hours, sometimes stretching up to 30 hours. The alternate VSR technique, based on vibrating the structure at its natural frequencies, is carried out for about an hour and the stress relieving is carried out by mechanical means. This process drastically reduces the consumption of energy and leaves no pollutant in the environment, whereas pollution created by burning fuel in oil fired furnace creates enormous pollutant gases.

3. Silicon Wafers for Solar Cells

Silicon Wafers are one of the main constituents of solar cell which are used to tap solar energy. The silicon wafers used in these solar cells are of 0.5mm thick and 73mm to 84.5mm diameter. As silicon wafers are very brittle machining them to required size is a very difficult task. Laser machining is best suited for this purpose as no cutting force is exerted on the component during machining. CMTI has carried out Silicon Wafer cutting operation in Laser Machine.

4. Conversion of Agro and Forest-Waste to Fuel Briquettes

There exist large stock of agro and forest waste like paddy husk, groundnut shell, sugar bagasse and cane trash etc. in the country. Generally, these wastes are burnt in the field causing lot of pollution in the atmosphere and at the same time no useful output is obtained. This assumes larger proportions if the fields are cultivated on modern lines with minimum manpower. The quantity of waste to be disposed will be more. This is especially so in developed countries, where very large tracts of land are cultivated and crop harvested by mechanical means. These agro wastes could be effectively used if they can be compacted into small and dense briquettes. They have high calorific value and can be a source of energy. Not only the pollution is avoided but the energy could be made available at the right places, where it will be costly to bring through conventional sources of energy. Availability of energy in villages and distant areas, and improperly connected places could change the very complexion of the region.

The conversion of the wastes into solid briquettes is carried out by a briquetting machine where it is compacted under high pressures. It is essential to design a machine to take up such high stresses in their structures and the briquetting die which is subjected to high structural stresses and wear. CMTI is working in this area.

5. Extrusion Die for the Manufacture of ceramic Honeycombs used in catalytic converters

Air pollution due to emission of harmful gases from automobiles is increasing at alarming rates in Indian cities. The government is trying to minimise the pollution by enforcing emission norms and making the use of 'Catalytic converters' compulsory in automobiles. Catalytic converter is a pollution control equipment which when fitted to the exhaust system of a vehicle, removes injurious gases like Carbon monoxide, Hydrocarbon and Oxides of Nitrogen by converting them into harmless gases like Carbon Dioxide, Water and Nitrogen.

The Catalytic converter consists of a 'Ceramic Honeycomb' which is prepared by extruding ceramic using a die made of steel. A number of small diameter holes are drilled on one side of the die and on the other, narrow slots are made resembling a honeycomb structure. During extrusion ceramic is forced from one side through the holes and when it comes out on the other side it assumes the shape of a honeycomb. After extrusion the honeycomb is dried and fired to impart sufficient strength.

The dies used for honeycomb extrusion are developed at CMTI along with M/s. Ceramic Technological Institute, BHEL, (Electroporcelains Division), Bangalore, successfully. Several dies have been manufactured and used in extrusion process. Catalytic converters prepared using these honeycombs have been tested on automobiles and excellent results have been obtained

6. Internal Combustion engine

Two wheeler automobiles are fitted with either 2 stroke or 4 stroke I.C. Engines. The atmospheric pollution is more with 2 stroke engines than with 4 stroke engines.

The reason is that in 2 stroke engines, the fresh charge is injected in to the engine cylinder before the exhaust gases are fully evacuated. The design of I.C. Engine is such that the evacuation of exhaust gases is done by fresh charge only. This causes a loss of nearly 15 to 30% of fresh charge. This fresh charge left unburnt to atmosphere, in the presence of sun light slowly gets transformed into certain polycyclic aromatic compounds which are known to be carcinogenic and thus not only polluting the environment but also wasting our resources.

CMTI is jointly working with Indian Institute of Science, Bangalore, in developing 2 stroke I.C. Engine. The design feature is such that the exhaust gases are thrown out by fresh air and then only the fresh charge is injected by an auxiliary piston, cylinder arrangement. This not only increases engine efficiency but also is more environment friendly as no fresh charge is left to atmosphere unburnt.

7. Other areas and Future activities

CMTI has also worked in the area of safe machining technique of pyrophoric material like magnesium, techniques of elimination of hazardous fumes in ECM etc.

It is also contemplating to work in the area of control of coolant contamination and deterioration of machine shop environment. Coolants are used in almost all machine tools to reduce the temperature generated during machining operations.

These coolants are stored and re-circulated regularly. With long usage, the coolants get contaminated with swarfs, machine oils and other dirts. The bacteria are also developed, creating a foul smell all around the machine shop. Moreover this contamination along with bacteria causes health hazards like skin eruption and irritation to the operator.

In order to control the deleterious effect a system containing process of filtratian of swarfs and dirt, separation of mineral oils and destruction of bacteria by either pumping ozone gas or ultra violet rays is being considered.

Dry Machining is another area of interest in Manufacturing at CMTI. The present status of Dry machining is given in fig 21.4.

GREEN MANUFACTURING AND SUSTAINABLE DEVELOPMENT

While developing new technology, processes and manufacturing techniques, it becomes important that they are sustainable in nature. It becomes necessary to have the clarity on the goal and vision that drive the concept of sustainable development.

Developing a system or process, that meets the requirement of today, without compromising on requirements of tomorrow, alone can be accepted as sustainable development. Technological solution to today's problem should not have detrimental effect on

Work Material	*Geometrically Defined Machining Process*					
	Turning	*Milling*	*Boring*	*Tapping*	*Reaming*	*Deep drilling*
Heat-treated steel	●	●	▲	▲	▲	▲
Cast Al Alloy	●	●	▲	▲	▲	▲
Construction-al Steel	●	●	▲	▲	▲	▲
Non ferrous Alloys	●	●	●	●	●	■
Cast/Ductile Iron	●	●	●	●	▲	▲

● *Possible* ▲ *Under Development* ■ *Not Yet Possible*

Fig. 21.4 : Technical capabilities of Dry Machining.

tomorrow's environment. We cannot live today on borrowed environment of future generation. The technology we practice or develop should have either neutral or positive effect on the environment. If the effect is negative, such development being non-sustainable is bound to collapse even if it is pushed hard. However, actual practice of sustainable development may vary in different technology areas in different countries, but the goal remains the same and this is going to be the challenge for future technological innovations.

CONCLUSION

In developing a product or a process, a new challenge is now posed from the environmental, safety, health and waste points of view. It, then becomes imperative that newer technology should be able to address these factors. Performing a difficult task or achieving high efficiency and high accuracy cannot remain the only goal today. The impact of product and process should also have to be looked into.

Without such consideration the new technologies will not be sustainable and will hardly find any takers for commercial exploitation. Moreover, the existing technologies are also to be given a fresh look as to whether we can continue with the present way of working, without hurting the environment of our present and future generation. This growing world wide consciousness on environment is making Scientists and Engineers to look for Green Manufacturing.

REFERENCES

Green Manufacturing, SME Blue Book Series-By Steven A Melnyk & Richard T. Smith.

Manufacturing and the environment- By P. Young et al. The International Journal of Advanced Manufacturing Technology (1997)13:488-493.

Green Automation : Environmentally conscious manufacturing By Cheryl Pellerin. Assembly Automation Vol. 15, No 1, 1995.

Indian Environmental Legislation, Guide for Industry and Business. Environment Management Division. Confederation of Indian Industry.

Picking a Coolant - By Jean V. Owen. Manufacturing Engineering, May 1998.

Canola Oil - based fluid is gentle on environment, By R. Adams et al. Hydraulics & Pneumatics April 1999.

22

Compost and Nitrogen Effect on Broccoli Yield

AZIZ SHIRALIPOUR*

To evaluate the effect of various compost products and nitrogen (N) applications on a broccoli (*Brassica oleracea botrytis*, var. (Brigadeer) crop yield, six compost products and N from ammonium nitrate fertilizer were utilized in field experiments. The basic design of the site for each compost type was a replicated split-split plot with compost type on the main plot, three rates of compost application (0, 37, 74 Mg ha^{-1}) on the split plots and three supplemental N rates (0, 75, 150 kg ha^{-1}) applied to split-split plots. Application rates of 37 Mg ha^{-1} and 74 Mg ha^{-1} for all compost types resulted in significant yield increases for broccoli. Broccoli yields were affected by the combination of compost type, compost rate, and N rate applied. The higher yields were obtained for higher compost rates, with the effect of supplemental N rate less important at these high compost rates.

INTRODUCTION

The major benefit from the application of compost to soil is the improved physical properties related to the increased organic matter content (Gallardo-Lora and Nogales, 1987; Shiralipour et

* Center for Biomass Programs, University of Florida, Gainesville, Florida.

Table 22.1 : Golita Loam Soil Analysis

Constituents Tested	*Unite*
(Total)	mg/kg dw
As	5
Cd	1.6
Cr	41
Cu	31
Zn	130
Fe	14,000
Mn	730
Pb	26
B	4
Mo	1.1
Ni	35
P	1,000
Se	<10
Ca	90,000
Mg	3,400
Na	130
K	5,300
Hg	<0.3
Water Soluble (saturated paste)	*% saturation*
Ca	0.8 meq/100 gms
Mg	1.67 meq/100 gms
Na	0.36 meq/100 gms
K	138 meq/100 gms
Cl	0.18 meq/100 gms
NO3	0.57 meq/100 gms
PO4	0.35 meq/100 gms
SO4	0.59 meq/100 gms
Electrical Conductivity (EC 10 mmhos/cm)	0.64 meq/1
Exchangeable-cations	
Ca^{++}	5.4 meq/100 gms
Mg^{++}	0.92 meq/100 gms
Na^{+}	0.07 meq/100 gms
K^{+}	0.26 meq/100 gms
Cation Exchange Capacity	16.5 meq/100 gms
Exchangeable Sodium Percentage	0.4 percent
pH on saturated paste	6.7

al., 1992). These improved physical properties include, but are not limited to, increased water holding capacity (Mays et al., 1973; Hernando et al., 1989; Shiralipour and Zachary, 1994) and erosion prevention (Watkins & Winch, 1974). In addition, compost application adds nutrients to soil (Sommers and Giordano, 1984; Gallardo-Lora & Nogales, 1987; Shiralipour et al., 1992) and suppresses soil-borne diseases (Hoitink & Fahy, 1986; Hoitink et al., 1991).

The long-term feasibility of composting depends largely upon building a market for the compost by demonstrating the benefits resulting from compost application. To remove barriers to compost acceptance, definitive information must be generated for end-users to show the safe use of compost products for public health, safety and the environment. Compost users must distinguish between compost attributes that provide measurable benefits, whereas compost producers need to understand how their choices for compost feedstock and composting process management, determine these desired attributes in the final compost products.

Often end-users (e.g., agricultural, community, landscapers, etc.) are unaware of municipal plans to build compost facilities, while local governments are unaware of the market demands of potential end-users. Field experiments can assist with the development of communication between end-users and local government officials, thus accomplishing the first step in long-term market development.

Studies are needed to measure compost attributes that give measurable benefits under field conditions. To demonstrate the safe use and benefits of compost application in a coastal Southern California climate, greenhouse experiments and field trials were conducted in Santa Barbara County. The main objective of the greenhouse studies was to determine the optimal rates of compost application for the field trial with cash crops. The results of the greenhouse studies were reported earlier (Shiralipour et al., 1996). The objectives of the field studies were to evaluate the effect of compost and N application on yield increases of broccoli heads under field conditions.

Table 22.2 : Nutrient Contents (Total) of Composts[a]

Constituents Tested	Compost Type[b]					
	A1	*A2*	*A3*	*A4*	*B1*	*B3*
	% Dry Weight (dw)					
Total N	1.7	1.5	1.7	1.6	1.3	1.2
Organic N	1.6	1.4	1.6	1.4	1.3	1.2
NH_3-N	0.002	0.0015	0.001	0.0011	0.0018	0.0016
NO_3-N	0.15	0.098	0.16	0.11	0.064	0.041
P_2O_5	0.65	0.63	0.61	0.61	0.63	0.61
K20	1.1	1.8	1.1	1.6	1.6	2.5
Ca	2.4	2.7	4.0	2.7	2.0	2.4
Mg	0.6	0.78	0.9	0.84	0.6	0.99
SO_4-S	0.11	0.083	0.1	0.079	0.079	0.056
Cl	0.37	0.32	0.38	0.35	0.31	0.25

[a] Soil analysis was performed by Soil Control Lab in Watsonville, CA.

[b] Key :

A1 - 100% MSW[c]
A2 - 53% MSW, 47% UPD
A3 - 96% MSW, 4% Bio
A4 - 52% MSW, 46% UPD, 2% Bio
B1 - 96% UPD, 4% Bio
B3 - 100% UPD

[c] MSW = municipal solid waste; UPD = urban plant debris; Bio = biosolids/UPD.

MATERIALS AND METHODS

Plant Type - Broccoli (*Brassica oleracea botrytis*, var. 'Brigadeer') was chosen for the field experiments, the same cash crop used in the greenhouse studies (Shiralipour et al., 1996). This vegetable was chosen because it represents one of the most important crops in terms of acreage and cash value (personal communication with the Santa Barbara County Agricultural Commissioner).

Table 22.3 : Trace Element Contents (Total) of Composts[a]

	Compost Type[b]					
Constituents Tested	*A1*	*A2*	*A3*	*A4*	*B1*	*B3*
	mg/kg (dw)					
As	10	<10	24	22	20	22
Cd	6.6	3	8	5.6	2.2	1.5
Cr	52	26	64	44	22	24
Cu	200	98	260	172	70	58
Zn	800	380	940	680	220	240
Fe	17000	18000	17000	12000	13000	11000
Mn	460	240	500	420	220	280
Pb	240	100	280	200	30	42
B	4	50	32	32	30	50
Mo	4.4	4	5.4	8	13	9
Ni	56	26	74	56	26	28
Se	<10	<10	<10	<10	<10	<10
Hg	0.55	0.45	0.33	0.4	0.61	0.45

[a] Soil analysis was performed by Soil Control Lab in Watsonville, CA.

[b] Key :

A1 - 100% MSW[c]
A2 - 53% MSW, 47% UPD
A3 - 96% MSW, 4% Bio
A4 - 52% MSW, 46% UPD, 2% Bio
B1 - 96% UPD, 4% Bio
B3 - 100% UPD

[c] MSW = municipal solid waste; UPD = urban plant debris; Bio = biosolids/UPD.

Soil Type - Soil type was Goleta loam, a well-drained soil on flood plains and alluvial fans. The soil formed in alluvium from sedimentary rock and is one of the most dominant soil types in Santa Barbara County. The chemical analysis results for this soil are shown in Table 22.1. The soil analyses were conducted by the Soil Control Lab in Watsonville, California and the methods they used were as follows:

Dried samples were stored in a 4°C cold room until ready for analysis, re-dried at 60°C, before grinding by mortar and pestle, then sampled for analysis.

Compost Types - Six compost products were selected for this experiment which included three feedstocks, used alone or in various ratios: urban plant debris (UPD), biosolids/UPD (Bio), and municipal solid waste (MSW). The UPD was composed of grass clippings and local plant leaves and branches. Bio was made from biosolids mixed with UPD at 1:1 ratio (v:v). MSW was an organic fraction of the local municipal solid waste. These three compost products, UPD, MSW and Bio, were obtained from facilities in Santa Barbara, California, Truman, Minnesota, and Carson, California, respectively. Six compost types were obtained

Table 22.4 : Statistical Test and Variance Components estimates for Broccoli Yield Data.

Source	*NDF*[1]	*DDF*[2]	*Broccoli*[3] *F-statistic*
Compost Type (Type)	5	12	197.34**
Compost Rate (C_Rate)	2	24	2531.03**
Type x C_Rate IA	10	24	84.99**
Nitrogen Rate (N_rate)	2	72	595.42**
Type x N_rate IA	10	72	2.55*
C_rate x N_rate IA	4	72	44.83**
Type x C_rate x N_rae IA	20	72	2.07*
	Main Plot		0.345
Variance Components	Split Plot		0.367
	Split-Split Plot		10.75

[1] Numerator degrees of freedom.

[2] Denominator degrees of freedom.

[3] F-statistic for the overall test of the effect. Significance indicated by *=p<0.05, **=0.05<p<0.01.

Table 22.5 : Broccoli Yields by Compost Type for Combinations of Compost Rate and N Rate

		Compost Product					
		100% MSW[1]	*53%MSW 47%UPD*	*96%MSW 4% Bio*	*52%MSW 46%UPD 2% Bio*	*96%UPD 4%Bi*	*100% UPD*
		(A1)	*(A2)*	*(A3)*	*(A4)*	*(B1)*	*(B3)*
Average Yield Dry Weight (g/plant)		*88.1*	*76.9*	*96.4*	*80.3*	*75.0*	*66.7*
C_Rate Mg ha^{-1}	*N_Rate kg ha^{-1}*						
	0	27 a	37 a	40 a	37 a	41 a	37 a
0	75	65 b	62 b	58 b	64 b	58 b	59 b
	150	77 c	70 c	65 c	75 c	67 c	69 c
	0	94 d	72 c	92 d	65 b	62 bc	54 b
37	75	99 d	78 d	99 e	80 c	73 c	65 bc
	150	103 d	84 de	102 e	87 d	82 d	76 cd
	0	100 d	90 ef	129 f	95 e	88 d	72 cd
74	75	102 d	95 f	138 g	105 f	98 e	80 de
	150	116 e	104 g	145 h	115 g	106 f	88 e
Plot variance		0.0	0.00	1.79	0.56	0.66	0.00
Split plot variance		1.56	2.28	0.00	2.22	0.00	0.00
Split-split plot var		17.22	9.72	6.21	7.44	8.78	10.33

[1] Means in the same column with similar letters are not significantly different via LSD at a Type I error of 0.01.

by mixing UPD, MSW or Bio in specific proportions to produce compost blends with the following radios:

A1 = 100% MSW
A2 = 53% MSW and 47% UPD
A3 = 96% MSW and 4% Bio
A4 = 52% MSW and 46% UPD and 2% Bio
B1 = 96% UPD and 4% Bio
B3 = 100% UPD

These represented the potential compost types expected to be produced at a Santa Barbara facility. Nutrient and trace element contents of the six compost types are shown in Tables 22.2 and 22.3. The trace element contents of all compost products were much lower than the Pollutant Concentration Limits (PCL) for sludge in the U.S. (U.S. EPA, 1990).

Table 22.6 : Various Forms of Nitrogen Added to Soil by Different Compost Types and Rates

Compost Rate ($Mg\ ha^{-1}$)	*Compost Types*	*Total N*	*Organic N*	*Estimated[1] Mineralised N*
		$Kg\ ha^{-1}$		
	A1	629	592	59-237
	A2	555	518	52-208
37	A3	629	592	59-237
	A4	592	518	52-208
	B1	481	481	48-194
	B3	444	444	44-177
	A1	1258	1184	118-472
	A2	1110	814	104-416
74	A3	1258	1184	118-472
	A4	1184	1036	104-416
	B1	962	962	96-385
	B3	888	888	89-355

[1] Based on estimated 10-40% mineralization during the first year of application.

Compost and Fertilizer N Application Rates - The compost was mixed with upper 15-20 cm of the soil by small garden shovels when used alone, or followed by N application when used in combination. The optimal application rates for broccoli was developed based on the results of greenhouse trials (Shiralipour et al., 1996). Rates of 0, 37 and 74 dry Mg ha^{-1} were used in these experiments for all compost products. The N source was ammonium nitrate fertilizer and was applied at N rates of 0, 75 or 150 kg ha^{-1}, either alone or in combination with compost.

Field Site and Design - A one-acre field site was located on an agricultural ranch (Rancho Dos Pueblos) in Goleta, California, about 18 miles north of Santa Barbara. The site was leveled and disked and then divided into three replicates of nine 1.5m x 1m plots for each compost type and separated by a 0.6 meter-wide alley to provide 27 plots for each compost type, giving a total of 162 plots (Fig. 22.1). The basic design of the site for each compost type was a replicated split plot with the three rates of compost application as main plots, and the three supplemental N rates as subplots.

Planting, Irrigation and Harvest - From uniformly grown seedling trays, 8 broccoli seedlings, approximately five cm in height, were transferred and planted in each plot with the immediate addition of water. Plants were irrigated by installation of drip tapes at the site. Using this method of irrigation, the nitrate movement from one plct to the next was minimized. Crops were harvested by hand at the end of the growing season and placed in paper bags. From each treatment 9 plants (3 plants from each plot) were randomly harvested. The bags were dried at 70oC for 48 h and weighed as a parameter of crop yield.

Statistical analyses - The field plot layout was a completely randomized design in compost type with compost rate applied as a split of the main compost unit and N fertilizer rate applied as split of the compost rate, thus leading to a split-split plot analysis in a completely randomized design with three replicates. A mixed effect general linear model (Littell et al., 1996) was used to examine main and interaction effects for the three factors (compost type, compost rate, and N rate), as well as estimate main plot, split plot, and split-split plot variance components. Least squares

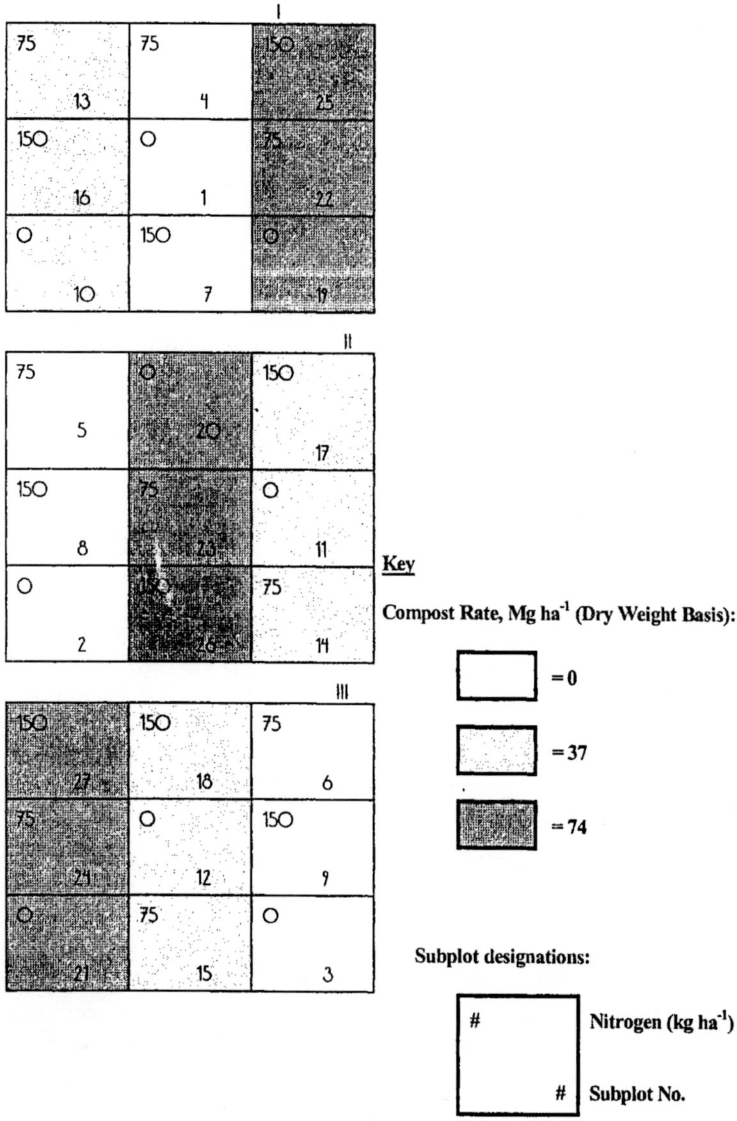

Fig. 22.1 : Compost Field Experiment Sample Plot Design

means were computed for appropriate effects and post-hoc means comparisons were performed using the LSD procedure at a comparisonwise Type I error rate of 0.01.

RESULTS AND DISCUSSION

Compost and N Effect on Broccoli Yield - The statistical analysis of the field test results on broccoli yields indicated all main, two-way and three-way interaction effects of compost type, compost rate and nitrogen rate were significant ($p < 0.05$) (Table 22.4). The significant interactions of compost type with the rate factors and concerns that the variability in yield was different for different compost types (Table 22.5) lead to a separate statistical analysis for each compost type. For each compost type, highest yields were obtained for the combination of high compost rate and high nitrogen rate. The general trend across all composts shows a significant difference in broccoli head yields at compost application rates of 37 Mg ha^{-1} and 74 Mg ha^{-1} compared to control (with no fertilizer or compost added), and between compost application rates of 37 mg ha^{-1} and 74 Mg ha^{-1} (Table 22.5). High nitrogen rates in the absence of compost still produced lower yields than in the presence of compost. The analysis indicates that broccoli head yields will be increased by a combination of type of compost, amount of compost and amount of N applied.

These results clearly indicate that the compost products utilized in these experiments can increase the yields of broccoli heads, regardless of whether N supplementation is used. The broccoli data showed a strong N response, demonstrating that 150 kg ha^{-1} inorganic N alone is insufficient to maximize head yield (commercial application rates of 150-350 kg ha^{-1} - personal communication with Dr. Ben Faber, University of California at Ventura). For all compost types, the addition of 74 Mg ha^{-1} of compost significantly increased broccoli yields. Those composts with more available N, namely A3 and A1, produced highest overall yields suggesting that the broccoli was responding primarily to the added N.

Urban plant debris, because of its high carbon to nitrogen ratio and lower N mineralization rate, produced the smallest increases in yields. Adding even a small amount of biosolids to yard waste improves the C/N ratio and hence improves yield. When comparing compost products, what seems to be the most important is the amount of available N (Table 22.6), since all will improve soil physical condition.

An additional benefit of compost over commercial N fertilization is the slow release of the N in compost to the plants through mineralization. Local rate of N mineralization of composted products ranges from 10 per cent to 40 per cent during the first year (personal communication with Dr. Ben Faber and Dr. Ben Crowley, University of California at Ventura and Riverside, respectively). However, it appears that the rate for this particular area falls at the lower end of the range. This process, in addition to the increased cation exchange capacity of the soil by compost application (Shiralipour et al., 1992), allows the plant roots time to utilize more of the released N, hence minimizing the loss of excess NO_3-N through leaching into the groundwater. This research suggests that at least for some crops, utilization of an appropriate compost type and rate of application may lead to reduction and possible elimination of commercial N supplementation.

ACKNOWLEDGEMENTS

This report is part of a three-year compost markets research project supported by The California Integrated Waste Management Board, the County of Santa Barbara, the Pew Charitable Trusts and The Procter & Gamble Company. The statements and conclusions of this report are those of the author and not necessarily those of the funding agencies.

REFERENCES

Gallardo-Loro, F. and R. Nogales. 1987. Effect of the application of town refuse compost on the soil-plant system: A review. *Biol. waste*. 19: 35-62.

Hernando, S., M. C. Lobo, and A. Polo. 1989. Effect of the application of a municipal refuse compost on the physical and chemical properties of a soil. *The Science of th. Total Environment*. 81/82: 589-596.

Hoitink, H. A. J. and P. C. Fahy. 1986. Basis for the control of the soilborne plant pathogens with composts. *Ann. Rev. Phytopathol.* 24: 93-114.

Hoitink, H. A. J., Y. Inbar, and M. J. Boehm. 1991. Status of compost-amended potting mixes naturally suppressive to soilborne diseases of floricultural crops. *Plant Disease*/September. 869-873.

Littell, R. C., G. A. Milliken, W. W. Stroup, and R. D. Wolfinger. 1996. SAS System for Mixed Linear Models. Cary, NC: SAS Institute Inc., 633 pp.

Mays, D. A. , G. L. Terman, and J. C. Duggan. 1973. Municipal compost: effect on crop yields and soil properties. *J. Environ. Qual.* 2(1): 89-92.

Shiralipour, A., D. B. McConnell, and W. H. Smith. 1992. Compost: Physical and chemical properties of soils as affected by MSW compost application. *Biomass and Bioenergy.* Vol. 3, Nos. 3-4, pp. 261-266.

Shiralipour, A. And J. Zachary. 1994. Compost field experiment guide for California Communities. Prepared for the Santa Barbara County Solid Waste Management Division and the California Integrated Waste Management Board. 89 pp.

Shiralipour, A., B. Faber, and M Chrowstowski. 1996. Greenhouse broccoli and lettuce growth using composted biosolids. Compost Science & Utilization. Vol. 4, No. 3, pp. 38-43.

Sommers, L. E. and P. M. Giordano. 1984. Use of nitrogen from agricultural, industrial and municipal wastes. ASA-SSSA, 677 South Segoe Road, Madison, WI 53711, USA.

U.S. Environmental Protection Agency. 1990. National sewage sludge survey: Availability of information and data anticipated impacts on proposed regulations. 55FR47210.

Watkins, E. M. and J. E. Winch. 1974. Composted organic waste, sewage sludges, and rock phosphate for the amelioration of acid uranium mine-tailings. *In Proceedings of the International Conference of Land Waste Management.* J. Tomlinson (ed). Ottawa, Ontario: Agricultural Institute of Canada.

23

Immunocontraceptive Biotechnology as Respite to Environmental Degradation Caused by the Pest Species. A Lesson from Brushtail Possum Overabundance in New Zealand

K.S. Sidhu*. K.E. Mate*, M.S. Harris*, F.C. Molinia**, J. A. Duckworth**, P.E. Cowan** and J.C. Rodger*

Problems posed by an introduced marsupial, the brushtail possum in New Zealand are such that they are considered number one mammal pest in that country. There are 60-70 million possums in New Zealand and they cause extensive damage to native ecosystems, agriculture and pastoral land and also spread bovine tuberculosis. The total cost and direct economic losses due to possums in the New Zealand economy amounts to US$ 30-40 (NZ$ 80-100) million per year with potential further costs of US$ 0.5 billion if threats of trade barriers due to bovine tuberculosis

* Cooperative Research Centre for Conservation and Management of Marsupials, Macquarie University, NSW Australia 2109.

** Landcare Research, Lincoln, New Zealand 8152.

(TB) were to eventuate. Overabundance of possums poses a difficult challenge for pest management because of negative economic impacts. The conventional methods of possum control like poison baits, traps, and shooting are partially effective but are increasingly considered inhumane and ecologically unsustainable. Therefore the only cost-effective, long-term acceptable way to achieve a sustainable reduction in possum population is by fertility biocontrol, using immunocontraception in combination with conventional methods. The major debate evolving in biocontrol at the present moment is whether biocontrol strategies should be delivered via disseminating (using an engineered living virus or bacteria) or non-disseminating (an engineered vector in plants or an attenuated species-specific pathogen) systems. Marsupials differ from eutherian mammals in many aspects of their reproduction, which pose unique problems in the application of reproductive technologies. The Cooperative Research Centre (CRC) for Conservation and Management of Marsupials has evolved various assisted reproductive technologies (ART) for marsupials, such as gamete maturation, superovulation, IVF, ICSI and AI. The ART have been used strategically to develop and test the first generation of non-disseminating biocontrol methods using gamete-specific immunocontraceptive vaccines. The genes for major egg zona pellucida (ZP) proteins (ZP I, ZP2 and ZP3) and several sperm proteins (e.g., SP-17) have been identified, sequenced, cloned and expressed in bacteria. The recombinant gamete proteins used as vaccines have been found to be highly effective in reducing the fertility of possums in laboratory trials. Currently efforts are being made to express these genes in carrots and other plants as the basis for an effective bait-delivered gamete antigen vaccine to regulate possum reproduction. The paper discusses the strategies being used in developing non-disseminating immunocontraception technology for the biocontrol of possums.

INTRODUCTION AND PERSPECTIVES

The common brushtail possum (*Trichosurus vulpecula*) is a cat-sized omnivorous Australian marsupial introduced in New Zealand in the mid-nineteenth century. It is a nocturnal, solitary, herbivorous and arboreal species inhabits mostly eucalyptus woodland in Australia. The possums were imported from Australia in order to establish a fur trade in New Zealand. The possum numbers in New Zealand grew at a much faster rate than the fur trade and soon it was realized that this exotic species was overgrowing its potential use with its current population at 60-70 millions. Possums now occupy about 95% of New Zealand at densities of up to 20 per ha, far in excess of those in Australia. Despite the presence of possums in New Zealand as early as 1840, their detrimental impacts were not recognized until much later and were the subject of considerable controversy (Pracy, 1974: Cowan, 1990, 2000). Possums are now recognized as the country's major mammal pest because they cause serious damage to native flora, fauna, and forest communities, and because they are the principal wildlife vectors of bovine tuberculosis (Cowan, 1990, Montague, 2000).

Direct costs of possum damage are estimated at US$ 30-40 million per annum, but non-tariff trade barriers because of the high levels of TB have the potential to cost the country up to US$ 0.5 billion (Hackwell & Bertram, 1999; Montague, 2000). Possum control mostly uses toxic baits broadcast aerially or deployed in bait stations costs about US$ 20 million per year. Increasing pressure for alternative approaches comes from national and international concerns about killing as a means of management, and about environmental contamination and risks to non-target wildlife (Fitzgerald et al., 2000).

WHY IS BIOLOGICAL CONTROL OF POSSUMS NECESSARY?

Possum numbers and areas of TB infection continue to grow in New Zealand despite of all attempts to control them. Overabundant possums pose serious threats to both conservation values and agricultural production. The current conventional methods used for possum control are only effective locally but the problems occur on a national scale. Therefore, integrated management of possums, using a combination of conventional control and biological control is needed for a long-term cost-effective solution to the national possum problem.

A wide range of factors drives the development of biological control for possums. Public opinion against killing as a means of controlling wildlife pests is mounting, not just in New Zealand but also internationally, and where these attitudes influence trade in New Zealand's products they become particularly important (Williams 1994). This effect is compounded by increased national and international concern about environment contamination with non-selective poisons used for pest control (Stewart et al 1994). New Zealand and Australia are the only countries that make extensive use of 1080 poison (sodium monofluoroacetate) for vertebrate pest management, and New Zealand's annual use of 2500 kg is an order of magnitude higher than that of Australia despite of much smaller land area. Biological control has the potential to reduce toxin use for possum control and hence concerns about environmental contamination, particularly of water supplies.

Despite of small number of non-target deaths that accompany 1080 poison operations, there is repeated public concern about it. Biological control also has the potential to reduce threats to non-target species. The expansion in conventional control over the last 5-10 years has produced a number of technical problems. The repeated use of 1080 and cyanide poisons has resulted in the development of possum that are bait and poison shy. Much effort is currently going into developing ways of overcoming shyness in field operations by changing operational procedures. New Zealand is also heavily dependent on a single poison, 1080, for most of its possum control. A major problem could arise if this poison becomes unacceptable to New Zealand's major trading partners. Biological control could alleviate both these problems. The Marsupial CRC is making attempts to develop biological control that will attack systems specific to marsupials, and possibly only possums (Cowan 1996, 2000).

Biological control by itself will not eradicate possums from New Zealand - the hope is either that it will reduce the possum problem to a scale manageable by sustained conventional control, or that in combination with conventional control it will provide opportunities for local or national eradication (Jolly 1993, Cowan 1996, 2000).

WHAT IS A BIOLOGICAL CONTROL FOR POSSUM?

Traditionally the biological control implies introducing a natural enemy of a pest species, such as disease or predators, to control the pest. The introduction of stoats to New Zealand to control rabbits is such an example though that was an ill-conceived and unsuccessful project. With the recent advances in molecular biology and genetic engineering, the concept of biological control has widened to include the use of genetically modified organism and biologically based substances such as immunotoxins and immunogens that interfere with normal biological processes. The application of these technologies to solve the problem of wild life population has increasingly been realized. Control of the European wild rabbit, red fox, feral cat, pig, goat, possum, horse and elephant using these technologies is currently being investigated.

DEVELOPMENT OF AN IMMUNOCONTRACEPTIVE VACCINE FOR MARSUPIALS

One of the major aims of the Marsupial CRC was to develop humane methods for fertility regulation of marsupials in captive and free-living populations with emphasis on non-disseminating immunocontraceptive delivery systems. The main requirements in developing a successful field delivered contraceptive vaccine are:

- Identification of target gamete antigen(s) neutralization of which will either block gamete transport, maturation, fertilization or early embryonic development.
- Species specificity to protect non-target species.
- Lack of interference with other biological processes, particularly the social functioning of the animal.
- A strong mucosal as well as humoral response at the appropriate target site.
- A delivery system restricted to target species with application in both captive and field situations.
- Comprehensive modeling to ensure feasibility of the system keeping in view species population dynamics and behaviour.
- Public and political concern about the technology.

The concept of regulating fertility in marsupials is based on three assumptions (1) contraception or interference with very early preimplantation stages of development produces effects at the population level, (2) the contraceptive vaccine interferes with a critical or regulatory process in reproduction, and (3) the molecular target will be species- or marsupial-specific.

The present paper reviews some of the current likely targets for immunocontraception in marsupials, progress made and problems encountered in developing such a strategy. The pattern of general marsupial reproduction-gamete production, maturation and endocrine regulation is generally similar to that of other mammalian species but is very peculiar in that the late embryo development occurs in pouch. Although marsupial reproduction and development has been the subject of much study (Tyndale-Biscoe and Renfree, 1987, Tyndale-Biscoe and Janssens, 1988), considerable basic research is required before generic concepts of immunocontraception can be applied effectively to this relatively poorly understood group of mammals.

ASSISTED REPRODUCTIVE TECHNOLOGY (ART) AND IMMUNOCONTRACEPTION

Because the knowledge base in reproduction in marsupials is limited, the Marsupial CRC has focussed research in this area, so that fertility control strategies could be evolved and applied effectively (for review see Rodger, 1995). This knowledge helped in developing an extensive suite of ART for possum biocontrol (Molinia et al., 1999). Table 23.1 summarizes the type, current status, and application of ART developed for possums.

The two key features of possum ART are that they permit control of reproduction in captivity, and increased productivity of animals all year round in a species that is naturally a seasonal breeder. This means that many aspects of possum reproduction can be intensively studied, and that large amounts of material can be generated for research with minimal animal numbers. Since ovarian follicle development, ovulation, fertilization and embryogenesis represent prime targets for immunocontraception, the use of possum ART like superovulation, AI, IVF and ET, has rapidly progressed research directed at these events. ART have also facilitated the screening of emerging vaccines. Most of this work to-date has been undertaken using natural breeding trials, which are time consuming, costly, and subject to irregularities in breeding in a captive environment. Using ART, immunized animals can be induced to superovulate and then AI can be used to determine their fertility in relatively short time. Moreover, with the development of IVF-ET technology, many potential antigens could be screened "in a petri-dish" for their ability to disrupt fertilization. The Marsupial CRC has successfully conducted the first laboratory trials with ZP vaccines using ART in two marsupial species, the brushtail possum and the tammar wallaby. The Marsupial CRC has also developed an IVF system for the brushtail possum using oviductal epithelial cell culture and co-culture systems, the first ever for any Australian marsupial (Sidhu et al., 1997, 1998, 1999a, b. Mate et al., 2000).

Current Status of Immunocontraceptive antigens and Fertility Trials for Marsupials

There are three approaches to manipulating fertility in mammals–surgical, chemical and immunological. Surgery is traumatic, expensive and labor intensive and useful only for confined small populations. Chemical methods can be effective but lack specificity, have limitations in delivery and rouse public concern about environmental contamination. Immunological methods are emerging as highly acceptable as they are able to address concerns associated with other methods.

Zona pellucida antigens

The zona pellucida (ZP) is an extracellular coat composed of three glycoproteins (ZP1, ZP2 and ZP3) present around all mammalian eggs that plays an essential role during fertilization and early embryo development. An increased understanding of the molecular and functional aspects of the marsupial zona pellucida is, therefore, vital to the development of ZP based marsupial immunocontraception. Like that of eutherian mammals, the brushtail possum ZP is composed of three glycosylated proteins which are orthologues of ZP1, ZP3 and ZP3. The overall amino acid identity between brushtail possum and eutherian ZP proteins is approximately 46% for ZP3 and 48-55% for ZP2. Wallaby ZP3 has an 86% amino acid identity with possum. Expression of possum ZP2 and ZP3 genes is restricted to the ovary. Analysis of ZP gene expression throughout pouch young development found that in the tammar wallaby both ZPI and ZP2 transcripts were present at 51-52 days and ZP3 expression was not detected until 75 days. Immunofluorescence and electron microscopic studies of ZP formation in the possum and wallaby found that the bulk of

Table 23.1 : Current status of possum-assisted reproductive technology (ART)

Type	*Status of development*	*Application*
Sperm Cryopreservation	Established	Indefinite storage of sperm for easy transportation
Superovulation	PMSG/LH protocol for breeding seasons established and FSH/LH protocol for seasonal anoestrus period being established	Generate multiple numbers of eggs for n vitro embryo production
Artificial insemination	Laparoscopic uterine and vaginal insemination methods established	Artificial breeding and fertility trials
Synchronisation	Commenced	Synchronisation of estrous cycles for embryo transfers
In vitro fertilization	Established in possum	Artificial breeding and fertility trials
Embryo transfer	Commenced	Artificial breeding
In vitro oviduct epithelial cell-sperm co-culture system	Established	In vitro production of capacitated sperm
Intra-cytoplasmic sperm injection (ICSI)	Established	Egg maturation and embryo production

Modified from Molinia et al. (1999).

ZP protein synthesis occurs within the oocyte (Mate & McCartney, 1998). Antibodies against ZP are ovary-specific and act by preventing sperm from binding and penetrating the ova and/or by disrupting the development of follicles in the ovary (Dunbar et al., 1994). The concept that immunization of possums against a ZP antigen would impair their fertility was proven when possums were immunized with porcine ZP and their immune responses and fertility assessed. Treated females produced serum antibodies that cross-reacted with possum ZP and reduced the likelihood of a fertile mating by 80% and significantly reduced the number of treated females producing pouch young during the breeding season (Duckworth et al., 1999a). Assessment of daily urine samples for cell changes and presence of sperm (Duckworth et al., 1999b) showed that treatment had no effect on the number of animals displaying oestrus or that mated. Treatment did not cause obvious ovarian dysfunction (Duckworth et al., 1999a). These results demonstrated that an immune response targeting the zona pellucida of the brushtail possum significantly reduced female fertility.

Possum ZP3, ZP2 and ZP1 have been characterized (Mate & McCartney 1998, McCartney & Mate 1999, Haines et al., 1999). The internal fragments of proteins have been expressed in *E. coli* and the purified recombinant possum proteins used for immunogenicity and fertility trials in possums (Duckworth et al., 1999a). When female possums were alloimmunised against recombinant possum zona pellucida 3 antigen (rZP3), anti-rZP3 antibodies were detected in serum of 14 or 15 immunised females. The effects of rZP3 immunization on fertility of possums was assessed following both natural matings (natural breeding trial) or after oestrous synchronisation and artificial insemination (AI trial) (Molinia et al., 1999). In the natural breeding trial, fewer treated animals were mated, conception rates were lower in immunized animals (17%) than in control animals (73%) and fewer immunized females produced offspring (6% immunized Vs 53% control). In the AI trial, fewer eggs were recovered following immunization and the proportion of eggs that were fertilized was lower in the immunized animals (23%) than in control animals (75%). Alloimmunisation with possum rZP3 elicited a strong humoral immune response and significantly reduced the fertility of treated female possums.

One of the most divergent regions of the ZP3 protein is an epitope thought to be involved in fertilization. This epitope, located immediately upstream of the furin cleavage site, was identified in the mouse using a monoclonal antibody that blocked fertilization (Millar et al., 1989). A potentially possum-specific ZP3 peptide was identified from the region in the possum ZP3 that corresponds to this epitope in the mouse (Duckworth et al., 1999a). Two vaccine formulations were tested in male and female possums. One vaccine formulation consisted of the ZP3 peptide conjugated to the carrier protein, keyhole limpet hemacyanin (KLH), and the other was a chimeric peptide consisting of the ZP3 peptide linked to a promiscuous T-cell epitope from tetanus toxin (TT). Both vaccine formulations elicited antibodies in serum and resulted in the accumulation of antibodies around the ZP layer in the ovaries of treated females. The antibodies bound to recombinant possum ZP3. Fertility assessment of the ZP3 peptide is in progress. The possum ZP3 epitope has also been expressed in transgenic plants for evaluation in immunogenicity and fertility trials.

In recent trials, two overlapping constructs of possum ZP2 were generated in *E coli* and the purified recombinant proteins used to immunize female's possums. Both N- and C-terminal portions of recombinant ZP2 constructs were immunogenic and stimulated high serum titers. When the fertility of possums was assessed following superovulation and artificial insemination, there was a significantly fewer eggs recovered from females immunized against either the C- or N-terminal region of the rZP2 protein compared to the controls, and the number of the eggs that were fertilized was reduced by 70-75%. Possum ZP2 is immunogenic and shows immunocontraceptive potential.

SPERM ANTIGENS

Immunization with sperm proteins causes a species-specific reduction of fertility in both male and female of a number of mammalian species (reviewed by Kerr, 1995). Sperm antigen targets suitable for immunocontraception need to be accessible to antibodies during some stage of gamete maturation and/or interaction and these antibodies must inhibit some aspect of gamete maturation, transport, or fertilization. When female possums were immunized with whole possum sperm the number of treated females producing offspring was reduced fertility by 80% indicating that sperm antigens may be effective immunocontraceptive targets for marsupials (Duckworth et al., 1998). Hence several marsupial sperm antigens are currently being investigated A number of novel marsupial sperm proteins have been characterized which display characteristics consistent with being suitable immunocontraceptive targets such as localization to the marsupial sperm surface and association with infertility (Harris, unpublished). Ongoing work is concentrating on identifying the genes encoding these proteins. Marsupial homologues of proteins known to participate in eutherian fertilization have also been identified and their genes cloned.

Sp 17 AS POSSUM SPERM TARGET ANTIGEN

The sperm antigen Sp 17 is the 17-kDa member of an autoimmunogenic family of proteins (RSA proteins 14-19 kDa) originally identified in the rabbit (O 'Rand and Irons, 1984). Sp 17 is a zona binding protein not available on the surface of live acrosome-intact rabbit and mouse sperm, but exposed upon initiation of the acrosome reaction (Richardson et.al, 1994, Wen et.al, 1999). Monoclonal antibodies to the RSA antigens blocked rabbit fertilization in vitro and in vivo and the penetration of zona-free hamster eggs by human sperm (reviewed O'Rand et.al, 1990). A synthetic peptide comprised of the dominant B-cell epitope and a promiscuous T cell epitope bovine RNase was then constructed and in female Balb/c mice induced a dose - dependent reduction in fertility of up to 60% (Lea et.al, 1998). The gene encoding brushtail possum Spl7 has been cloned, a bacterial recombinant protein produced and shown to be immunogenic in possums (Harris and O'Rand, unpublished). The effect of recombinant possum Spl7 on the fertility of female possums is currently being assessed.

DEVELOPMENT OF AN EFFECTIVE DELIVERY SYSTEM

Currently one major problem confronting scientists is the development of delivery systems for immunocontraceptive vaccines

(Rodger, 1999). A major debate evolving in biocontrol at the present moment is whether biocontrol strategies should be delivered via disseminating (using an engineered living virus or bacteria) or non-disseminating (an engineered vector in plants or an attenuated species-specific pathogen) system. Initially, a non-disseminating oral bait system for the target species like possum is likely to be most acceptable. However, in New Zealand such a system could be the first step in the development of a disseminating system. Non-disseminating systems being evaluated for oral delivery include non-replicating viruses, bacteria and plants with the genes for the antigenic epitopes and immune potentiators transgenically engineered into their genomes. A range of viral vectors i.e. proxviruses, adenoviruses, and herpesviruses and bacterial vectors i.e. attenuated Salmonella *typhimurium* have been evaluated as delivery system for heterologous antigens. An excellent example is the successful use of recombinant Vaccinia (cowpox) virus expressing a part of the rabies virus genome to immunize populations of wild foxes (Brochier et al., 1990). Synthetic oral particular delivery system i.e. microspheres, liposomes etc. are also available, although the cost of production are currently high (Miller, 1997).

Recent developments in delivery systems include bacterial ghosts, who are killed and evacuated bacterial membranes with associated recombinant antigen (Szostak et al., 1996). Transgenic plants have been developed that express mammalian contraceptive antigens capable of inducing an immune response (Smith et al., 1997) and a second approach to plant-driven vaccine uses modified plant viruses to carry and express the gene of interest in infected plants and the latter could be fed to the target animals (Cowan et al., 1999). The Marsupial CRC in collaboration with Boyce Thompson Institute (BTI) for Plant Research in New York and University of Vienna, Vienna and Queensland Department of Primary Industries, Australia is actively involved in developing plant-derived and bacterial ghost-derived vaccines respectively expressing zona pellucida proteins or sperm proteins for marsupial biocontrol. The Marsupial CRC and the BTI are also evolving strategies for developing transgenic plants i.e. carrots and potatoes and the use of modified plant viruses to express foreign marsupial

ZP genes for possum biocontrol. Transgenic potato expressing LT-B gene as a model antigen when fed to possums has been found to generate systemic as well as 10 mucosal immune responses in possum. This demonstrates the potential of plant vaccine systems for possums.

CONCLUSION

Fertility control in wild populations using immunocontraception is a relatively new concept and is emerging as a highly acceptable means of controlling vertebrate pest species. A major challenge is generating the basic knowledge in reproductive biology, immunology and ecology in wild pest species where not much is known. This knowledge plays a key role in developing this technology. Nevertheless, some headway in this regard has already been made in developing fertility control for such wild species as red foxes, rabbits, wild cats, and horses using immunocontraceptive technology. Encouraged by the fact that ZP vaccine evolved for brushtail possum has been found successful in curtailing reproduction in laboratory trials, the other overabundant marsupial species such as Eastern Grey kangaroos, are also being targeted. Future challenges lie in: (1) evolving combination vaccines, as immuno-neutralization of a single gamete-specific antigen may not be sufficient, and (2) evolving effective delivery systems for such vaccines. Equally important is building public opinion towards acceptability of such technology by effective communication, discussions, and participatory decision-making.

ACKNOWLEDGEMENTS

The research work of the authors cited in this paper was supported by funds from the Australian Government's Cooperative Research Centre Program, New Zealand Ministry of Agriculture and Forestry, Policy Division and New Zealand Foundation for Research, Science and Technology.

REFERENCE

Brochier, B, Thomas, I., Baudin, B., Leveau, T., Pastoret, P., Languet, B., Chappuis, G., Desmettre, P., Blancou, J., and Artois, M. (1990), Use of vaccinia-rabies recombinant virus for the oral vaccination of foxes against rabies. Vaccine 8, 101-104.

Cowan, P. E. (1990), Brushtail possum in: King, C.M. Ed. The Handbook of New Zealand mammals. Auckland, Oxford University Press. pp. 68-98.

Cowan, P. E. (1996) Possum biocontrol: prospects for fertility regulation. Reprod. Fertil. Dev. 8, 655-660.

Cowan, P. E. (2000), Biological control of possums: prospects for the future. *In*: Montogue, T.L. Ed. The Brushtail Possum: Biology, Impact and Management of an Introduced Marsupial. Manaaki Whenua Press. pp. 262-270.

Cowan, P.E., Wairnsley, A., Kirk, D., Lee, S.M., and Young, P. (1999), Plant-derived vaccines for possum fertility control. Proc. Royal. Soc. New Zealand 56, 24-27.

Duckworth, J.A., Buddle, B.M., Scobie, S. (1998), Fertility of brushtail possums (Trichosurus Vulpecula) immunised against sperm. J. Reprod. Immunol. 37: 125-138.

Duckworth, J., Harris, M., Mate, K., McCartney, C., Buist, J., Scobie, S., Jones, D. and Jones, S. (1999a) Development of brushtail possum immunocontraception: targeting sperm and egg antigens. Proc. Royal. Soc. New Zealand 56, 72-75.

Duckworth, J.A., Scobie, S.S., Jones, D.E. and Selwood, L. (1999b). Determination of oestrus and mating in captive female brushtail possums Trichosurus Vulpecula (Marsupialia: Phalangeridae) from urine samples, Aus. J. Zool. 46: 547-555.

Dunbar, B.S., Avery, S., Lee, V., Prasad, S., Schwahn, D., Schwoebel, E., Skinner, S. and Wilkins, B. (1994), The mammalian zona pellucida: its biochemistry, immunochemistry, molecular biology and development expression. Reprod. Fertil. Dev. 6:331-347.

Fitzgerald, G., Wilkinson, R. and Saunders L. (2000) 'Public Perceptions and Issues in Possum Control'. *In*: Montogue, T.L. Ed. The Brushtail Possum: Biology, Impact and Management of an Introduced Marsupial. Manaaki Whenua Press. p. 292.

Hackwell, K., and Bertram, K. (1999), Pests and Weeds. The cost of restoring an indigenous dawn chorus. Wellington, New Zealand Conservation Authority. p. 71.

Haines, B.P., Rathjen, P.D., Hope, P.D., Whyatt, L.M., Holland, M.K. and Breed, W.G. (1999), Isolation and characterisation of a cDNA encoding a zona pellucida protein (ZPB) from the marsupial *Trichosurus vulpecula* (brushtail possum). Mol. Reprod. Dev. 52:174-82.

Jolly, S.E. (1993), Biological control of possum. N.Z. J.Zool. 20, 335-340.

Kerr, L.E. (1995), Sperm antigens and immunocontraception. Reproduction, Fertility and Development 7: 825-830.

Lea, I.A, van Lierop, M.J.C., Widgren, E.E, Grootenhuis, A., Wen, Y., van Duin, M. and O'Rand, M.G. (1998), A chimeric sperm peptide induces antibodies and strain- specific reversible infertility in mice. Biol. Reprod. 59, 527-536.

Mate, K., Sidhu, K.S., Molinia, F.C., Glazier, A.M. and Rodger, J.C. (2000) Sperm Binding and Penetration of the Zona Pellucida *In vitro* but not Sperm-Egg Fusion in an Australian marsupial, the Brushtail Possum (*Trichosurus vulpecula*). Zygote 8, 189-196.

Mate, K.E., McCartney, C.A. (1998), Sequence and analysis of the zona pellucida 2 (ZP2) cDNA from a marsupial, the brushtail possum, *Trichosurus vulpecula.* Mol. Reprod. Dev. 51:322-329.

McCartney, C.A. and Mate, K.E. (1999), Cloning and characterisation of a zona pellucida 3 cDNA from a marsupial, the brushtail possum, *Trichosurus vulpecula.* Zygote 7: 1-9.

Millar, S.E., Chamow, S.M., Baur, A.W., Oliver, C., Robey, F. and Dean, J. (1989), Vaccination with a synthetic zona pellucida peptide produces long-term contraception in female mice. Science 246: 935-938.

Miller, L.A. (1997), Delivery of immunocontraceptive vaccines for wildlife management. *In* : Kreeger, T.J. ed. Contraception in wildlife management. Technical Bulletin No. 1853, Animal and Plant Health Inspection Service, Washington DC, USDA. pp. 49-58.

Molinia, F.C., Glazier, A., Sidhu, K., Mate, K., Rodger, J., Smith, J. and Berg, D. (1999), Assisted reproductive technologies for possum. Proc. Royal. Soc. New Zealand 56, 50-53.

Montogue, T.L. (2000), The Brushtail Possum: Biology, Impact and Management of an Introduced Marsupial. Manaaki Whenua Press. pp. 292.

O'Rand, M.G. and Irons, G.P. (1984), Monoclonal antibodies to rabbit sperm autoantigens. II. Inhibition of human sperm penetration of zona-free hamster eggs. Biol. Reprod. 30, 731-6.

O'Rand, M.G., Widgren, E.E., Nikolajczyk, B.S., Richardson, R.T., and Shabanowitz, R.B. (1990), Receptors for zona pellucida on human spermatozoa. *In*: Alexander, N.J., Griffin, D., Spieler, J and Waites, G. eds. Gamete interaction: prospects for immunocontraception. Wiley-Liss, New York. pp. 213-224.

Pracy, L.T. (1974), Introduction and liberation of possum (*Trichosurus vulpecula*) into New Zealand. New Zealand Forest Service Information Series No. 45. 2nd ed. p. 28.

Richardson, R.T., Yamasaki, N. and O'Rand, M.G. (1994), Sequence of rabbit sperm zona pellucida binding protein and localization during the acrosome reaction. Dev Biol 165, 688-701.

Rodger, J.C. (1995) Likely targets for immunocontraception in marsupials. Reprod. Fertil. Dev. 9, 131-136.

Rodger, J.C., (1999), Delivery, arguably the major challenge for possum biocontrol. In Advances in the biological control of possums, The Royal Society of New Zealand, Miscellaneous series 56:16-18.

Sidhu, K.S., Molinia, F.C. and Rodger, J.C. (1997), Sperm-oviduct epithelial cell monolayer coculture : an *in vitro* model for studying sperm-female tract interactions in the possum. Proc. Royal. Soc. New Zealand 45, 79-81.

Sidhu, K.S., Mate, K. E. and Rodger, J.C. (1998), Sperm-oviduct epithelial cell monolayer coculture : an *in vitro* model for studying sperm-female tract interactions in a marsupial, the tammar wallaby (*Macropus eugenii*). J. Reprod. Fertil. 114, 56-61.

Sidhu, K.S., Mate, K.E., Molinia, F.C. and Rodger, J.C. (1999a), Induction of thumbtack sperm during co-culture with oviduct epithelial cell monolayers in a marsupial the brushtail possum. Biol. Reprod. 61, 1356-1361.

Sidhu, K.S., Mate, K.E., Molinia, F.C., Glazier, A.M. and Rodger, J.C. (1999b), Secretory proteins from the female reproductive tract of the brushtail possum (*Trichosurus vulpecula*): binding to sperm and effects on sperm survival *in vitro*. Reprod. Fertil. Dev. 112, 329-336.

Smith, G.L., Wairnsley, A. and Polkinghome, I. (1997), Plant-derived immunocontraceptive vaccines. Reprod. Fert. Dev. 9, 85-89.

Stewart, H., Jenken, B., McCarthy, R., and Flinn, D. (1994), Public versus user perceptions of risks and benefits of plantation forestry in Victoria, Australia. *In*: Seawright, A.A., Eason, C.T. ed Proc. Sci Workshop on 1080. The Royal Society of New Zealand Miscellaneous Series 28, 10-19.

Szostak, M.P., Hensel, A., Eko, F.O., Klein, R., Auer, T., Mader, H., Hasiberger, A., Bunka, S., Wamner, G. and Lubitz, W. (1996), Bacterial ghost, non-living candidate vaccines. J. Biotechnology 44, 161-170.

Tyndale-Biscoe, C. H. and Renfree, M.B. (1987), 'Reproductive Physiology of Marsupials'. (Cambridge University Press, Cambridge).

Tyndale-Biscoe.C.H. and Janssens, P.A. (1988), 'The Developing Marsupial'. (Springer-Verlag, Berlin).

Wen, Y., Richardson, R.T. and O'Rand, M.G. (1999), Processing of the sperm protein Spl7 during the acrosome reaction and characterization as a calmodulin binding protein. Dev. Biol. 206, 113-22.

Williams, J.M. (1994), Food and fiber markets and social trends: implications for pest management. *In*: Seawright, A.A., Eason, and C.T. Ed Proc. Sci Workshop on 1080. The Royal Society of New Zealand Miscellaneous Series 28, 20-32.

24

Agrotechnologies and Tissue Culture in India

DR. ANIL KUSH*

Like in most of the developing countries. Agriculture is the backbone of Indian economy, fabric of Indian society and the nerve of the nation. Despite the significant increase in our food productivity, specially because of Green and White Revolution, Indian agriculture has a very long way to go to feed this exponentially growing population. The human population is likely to double in next two-three decades. More than 10 billion people will have to be fed, clothed and provided with jobs under conditions of shrinking land and water resources for .agriculture, expanding biotic and abiotic stresses increasing genetic erosion and raising cost of fuel energy reserves. India and China, the two popular nations will have to import 40 million tonnes and 140 million tonnes respectively by the year 2030. Interestingly the entire world foodgrain is presently estimated to be only 200 million tonnes. In Indian scenario whereas rice production increased by 7.76% [74.7 million tonnes to 80.5 million tonnes] between 1991-92 to 1996-97, wheat production was considerably better by 23.56 per cent [an increase from 55.6 million tonnes to 68.7 million tonnes] in the same period. The overall increase in the production of all foodgrains by 17.97 per cent [from 168 million tonnes to 198.2 million tonnes]

* Vice President, Indo American Hybrid Seeds (India) Pvt. Ltd., Bangalore-660 070, India.

in five year duration is by no means a small achievement. Even so, there is little doubt that an additional 40 million tonnes of foodgrains that is estimated to be required in the medium term cannot be harvested from existing farms, technologies and system. We need technologies which can improve continuously the productivity, profitability and sustainability of our farming system. The cultivable land is fast depleting and demand from per unit of this is shown in the other way trend. About 80-90% of the total foodgrain production in this country comes from top 15-20 cm of soil. Unfortunately this soil too has become a heap of polluting chemicals with little organic matter left. We need to tap the waste land and follow lands which amount to about 25 and 60 million hectare respectively. Most of our agricultural production comes from rain fed areas. Hence development of our varieties which give better yield even under low irrigation and at the same time optimal use of water resources would be the priority in near future. New technologies like hybrid seeds, disease free tissue culture plants, biotechnology, integrated pest management, biofertilisers and eco-friendly approaches to maintain sustainable increase in productivity will be required. Agriculture as such can no more be looked at as an only provision which provides product for the kitchen alone.

Diversifying Agriculture

There is a need for diversification and relatively clear examples in this direction we had in India points to White Revolution, Floriculture and Aquaculture - just to name a few. Just to mention about floriculture – though we have got big potential of floriculture development and its experts, our world share of floriculture market is less than 1% of the total which is about 50 billion US$. Development of right type of plant material, proper cultural practices and the proper marketing are few of the events in the chain reaction of a successful floriculture venture. The top 5 flower growing states in the country include Karnataka, Tamil Nadu, Andhra Pradesh, West Bengal and Maharashtra which together cover about 86% of the total floriculture area in this country. However only about 20% of the flowers are grown as per international standards, making it tough quality competition in the international scenario.

Hybrid Seeds

The exceptional agronomic performance of hybrid seed in terms of yield, vigor, adaptability and uniformity represents a superior product for farmers. The continued development of new and hybrid seed varieties is the cornerstone of a huge industry. Developing hybrids by conventional methods is a long process of developing suitable germplasm. In order to produce hybrid seeds uncontaminated with selfed seed, some sort of pollination control system must be used to block self pollination. These systems can be mechanical, chemical or genetical. Typically this process is hampered because the components of the hybridization system must be included during germplasm development thereby restricting the range of genetic material that can be incorporated. The commercial limitations and cost associated with producing hybrid seed by conventional methods have generated considerable interest in developing alternate system based on molecular biology. Much of the research efforts is based on the fundamental concept of disabling or ablating, pollen or anthers by selectively expressing some sort of toxic agent. The practical success of these methods hinges on the effectiveness of the toxic agent and the function of the promoter driving the expression.

Though North America and Europe are the largest producer of seeds on the globe, Asia is largest consumer. Hence there is a big gap between North and South in terms of production and utility. Hybrid seeds, though now established in almost all the major cereal and vegetable crops still does not make more than 10% of the total seed used by the farmers on the globe. Hence there is large potential increase in productivity through these seeds. However, hybrid seeds are more demanding in terms of inputs. Hence the environment affect if any of these seeds cannot be completely ignored. One need to have a balanced look at productivity and environment.

Global seed exports (million $)

Crop	*Value*
Maize	400
Herbage Species	350
Potatoes	340
Beets	260
Wheat	70
Other agricultural seeds	480
Small-seed vegetable and Flowers	750
Broad seed veg. legumes	250

Major countries for domestic seed market (million $)

Country	*Value*
U.S.A.	4500
CIS	3400
Japan	2500
China	2000
France	1800
Germany	1500
Brazil	1200
India	900
U.K.	570
Spain	550
Argentina	400
Poland	400
Hungary	400
Canada	350
Mexico	350
Netherlands	300
Australia	280
Belgium	200
Austria	170

Annual Seed consumption (million tonnes) — Geographic distribution

Region	*1994*
Asia	43.2
CIS	39.2
Europe	19.2
North & Central America	10.4
Africa	10.1
South America	4.4
Oceania	1.0

Major seed-exporting countries [million $]

Country	*Agric.*	*Veg.&f.*	*Total*
U.S.A.	470	150	620
Netherlands	400	200	600
France	375	75	450
Germany	190	35	225
Denmark	110	20	130
Italy	90	30	120
Others	265	490	755
Total	1900	1000	2900

Protected Cultivation

Protected cultivation is usually undertaken in glass houses or green houses or even simpler polyhouse with arrangements for regulation of light, relative humidity, temperature and irrigation. GH technology had been very successful in production of all the year round or in so called off season various vegetables, ornamental plants and even fruits.

Some of the crops that can be raised in Green House

Vegetables	*Ornamentals*	*Fruits*
Tomato	Anthurium	Strawberry
Capsicum	Chrysanthemum	
Cucumber	Gerbera	
French Beans	Marigold	
Bitter Gourd		
Okara		

Spices of Indian Agriculture

In the world of spices and herbs, India plays a pivotal role. More than 52 spices and herbs are grown in our country. Our annual production is over 2 million tons. Currently, India's spice export amounts to 40 per cent of global spice trade in quantity and 19.5 per cent in value equivalent to 2,20,000 tons in quantity and US$ 340 million in value. Our export spectrum is led by pepper followed in the order by Chilli, Spice Oil and Oleiresins and other spices. By the year 2000 AD, Indian industry is poised to capture 25 per cent share in value of the world trade in spice by adding value and quality. In future spice industry shall be more focussing on the major concerns of trade such as pesticide residues, mycotoxins, heavy metals, microbial contamination etc. Tissue

Spice Exports

	1996-97		*1997-98[E}*	
Item	*Qty. (in MT)*	*Value (in lakh)*	*Qty. (in MT)*	*Value (in lakh)*
Pepper	47893	41231.84	34250	47956.75
Cardamon (Small)	226	869.67	260	892.30
Cardamom (Large)	1628	1209.53	1450	1035.90
Chilli	50051	20145.15	43450	14013.25
Ginger	29737	5924.41	28550	6781.25
Turmeric	23019	5844.61	24900	7510.00
Coriander	12574	3136.58	20100	5441.25
Cumin	6375	3437.80	12150	6089.50
Celery	3780	801.76	3000	690.00
Fennel	4850	1788.60	11100	3206.10
Fenugreek	8891	1204.57	5000	795.00
Other seeds (1)	3059	842.60	4170	855.75
Garlic	4889	797.74	4000	760.20
Other Spices (2)	18955	4429.68	16450	5526.45
Curry Powder	4639	2056.52	5125	2225.55
Mint Oil	2371	13449.50	2525	8788.00
Oleoresins	2358	15901.21	2270	22648.00
Total	225295	123071.77	218750	135215.25
Value in Mn US$	197.86	241.43	346.97	363.60

culture technology [micropropagation] is now available at commercial level for over 35 spices including small cardamom, large cardamom, vanilla, black pepper, ginger, turmeric and certain herbal spices. The Spices Board is implementing two important projects in popularization T.C. Vanilla and Large Cardamom using about four lakh T.C. produced plantlets.

Horticultural Scenario

Vegetables and horticulture crops have an important role to play specially in terms of increasing the net return of the farm land, increase export potential and provide balanced nutrition to the Third World countries. India is the second largest producer of vegetable crops on this globe. The net income per hectare on an average from different vegetable/horticulture crops range from Rs. 20,000 to Rs. 50,000 which is 3-4 times the average returns from cereals. During the past 50 years, we have been able to increase our production of fruits and vegetables by three times. However to meet the growing demand, we need to double it again in next five years. It is here we have a hard nut to crack with new technologies and bettor management. Post harvest technology is another area which requires immediate attention specially because of its export potential, seasonal nature of the harvest and poor infrastructure available in the country for transport.

Following is the summary of the fruit scenario in India :

Crop	*Area [lakh ha]*	*Production [lakh tonnes]*	*Productivity [t/ha]*
Mango	11.36	92.23	8.11
Banana	3.96	104.59	26.40
Citrus	3.69	29.79	8.06
Apple	1.91	11.68	6.11
Guava	1.12	12.04	10.77
Pineapple	0.59	8.58	14.45
Litchi	0.53	2.60	4.87
Papaya	0.47	8.03	16.94
Grapes	0.34	6.53	19.20
Sapota	0.30	4.23	13.75
Others	7.74	49.20	6.35
Total	32.05	329.55	135.01

Oil and Pulses

Oil is an important human requirement for both edible and non-edible applications. Based on uses, oils and fats are divided into three broad classes– human foods, animal feeds and other uses– especially oleochemicals which include soap and paint. The largest application is for human consumption which include cooking oil, margarine and processed oil. Oil bearing crops include *Brassica sp.* [rape seed], *Glycine max* [soyabean], *Gossypium hirsutum* [cotton], Zea mays [maize/corn], *Elaei's guineensis* and *E. oleifora* [oil plam], *Helianthus annus* [sunflower], *Arachis hypojea* [peanut], *Simmondsia chinesiss* [jojoba], *Carthamus tinctorius* [safflower], *Cocos nucifera* [coconut], *Olea eurupera* [Olive] and *Linus usitatissimum* [flax]. There are a number of genes that can be targeted in the existing steps of the pathway which might result in a change of oil composition. Acyl Carrier Protein [ACP], Acetyl Co-A Carboxylase [ACC], Ketoacyl - ACP synthase [KASI] are just to give some successful examples in genetic engineering of oil seed crops.

Pulses are the major source of vegetarian – protein specially in a society like India. Pulses are mostly grown in the rainfed conditions and are in fact poor farmer's crop. The productivity of pulses/unit has not increased significantly during past 20-30 years. In fact they have not been exposed to technological revolutions of recent past either. There are not many hybrids popular in pulses. They are most prone to biotic and abiotic stresses. But for Soybean and alfalfa, genetic engineering had not been seriously applied to them specially in terms of disease tolerance. Hence there is an urgent need for a focused efforts to increase their production.

Genetic variability and environment

The recent happening in Andhra Pradesh and Karnataka, about cotton cultivation again points to the few lessons we need to learn. Uncontrolled and unspecified use of pesticide not only kills the environment, it makes the pathogen more resistant to these chemicals. Also there is a need to develop genetic tolerance either through conventional breeding or through genetic engineering in the cultivars which are susceptible to various diseases and pests.

Genetic variability is the fountain head of all substance, innovation and improvements of plant and animal productivity. Yet the very primary of genetic variability is threatened by human avarice for the most part with a bit of ignorance to boot. Genes, species and habitats are being man handled by the North. Flexing their technological and financial powers as much as by the naivete' and ignorance of the South. The loss of every gene and species, either permanently or worse to the maze of modern patenting regimes, severely limit our options for the future. It is even more compelling now to augment our biodiversity than ever before through conservation, preservation and legislation because of the urgent need to better utilise this genomic variation to sustain the growing demands. The rapid deteriorating global scenario, as seen from changes in temperature, sea-level, rain, snowfall and uv radiation are the premonitions of events to follow. Bioethics, biosafety equitable sharing are centres to any just system which contribute significantly in pulling us through the tough times that are lurking round the corner. This may be a time we need to look in the context of Indian eco and bio systems. Biosafety aspects related to transfer of genes cannot be ignored when we are transferring some of this technologies. The new technologies and products needed have to be safe to both human well-being and to the entire eco system. Thus it must encompass agri-horticultural environment and energy under the following parameters:

Agriculture :
- Improved Pesticides
- Integrated Pest Management
- Stress Tolerant Plants
- Disease Tolerant Plants/Animals
- Plant Vaccines
- Improved Nutritional Quality
- Improved Keeping Quality
- Better Growth Characters

Environment :
- Bioremediation
- Biofilters
- Natural Plastics
- New Composite Materials
- Biosorption Processing
- Biosensors
- Biofertilisers

Energy . Cleaner fuel
Desulfurization Process
Enhanced Oil Recovery.

Genetic Engineering

Traditional breeding continues to be the principal source of new cultivars and varieties for the ornamental market. Inspite of its success saga, traditional breeding has several limitations, the most obvious being the limited gene pool contained within a species. Another limitation of both traditional and *in vitro* breeding is the need for successive back crosses to purify the desired trait in the original offspring. This process can take many years and often the final product is a compromise between desirable and undesirable characteristics.

Genetic transformation overcomes both the limited gene pool and the necessity for successive backcrossing. Since the DNA used in the transformation can be isolated from any having organism, gene pool barriers are eliminated. Genetic transformation involves the transfer into the host plant of at most a few genes. The result is a progress that is genetically identical to its parent with the addition or dilution [in the case of antisense technology] of one or two additional characters. As a result the desired horticultural properties of the host plant are preserved and no back crossing is necessary to purify the desired phenotype.

For genetic transformation, *Agrobacterium* mediated gene transfer remains the most popular technique though the success rate in the case of monocots is rather low. Microprojectile bombardment is the most common alternative. However, the most effective method of particle delivery for this technique is still getting optimised. Several of the newer gene transfer techniques such as electrophoresis, silicon carbide mediated gene transfer and ultra sonication, offer considerable promise as viable alternative techniques

It was in 1983 that the development of the first transgenic plant was announced in a scientific meeting at Miami. During past 15 years this technology has been picked up very fast specially in the field of herbicide resistance, plant protection, quality aspects and production of novo biomolecules. Initially dicots and oflate monocots have been successfully manipulated by this technology to introduce/regulate the expression of desired traits/character. More than 3000 field trials of more than 50 different plant species are being conducted all over the world – major clustering of that being still in North America and Europe. Some of the commercially important plants which have been transferred by various agronomic and/or disease tolerance or quality parameters include alfalfa, apple, barley, cotton, maize, pine, potato, rice, sorghum, soybean, sugarbeet, sunflower, tobacco, tomato and wheat.

Tissue Culture

In the context of Indian agro-horticulture scenario, the tissue culture techniques as such can help us in :

- Faster multiplication of disease free plants in orchids, anthurium, gerbera, begonia, gladiolus, chrysanthemum and many ornamental plants.
- Selection of somaclonal variation in flower colour as in chrysanthemum.
- Production of haploids in petunia, cuclamen and few others, for developing homozygous lines for production of hybrid seeds.
- Protoplast fusion for somatic hybridization, useful in petunia.
- Production of synthetic seeds through somatic embryogenesis.
- Multiplication of flower bulbs in liquid bioreactors has almost come to commercialization stage.
- *In vitro* preservation of germplasm, valuable commercial varieties or endangered species.

In the world during past decade there had been exponential increase in the production of T.C. plants. While in 1985-86 only 130 million plants were produced. Last year the total production of T.C. plants was about 10 times of it. In fact at present the demand of T.C. plants exceeds the production capacity. In India too, annually there is production capacity of about 50 million plants. The major items produced through T.C. include plantation crops like Banana, Papaya, elite clones of various spices, forestry crops like tea and sandalwood and large number of floriculture selected crops. Of late T.C. is also used in haploid breeding which enhances the pace with which new varieties and genotypes can be evolved.

Constrains of T.C. Production

However there are some constrains also in optimising the production of T.C. plants. Biotechnology is a highly capital intensive industry and since the delivery of large volumes at targeted time is not possible in most cases, the per plant cost becomes prohibitive. When the economics of an industry is under consideration, it becomes necessary to scrutinize the factors responsible for its performance over the past years, in order to set the priority for its objectives. Though an extensive range of techniques has been developed during the last 20 years, there are still some major difficulties, especially those associated with the expansion of this technology to a wider variety of crops. Some of these problems can be discussed under the following headings:

Production Costs

One of the key hurdles in commercialisation of tissue culture is the initial investment. Venture capitalists are very shy to look towards this technology as they are still not very convinced about the returns in a predicted time frame. Bank loans in India are too cumbersome and the interest rate is probably one of the highest in the world. It is really hard to produce tissue culture plants in an economically viable way if one has to pay 17-19% as the interest to the bank. The interest rate simply evaporates the benefits. Initially, in the case of selected crops, conventional planting materials were being sold at a much lower price in comparison to tissue cultured transplants or propagules. However, later, when it was proved that through tissue culture technology, uniform and disease free plants could be produced in large numbers all year round, the commercial growers accepted tissue culture transplants at a higher price. The present trend in the micropropagation industry gives us a picture that more and more companies are producing only a few selected varieties of plants in large numbers, thereby generating stiff competition in the market. For example, Vietnamese farmers are producing tissue cultured potato plants at a total cost of one US cent each, as compared to eight cents per plant in the USA. The reason is that the current procedure of plant tissue culture is labour-intensive. Consequently, the labour inputs [in the form of salaries] form a significant proportion of the total production cost. Another important factor which contributes towards higher cost is the imported chemicals specially the harmony. The quality of indigenous harmony is highly unpredictable. In order to extend the technology to a variety of high value crops, one must be able to cut down production costs to the minimum by implementing the technical advances that are made in the research and development field. Given technical support and low transportation costs, future production centers will shift to countries where labour is less expensive. The standard of living in every country is increasing day by day and the same trend, as can be seen in Vietnam, may not continue in the future. Hence, the future of the micropropagation industry clearly lies in the ability to develop new technologies such as the application of computer controlled handling systems. Automation of plant culture system too has met with only limited success because of the very nature of the system as well as the prohibitive cost vs. cheap labour.

Seasonal nature and heavy peak demands

Most of the crops are seasonal and are in very demand only at certain times of the year. To overcome this problem and to meet the requirements of extremely large number of plants, efficient number of plants, efficient methods must be developed. Also, most commercial laboratories produce a mixture of plants required at different times in order to maximize the use of the facilities throughout the year. These definitely cut down the production cost. Just to give an example of this, while banana is grown almost all the year round in Malaysia, Thailand, Indonesia, some part of Philippines and Southern India but its planting time in North India is limited to about three months i.e. July, Aug, Sept. only. Hence the production has to be tailored to meet the demand.

Personnel/Organisational problems

Since large-scale micropropagation is a labour intensive process, one must correspondingly increase the number of employees to increase the number of plants produced. As the activities in any commercial laboratory expand, organisational problems will arise and, at times, become critical. To reduce these problems, the management structure must be well planned.

Supporting information systems such as inventory control, production scheduling, space utilisation, daily targets and actuals should be well defined for better management. For the successful large scale micropropagation of quality products, there must be perfect co-ordination among the technicians, supporting staff, supervisors and the researchers. The job description and reporting system should be very clear. In addition to these, personnel training is a critical prerequisite for successful large-scale production, which needs to be periodically reviewed on the basis of the customers needs. As such human resource development plays a key role in the efficiency of a production laboratory. Also it is not easy to get well trained technicians either. The inventory control must be computerised. All these features will, however, add considerably to the management responsibilities.

Contamination control

Close attention has to be given to this aspect, to avoid most of the industrial effort going waste. For example, if in an Indian commercial laboratory, 10 per cent of the plants produced through micropropagation methods perish due to contamination. This loss of cultures during the culture period would increase the production cost of the remaining plants to a large extent.

Contamination in cultures are caused mainly due to the following:

a. Natural contamination : occurs due to any of the following agents – dust, air-borne salts, metal filings, chippings, spares, seeds, vegetable matter, fibers, animal hair etc.
b. Man-made contamination : occurs mainly from body and clothing. It was calculated that each operator generates a minimum of 500,000 to 1,000,000 particles [greater than 0.5 um diameter] per minute.

To reduce .contamination rates and increase productivity, a clean indoor room environment for culture work is essential. In the work-area (where aseptic manipulations are carried out], three sources contribute to work-area particle contamination of the room air. They are the outside air, the liberation of particles by equipment machinery and processes, and the liberation of particles by operator. The objective of clean room is to establish and maintain an internal atmosphere devoid of particulate contamination in the working environment. Air cleanliness is measured as particle counts such as Class I, Class 10, Class 100, Class 1000, Class 10000 etc. i.e. on a particle concentration [number of particles per unit volume]. For the high cleanliness area of Class 100 or better, low turbulence displacement flow, also called laminar flow, has to be employed. Here the air moves along parallel flow paths and permits the particles liberated in the room to be carried quickly and as directly as possible. This sort of facility may be established by introducing the supply of air into the room through walls or ceilings built up from HEPA or ULPA filters or through special clean air distributing elements. Since the installation and running cost of laminar flow cabinets is high, it is advisable to design it according to the plant species to be worked on, so that the working space is utilised effectively. To achieve this, a perfect understanding is needed between researchers, engineers and architects.

Modern high quality clean rooms have environmental control systems in the widest sense. Besides cleanliness, temperature and humidity control, they encompass advanced air handling technology, vibration protection and many other aspects. In addition to this, correct and well-defined working procedures are necessary. All this should be backed up by performance monitoring of the operation as a whole. Wearing gowns, which is mandatory in Class 100 clean rooms, reduces the contamination from clothing, skin and hair.

Maintaining cleanliness in the working area is equally important. Apart from contaminants appearing in the cultures and those carried along with the explant, there is a continuous entry of contaminants along with the worker and through air entering the room. A definite protocol should be developed in order to reduce contamination in the laboratory and this varies with the location of the laboratory.

Marketing Problems

Besides the problems associated with production technology, the marketing of biotech products adds another dimension to the problems, restricting the furtherance of agricultural-biotechnology in developing countries. The delicate perishable nature of the product itself is the first marketing hurdle. That means the shelf life of the product is highly limited. Hence to take care of the product better infrastructure and lurk with end user/farmer is required. That is one of the reasons that most of the production is customer tailored. One cannot produce the plants and then look for the customer. Another important factor is the insensitivity of the farmer to per plant returns and the ultimate potential of his crop, acts as a serious impediment to hastening the adoption of the technology in developing countries. He is less receptive to the concept and thus needs a gestation period before he is convinced.

On the export scene, the lack of variety of packaging material at affordable prices, high air freights, unsuitable air connections, lengthy phytosanitary, customs and export documentation formalities make this nontraditional, highly perishable export commodity much less interesting as compared to the potential it has. Also in some of the European markets we are at disadvantage due to the fact "Made in India". Products from India attract extra levy and taxation. The introduction of TRIP and IPR too is going to affect the T.C. market export business.

CONCLUSION

In summary, Indian Agro horticulture scenario is going through a dynamic phase where it is facing daunting constraints, challenges and uphill task amidst lot of new technologies, to choose from, to answer some of the question or to ease some of the pressure it has. The pressure is both from biotic as well as the abiotic factor. At the same time the technologies available too have to be analysed in correct perspective as we cannot afford to try out many things with repeated failures. Abiotic stress from water and climatic conditions had been major hurdle in increasing productivity significantly from 60 - 70% of the rainfed agriculture. Biotic stresses through pests and disease every year deplete our yield by about 30%. Another mounting biotic stress is our exponentially growing population. Population has put immense pressure not only on cultivable land productivity but also on fresh, eco-system and inputs like water and organic matter. Conventional breeding has brought us a long way and shall keep to play an important role in time to come to develop new lines/varieties to address some of these problems. And to our breeder new help and light has come through molecular breeding and genetic transformation. Inter specific genetic barrier started crumbling and now virtually one can transfer gene from any biotic to another. How logically and smartly we use this technology would depend on need and the human mind. Conservation and proper utilisation of our biodiversity could be an important key to survival of mankind on this planet. As it is there we need to hunt for novo biomolecules for disease tolerance or better product. Tissue culture has helped us in a big way in large scale production of elite clones which would have been difficult or impossible in some of the plant species.

Index

D

E

F